Molecular Catalysis

Volume 3

Series Editors

Carmen Claver, Departament de Química Física i Inorgànica, Universitat Rovira i Virgili, Tarragona, Spain

Nicholas Turner, School of Chemistry, University of Manchester, Manchester, UK

This book series publishes monographs and edited books on all areas of molecular catalysis, including heterogeneous catalysis, nanocatalysis, biocatalysis, and homogeneous catalysis. The series also explores the interfaces between these areas. The individual volumes may discuss new developments in catalytic conversions, new catalysts, addressing existing reactions and new reactions regarded as desirable from a societal viewpoint. The focus on molecular insight requires an appropriate attention for synthesis of catalytic materials, their characterization by all spectroscopic and other means available, and theoretical studies of materials and reaction mechanisms, provided the topic is strongly interwoven with catalysis. Thus the series covers topics of interest to a wide range of academic and industrial chemists, and biochemists.

Ming Bao · Jiasheng Wang · Xiujuan Feng ·
Jingjie Luo · Jian Sun

Nanometal Catalysis in Organic Synthesis

Springer

Ming Bao
Dalian University of Technology
Dalian, China

Xiujuan Feng
Dalian University of Technology
Dalian, China

Jian Sun
Jilin University of Chemical Technology
Jilin, China

Jiasheng Wang
Dalian University of Technology
Panjin, China

Jingjie Luo
Dalian University of Technology
Panjin, China

ISSN 2522-5081 ISSN 2522-509X (electronic)
Molecular Catalysis
ISBN 978-981-97-4571-5 ISBN 978-981-97-4573-9 (eBook)
https://doi.org/10.1007/978-981-97-4573-9

This Springer imprint is published by the registered company Springer Nature Singapore Pte Ltd.
The registered company address is: 152 Beach Road, #21-01/04 Gateway East, Singapore 189721, Singapore

Contents

Chapter 1
Introduction

Abstract Catalysts are an important part of organic reactions, they can reduce the reaction energy barrier, and thereby accelerate the reaction rate and make the reaction more efficient. Modern industrial organic synthesis and laboratory research are increasingly dependent on the development and utilization of high-performance catalysts. At a rough estimate, more than 75% of all existing industrial chemical processes and 90% of newly developed reactions involve the use of catalysts. In 1895, Ostwald came up with the definition that we use until today: A catalyst is a substance that changes the rate of a chemical reaction without itself appearing into the products.

1.1 Catalysts and Catalytic Reactions

Catalysts are an important part of organic reactions, they can reduce the reaction energy barrier, and thereby accelerate the reaction rate and make the reaction more efficient. Modern industrial organic synthesis and laboratory research are increasingly dependent on the development and utilization of high-performance catalysts. At a rough estimate, more than 75% of all existing industrial chemical processes and 90% of newly developed reactions involve the use of catalysts. In 1895, Ostwald came up with the definition that we use until today: A catalyst is a substance that changes the rate of a chemical reaction without itself appearing into the products. As shown in Fig. 1.1, during the catalytic cycle the catalyst interacts with the reactants to generate several intermediates, which are subsequently converted into a product and the catalyst is regenerated. Therefore, the modern definition of a catalyst is as follows: A catalyst is a substance which increases the rate at which a chemical reaction approaches equilibrium without becoming itself permanently involved, and its composition and quantity did not change before and after the catalytic reaction.

Catalysts used in industrial chemical processes and academic research are usually liquids or solids, which are soluble or insoluble in an organic solvent. And thereby the catalytic reactions can generally be divided into two categories: homogeneous catalytic reactions and heterogeneous catalytic reactions. A catalytic system in which the substrates for a reaction and the catalyst components are brought together in one

M. Bao et al., *Nanometal Catalysis in Organic Synthesis*, Molecular Catalysis 3,
https://doi.org/10.1007/978-981-97-4573-9_1

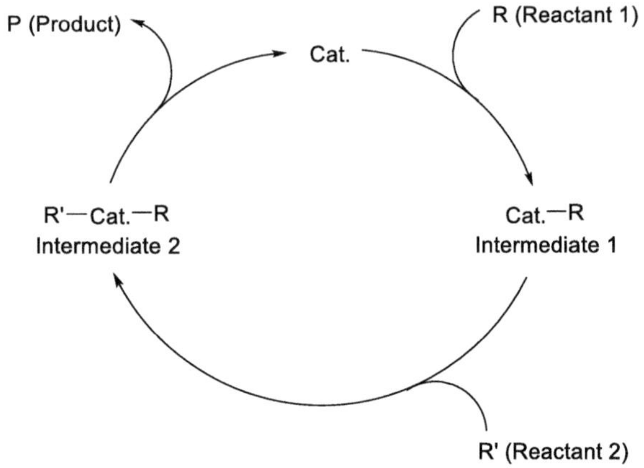

Fig. 1.1 Catalytic cycle

phase, most often the liquid phase, was called homogeneous catalysis. Various catalysts including acids, bases, Lewis acids, small organic molecules, (organo)metallic complexes, and enzymes were used in homogeneous catalysis. The reactant and the catalyst are not in the same phase, there is a phase interface between them, and the reaction is carried out on the two interfaces, which is called heterogeneous catalysis. The common catalysts are solid, and the reactants are gases or liquids in the heterogeneous catalysis.

1.2 Classification of Catalysts

Numerous catalysts known today can be classified according to various criteria: source, state of aggregation, area of application, primary and secondary functions, or phase of the reaction system.

(1) According to the source, the catalysts can be classified into biocatalysts and abiotic catalysts. Biocatalysts include enzymes and microorganisms. Abiotic catalysts include metal complexes, precious metals, sulfuric acid, etc.
(2) According to the state of aggregation, the catalysts can be classified into gas, liquid, and solid catalysts.
(3) According to the area of application, the catalysts can be classified into polymerization, condensation, esterification, acetylation, hydrogenation, dehydrogenation, oxidation, reduction, alkylation, isomerization, and other catalysts.
(4) According to the primary and secondary functions, the catalysts can be classified into main catalysts and co-catalysts. For example, in the Ziegler–Natta catalysts, $TiCl_4$ and $TiCl_3$ are the main catalysts, and $AlEt_3$ is the co-catalyst.
(5) According to the phase state of the reaction system, the catalysts can be classified into homogeneous catalysts and heterogeneous catalysts. Homogeneous catalysts include acids, bases, soluble transition-metal compounds, and peroxides.

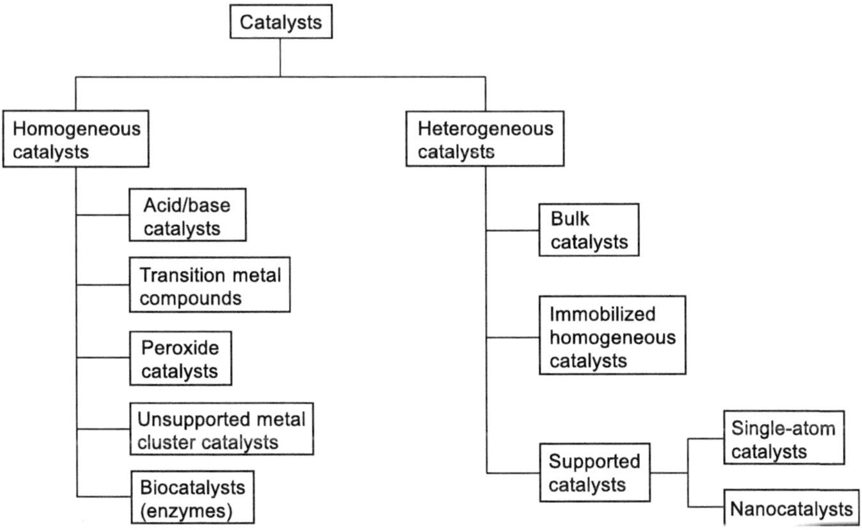

Fig. 1.2 Classification of catalysts classified according to the phase state of the reaction system

Heterogeneous catalysts include solid acids, organic bases, precious metals, metal oxides, metal complexes, rare earth compounds, molecular sieves, and nanomaterials and so on. Although the catalytic reaction mechanism of immobilized homogeneous catalysts is the same as that of homogeneous catalysts, the insoluble immobilized homogeneous catalysts were also classified as heterogeneous catalysts. Figure 1.2 shows the classification of catalysts classified according to the phase state of the reaction system. However, if the catalysts were classified according to the catalytic reaction mechanism, the immobilized homogeneous catalysts should belong to the homogeneous catalysts.

1.3 Homogeneous Catalysts

1.3.1 Effectivity

As classified above (Fig. 1.2), homogeneous catalysts include acid/base catalysts, transition metal compounds, peroxide catalysts, unsupported metal cluster catalysts, and enzymes. In homogeneous catalysis, catalysts, starting materials, and products are present in the same phase, usually the liquid phase, and thereby there is no mass transfer problem. Except for unsupported metal cluster catalysts, all molecules of the homogeneous catalyst act as active centers showing higher activity per unit mass than heterogenous catalysts. High selectivity including chemoselectivity, regioselectivity, diastereoselectivity, and enantioselectivity can be achieved by using homogeneous catalysts because homogeneous catalyzed reactions are controlled mainly by kinetics rather than by material transport.

The unsupported metal cluster catalysts have recently attracted considerable attention as they exhibited high activity and selectivity in organic molecule transformation. The reason behind this behavior could be attributed to their large specific surface area and several active centers on the crystallite surface.

The biocatalysts, namely, enzymes, exhibited extremely high activity and selectivity in organic synthesis. For the asymmetric synthesis, enzymes generally give 100% enantioselectivity. The enzyme-catalyzed reactions are usually carried out in aqueous media at ambient temperatures in line with the concept of green chemistry. In the near future, more enzyme-catalyzed reactions will be developed for organic synthesis.

1.3.2 Property

The structures of homogeneous catalysts can be well characterized by spectroscopy determination including infrared spectroscopy, nuclear magnetic resonance spectroscopy, single crystal X-ray diffraction, etc. The catalyst molecules can be modified by changing the central metal atoms and ligands to modify their electronic effects and stereoscopic effects for design and synthesis of new catalysts with various properties. Due to the intrinsic thermal stability of homogeneous catalysts, especially transition metal complexes, in the liquid phase, the homogeneous catalytic reaction is limited to temperatures below 200 °C. Although homogeneous catalysts are less stable under relatively high temperature, they are less susceptible to poisoning by sulfur-containing compounds. The homogeneous catalysts usually cannot be easily separated from the reaction mixture and reused for the target reaction. This is the major disadvantage of homogeneous catalysts. Therefore, many research studies focus on the synthesis and application of immobilized homogeneous catalysts. The supports used can be inorganic or organic materials.

1.3.3 Mechanism

The performance of the homogeneous catalysts can be explained and understood at the molecular level through spectroscopy determination. The mechanism of homogeneous catalysis is easy to propose by kinetic study, control experiments, intermediate separation, and density functional theory calculation. For a homogeneous catalysis using transition metal complex as the catalyst, the reaction mechanism may involve ligand dissociation, substrate coordination, oxidative addition, transmetalation, migration insertion, reductive elimination, β-elimination, etc. For example, a reasonable mechanism for the Stille reaction of acid chloride with phenyl tributylstannane is shown in Fig. 1.3.

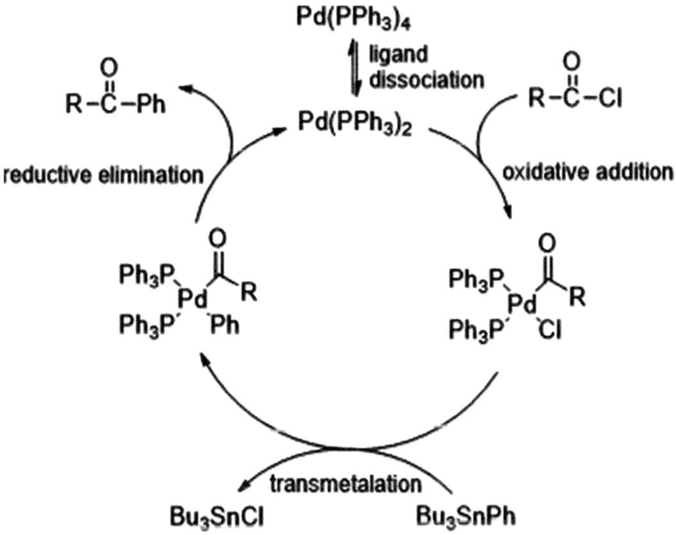

Fig. 1.3 Mechanism for the Stille reaction of acid chloride with phenyl tributylstannane

1.3.4 Application Fields

Owing to the high activity and selectivity of homogeneous catalysts, they cannot only be applied to produce bulk chemicals, such as acetic acid and methanol, but more importantly, they can be employed to provide high value-added fine chemicals, such as polymers, intermediates of high functional materials, pharmaceuticals, pesticides, food additives, etc.

1.4 Heterogeneous Catalysts

1.4.1 Effectivity

Only the surface atoms of heterogeneous catalysts act as active centers showing lower activity per unit mass than homogeneous catalysts. Owing to the heterogeneous catalytic reaction taking place on the surface of catalysts, macroscopic diffusion plays an important role. The heterogeneous catalysts were usually utilized at high temperature, and hence the high reaction rate can be achieved. However, the reaction selectivity is generally low.

The almost same high activity and selectivity can be obtained by using insoluble immobilized homogeneous catalysts in organic synthesis as heterogeneous catalysts were used. However, the leaching of the active species into the solvent makes the catalyst gradually loss activity, and cannot be recycled and used for a long time.

Therefore, the stability of the insoluble immobilized homogeneous catalysts become a point of concern.

1.4.2 Property

The structures of heterogeneous catalysts were less well defined. Many modern methods, such as X-ray diffraction (XRD), small-angle X-ray scattering (SAXS), infrared spectroscopy (IR), Raman spectroscopy (RS), solid-state nuclear magnetic resonance spectroscopy (SS-NMR), X-ray photoelectron spectroscope (XPS), scanning electron microscopy (SEM), and transmission electron microscope (TEM), were used for the characterization of heterogeneous catalysts. The thermal stable heterogeneous catalysts can be recycled and used at high temperatures, even above 200 ºC. The major advantage of heterogeneous catalysts is the ease of separating the catalyst from the product. In the liquid-phase reaction, the heterogeneous catalysts can be separated by simple methods such as filtration or decantation.

1.4.3 Mechanism

Except for using insoluble immobilized homogeneous catalysts, the mechanism of heterogeneous catalytic reactions is very complex and difficult to understand at the molecular level. The reaction mechanism may involve diffusion of the reactants to the catalyst surface, physical adsorption and chemical adsorption on the active sites, surface reaction, desorption of products, and product diffusion from the surface, etc.

1.4.4 Application Fields

Due to the wide scope and higher thermal stability of heterogeneous catalysts than homogeneous catalysts, they are more widely utilized in the chemical industry. Heterogeneous catalysis accounts for more than 90% by volume of all industrial catalytic processes. Heterogeneous catalysts were not only used for cracking and reformation, but also for producing bulk chemicals, such as ammonia and methanol.

1.5 Some Famous Industrial Catalysts

Catalysts played a very important role in the progress of human civilization and the development of the world economy. Every major breakthrough in a catalyst made major social change. Here are seven catalysts that have changed the world since the Industrial Revolution.

1.5.1 Catalyst for Synthesis of Sulfuric Acid

Sulfuric acid industry is the most important industry in modern times. In 1875, the first production line for fuming sulfuric acid was created in the world by a German chemist named E. Jacob using a very expensive platinum catalyst. A decade later, a more cost-effective alternative process was developed by chemists at the German company BASF using the currently widely used V_2O_5 catalyst.

Whether in the past or now, sulfuric acid is a very important raw material for almost all industries, such as chemical industry, metallurgy, machinery, textile and so on.

1.5.2 Catalyst for Synthesis of Ammonia

In 1909, German chemist Fritz Haber directly obtained ammonia from nitrogen and hydrogen with a yield of 6% at 17.5–20 MPa and 500–600 °C by using osmium as a catalyst. Two years later, Haber developed a more cheap and active iron catalyst and built an ammonia plant for producing 30 tons per day. This is a major victory for artificial nitrogen fixation, and also a major victory for catalysts in chemical industry and chemical research. In 1918, Haber received the Nobel Prize in Chemistry for his work on the direct synthesis of ammonia from nitrogen and hydrogen.

In 1912, Alwin Mittasch and Carl Bosch at the German company BASF, conducted 6,500 tests with 2,500 different catalysts, and finally developed a cheap and easily available iron catalyst with potassium and aluminum oxides as co-catalysts, which is also the prototype of catalyst components in the modern synthetic ammonia industry. This ammonia synthesis process is known as the Haber–Bosch process, and it marks the first milestone in the industrial realization of high-pressure catalytic reactions. In 1931, Bosch received the Nobel Prize in Chemistry for his development of high-pressure technique for synthesis of ammonia.

The synthetic ammonia industry is considered the greatest chemical invention in the twentieth century, and also known as the bellwether reaction in heterogeneous catalysis. Through the synthetic ammonia industry, a large amount of nitrogen fertilizer can be provided, thus greatly promoting the development of agriculture.

1.5.3 Catalyst for Fischer–Tropsch Synthesis

The process of catalytic synthesis of liquid hydrocarbons or alcohols with syngas (a mixture of carbon monoxide and hydrogen) as raw materials under appropriate conditions is called Fischer–Tropsch synthesis. In 1923, German chemists Franz Fischer and Hans Tropsch prepared hydrocarbons at 400–455 °C under 10–15 MPa pressure using CO and H_2 as raw materials in the presence of basic iron filings as the catalyst. This marks the birth of indirect coal liquefication technology. Subsequently, they developed Ni and Co based catalysts.

At present, high-quality oil and more than 20–30 kinds of high value-added chemicals were obtained by the most advanced high-temperature Fischer–Tropsch synthesis technology. This further extends the coal chemical industry chain and increases the types of downstream products.

1.5.4 Catalyst for Petroleum Refining

Almost all petroleum refining processes require catalysts, including catalytic cracking, catalytic reforming, catalytic hydrogenation, alkylation, etherization, catalytic dehydrogenation, and catalytic desulfuration. Petroleum refining catalysts accounted for one-third of the catalyst's market share. The earliest catalysts used were zeolite molecular sieves, such as X- and Y-zeolites. The Y-zeolite is still the main active component of catalysts in the current cracking process. At present, there are many kinds of catalysts with various properties can be used in petroleum refining. For example, ZSM-5, USY, Pt/Al_2O_3, and MoS_2.

Catalytic cracking is one of the most important processes in the petrochemical industry. Petroleum refining catalysts accounted for one-third of the catalyst market share becoming the most used catalysts in the world.

1.5.5 Catalyst for Olefin Polymerization: Ziegler–Natta Catalyst

In 1953, German chemist Karl Ziegler discovered that a catalytic system combining $TiCl_4$ and $AlEt_3$ could catalyze ethylene polymerization at room temperature under ordinary pressure to obtain polyethylene with high molecular weight.

In 1954, Italian chemist Giulio Natta used the $TiCl_3/AlEt_3$ catalyst system to realize propylene polymerization; polypropylene with high isotacticity was obtained. This is the pioneering research of isotactic polymerization. In 1963, Ziegler and Natta shared the Nobel Prize in Chemistry for their discoveries in the chemical properties and techniques of polymers.

After more than 70 years of development, Ziegler–Natta catalyst has become the most mature and most widely used catalyst for olefin polymerization today, which is used in more than 90% of polyolefin product preparation in the world. The role of Ziegler–Natta catalyst in the development of the entire human society is unparalleled.

1.5.6 Three-Way Catalyst for Treatment of Automobile Tail Gases

In 1974, the first Clean Air Act was implemented in the United States, when only CO and HC were controlled, and the catalyst used for automobile exhaust treatment was the Pt–Pd based oxidation catalyst. Later, with the damage of photochemical smog to the air becoming more and more serious, NOx was also regarded as an important indicator of exhaust emissions. In 1989, the Ford Motor Company in the United States tested Pd/Rh catalyst as a part of the three-way catalyst for the first time, while CO, HC, and NOx are three harmful substances which were catalytic purified. Since then, the three-way catalyst has become the classic catalyst in the automobile exhaust gas treatment.

1.5.7 Catalyst for Asymmetric Synthesis

In 1968, an asymmetric catalytic hydrogenation reaction was successfully realized by W. S. Knowles at Monsanto Company in the United States using the complex formed by the chiral phosphine ligand and metal rhodium as a catalyst for the first time in the world. This is the pioneering research of synthesis of chiral molecules by homogeneous asymmetric catalysis. In the early 1970s, Knowles achieved industrial synthesis of L-dopa for Parkinson's disease using asymmetric hydrogenation at the Monsanto Company. This was the first industrial example of asymmetric synthesis in the world.

In 1985, an asymmetric hydrogenation reaction was successfully conducted by the Japanese chemist Ryoji Noyori using the famous bisphosphine ligand, BINAP, prepared by Ryoji Noyori himself. The catalyst system (*S*)-BINAP-Rh^+ was applied to the industrial production of menthol flavoring.

In 1980, an oxidation reaction of allyl alcohol with 2-hydroperoxy-2-methylpropane in the presence of chiral titanate was reported by K. B. Sharpless in the United States. Then, the asymmetric epoxidation reaction of allyl alcohol was successfully achieved by K. B. Sharpless using the complex (5–10 mol%) formed by tetraisopropyl titanate and diethyl tartrate as the catalyst in the presence of molecular zeolites.

In 2001, Knowles, Noyori, and Sharpless shared the Nobel Prize in Chemistry for their work on asymmetric catalytic hydrogenation/oxidation reactions.

1.6 Nobel Prize in Chemistry for Catalytic Organic Synthesis

Since the last century, numerous scientists have won the Nobel Prize in Chemistry related to catalytic organic synthesis. The winners and their contributions are listed in Table 1.1.

Table 1.1 Nobel Prize in Chemistry for Catalytic Organic Synthesis

Year	Country	Name	Reason
1912	France	Paul Sabatier	For the development of a method for hydrogenation of organic compounds using metals as catalysts
1963	Germany	Karl Ziegler	For their discoveries in the field of the chemistry and technology of high polymers
	Italy	Giulio Natta	
1973	UK	Geoffrey Wilkinson	For their pioneering work, performed independently, on the chemistry of the organometallic, so called sandwich compounds
	Germany	Ernst Otto Fischer	
1975	Australia (worked in the UK)	John Warcup Cornforth	For research in the stereochemistry of enzyme-catalyzed reactions and the stereochemistry of organic molecules and reactions
	Switzerland	Vladimir Prelog	
2001	USA	William S. Knowles	For their work on chirally catalyzed hydrogenation reactions
	Japan	Ryoji Noyori	
	USA	K. Barry Sharpless	For his work on chirally catalyzed oxidation reactions
2005	France	Yves Chauvin	For the development of the metathesis method in organic synthesis
	USA	Robert H. Grubbs	
	USA	Richard R. Schrock	
2010	USA	Richard F. Heck	For palladium-catalyzed cross couplings in organic synthesis
	Japan (worked in the USA)	Ei-ichi Negishi	
	Japan	Akira Suzuki	
2021	Germany	Benjamin List	For the development of asymmetric organocatalysis
	USA	David MacMilan	
2022	USA	Carolyn R. Bertozzi	For the development of click chemistry and bio-orthogonal chemistry
	Denmark	Morten Meldal	
	USA	K. Barry Sharpless	

1.7 Conclusion

At present, the production of both bulk and fine chemicals is increasingly dependent on catalytic reactions. More than 60% of all chemicals are synthesized through catalytic reactions, and more than 90% of industrial chemical reactions are catalytic reactions. Especially, the synthesis of pharmaceutical and pesticide molecules is almost inseparable from catalysts. Therefore, the design and preparation of a new type of catalyst has attracted considerable attention. It was recently found that nanostructured catalysts, such as metal nanoparticles, metal nanoclusters, and nanoporous metals, exhibited high catalytic activity in organic transformation. The supported metal single-atom catalysts were also found to possess unique catalytic activities that differ from those of their cluster and particle counterparts. In this book, the preparation, characterization, and application of these nanostructured catalysts and supported metal single-atom catalysts in organic synthesis are introduced.

Chapter 2
Organic Synthesis Catalyzed by Metal Nanoparticles

Abstract Metal nanoparticles, which are particles with sizes in the range of 2–100 nm, have been widely applied as active catalysts in organic molecule transformation. In such systems, an appropriate support is required to prevent agglomeration of the fine metal nanoparticles. This chapter introduces the preparation, characterization, and application of supported metal nanoparticles in organic synthesis based on current research works and studies. Interactions between the support and metal nanoparticles are also discussed. The most significant applications of metal nanoparticles will be discussed in terms of the reaction types and essential metal elements. The structure-performance relationship in the view of metal-support interaction, size effect, formation of active metal centers will be analyzed according to the most typical research studies.

Metal nanoparticles are defined as particles smaller than 100 nm in at least one dimension, which can be fabricated via mature synthesis technology based on current chemical engineering. In organic synthesis, supported metal nanoparticles become the great range of catalysts for the heterogenous reaction system with excellent catalytic performances for a variety of reactions. Different from most homogeneous reactions using coordinated compounds, soluble metallic inorganic salts, strong base or acid, supported metal particles can be directly recycled by facile separation of solid powder from the liquid or gas reaction systems. There would not be tough problems by metal leaching or poisoning if the supported particles were deliberately designed with stability. It provides more possibility and perspectives for supported metal nanoparticles to be further applied in the industrial world involving organic synthesis, which is positively correlated with the national economy and global social development in all respects.

In this chapter, the synthesis method of metal nanoparticles and the supported catalysts were introduced in the first section. Common characterization and fine analysis based on current research works and studies will be put forward subsequently. The most significant applications of metal nanoparticles will be then discussed in terms of

M. Bao et al., *Nanometal Catalysis in Organic Synthesis*, Molecular Catalysis 3,
https://doi.org/10.1007/978-981-97-4573-9_2

the reaction types and essential metal elements. The structure-performance relationship in the view of metal-support interaction, size effect, formation of active metal centers will be analyzed according to the most typical research studies.

Metal nanoparticles on support can be fabricated by different methods, which can be divided into the physical preparation method, chemical method, and physicochemical mixing method. Considering the potential application of metal nanoparticles in the production of high-value added down-steam chemicals by organic synthesis, the mostly commonly used chemical methods by either ex-situ or in situ strategy will be discussed in this section.

2.1 Metal Colloid and Ex-Situ Synthesis of Supported Metal Particles

Among these supported materials, metal nanoparticles can be synthesized before depositing/anchoring onto the surface of the supporting materials. One of the well-utilized methods is the sol–gel process used to form colloidal metal particles. Such fabrication strategy is processed by growing of metal atoms into particles by forming metal seeds in the solution-phase, which is denoted as the "bottom-up method". Subdividing the bulk-phase metal using a physical method can also provide colloidal particles and named as the "top-down methods". In this section, the chemical strategy from bottom to up method will be discussed in detail. Colloidal chemical synthesis can be used to synthesize uniform nanoparticles. The size distribution and shape of nanoparticles can be controlled according to the synthesis conditions including the reducing agent, solvent, temperature, and coordination agent.

2.1.1 Synthesis and Characteristics of Metal Colloid

Nanoparticles can be formed and well dispersed in liquid solvent by chemical reduction of the metal salt. In such a system, the metal salt and reducing agent are indispensable to the geometric structure and the stability of the metal particles in nano-scale. The reducing agent can be an organic matter with $-NH_2$, $-CHO$ or $-OH$ groups, e.g., hydrazine, acetaldehyde, or polyethylene glycol; reducing gas such as H_2 and CO; inorganic salt ($NaBH_4$) or even a secondary metal element. A specific protecting agent is also beneficial and essential to keep a stable colloid with the dispersed metal particles. The protecting agent could be a donor ligand, polymer, and surfactant in many syntheses. The main steps for a typical synthesis are illustrated in Fig. 2.1. The metal nanoparticles were dispersed in the liquid solution with the particle surface encapsulated/protected by the protecting agent. One of the characteristics of the successful formation of colloidal nanoparticles is the change of color, which is a direct result of the localized surface plasmon resonance (LSPR) effect. If a laser

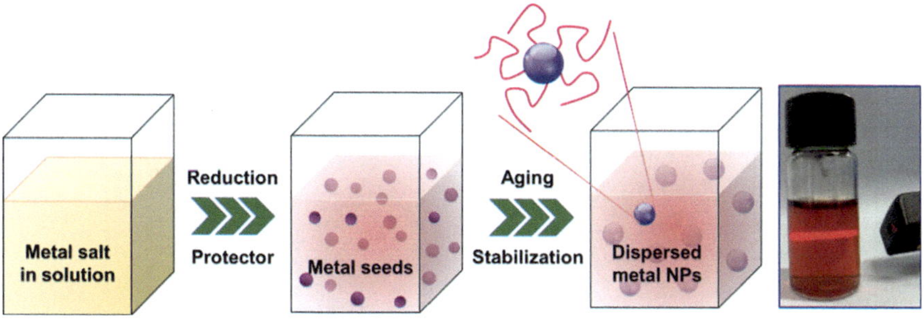

Fig. 2.1 Illustration of chemical reduction of metal salt in the presence of reducing agent and protector to produce metal NPs dispersed in liquid solution. The electronic photo on right illustrates the gold colloid displaying the Tyndall effect

beam was used to illuminate the colloid solution, Tyndall effect appears as revealed in the electronic photo in Fig. 2.1.

The chemical reduction strategy is becoming one of the most popular method for colloidal metal particles. It can be described by the following Eq. (2.1).

$$xM^{n+} + nxe^{-} + Protector \rightarrow M_n^0 \tag{2.1}$$

The synthesis of colloidal nanoparticles by chemical reduction was started in the mid-twentieth century [1, 2]. Metal colloids made by chemical reduction were for the first time synthesized and characterized by Turkevich et al. in 1951 [1], which was published in the *Transactions of the Faraday Society*. According to their reports, the particle sizes were estimated to be around 30 Å. After then, Leheny et al. [2] successfully oxidized the tetrathlafulvalene (TTF) to TTF^{+} using colloidal Pt particles. The aqueous solution of $PtCl_4^-$ was reduced by sodium citrate (0.2 wt%) after reflux for 4 h. The solution changed from transparent colorless at the beginning to a transparent blackish, suggesting the formation of small Pt colloidal particles. Since then, studies on colloidal nanoparticles used for homogeneous and heterogeneous catalysis were spontaneously carried out by research fellows.

A statistic of the published literature concerning the colloidal metal particles were illustrated in Fig. 2.2. Although the first reports engaged by Turkevich et al. was published in 1950s [1], it took almost three decades to attract people's attention until 1980s. During these years, only 1–3 publications per year were concerned about this subject. Most of the research studies were focused on the concept, characterization, applications in Physics and Spectroscopy field. The perception of colloidal metal particles in chemical field became a conscious awakening since 1990s.

The discussed subject includes the colloidal nanoparticles either synthesized by the chemical method or used for chemical applications. The related topics increased with the timeline accounting for about 30% of all publications on colloidal metal particles. Investigations based on colloidal metal particles for catalysis began at the

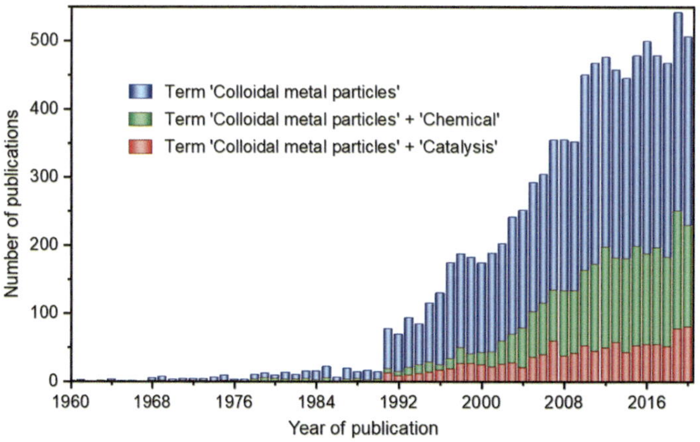

Fig. 2.2 Statistical data of the published research papers and reviews retrieved with the term of "Colloidal metal particles", "Colloidal metal particles" plus "Chemical", and "Colloidal metal particles" plus "Catalysis" according to the data from the Web of Science

same time and the number of current publications increased about ten-folds compared to the publications in the early years.

Different morphologies of the metal particles, ranging from spherical nanoparticles to anisotropic, core–shell and hetero-structured shapes can be demonstrated according to the detailed synthesis including the ratio between the raw materials, temperature, gas atmosphere, and the pair of reducer/protector [3]. In this section, we will focus more on those particles with spherical or other superior shapes, which would be suitable as subsequent heterogeneous catalysts for catalysis.

Compared to transition metals, precious metals more easily form a stable colloid in metallic states due to their higher reduction potential [4]. Herein, metal colloids including Au colloid, Pt colloid and silver colloid were frequently reported and used in early literature.

Taking the gold colloid as a special instance, it displays its own specificity in the catalysis field using gold as the key center for many organic reactions. Gold colloid can be simply obtained by reducing the mixture of the Au precursor (e.g., $HAuCl_4$ in many cases) and a protecting agent, where $NaBH_4$ or citrate can be used as the most common reducing agent [1]. Citrate can be used as both the reducing agent and protecting agent, which is very mild for the uniform growth of Au nanoparticles in the colloid. No matter how, one of the challenges is to keep the size distribution controllable with stability [6]. The Turkevich's method was herein promoted by optimizing the solvent, pH value, protecting/reducing agent to control the growing mechanism/steps of the gold seeds and particles [7, 8]. Gold nanoparticles in these colloids can be tuned with narrow size distributions from about 3 nm to 200 nm and even larger sizes as illustrated in Fig. 2.3. One of the amazing issues is the wonderful color change (e.g., steadily changed brownish-orange, pink, red, win-red, etc.) of the transparent solution due to the variance of the shapes and sizes of nanoparticles.

Piella et al. [5] studied the position and width of the LSPR peaks in different colloids by Uv–vis spectra, which was correlated with the shape, size distribution and color of Au nanoparticles.

With the advancement of technological and detection methods, bimetallic alloy nanoparticles or even high entropy alloy particles in colloids can be synthesized by modified chemical reduction. As listed in Table 2.1, bimetallic AuAg colloid is a kind of typical colloidal nanoparticle reported in literature, using common protecting agents such as PVA and PVP coupled with $NaBH_4$ or citrate as the reducing agent [10, 11]. A bimetallic Au-M colloid even displays a greatly shifting adsorption peak due to the obviously modulated particle morphology and LSPR effect as reported in the AuAg colloid systems. Different from the wine-red Au colloid, Ag colloid displays the bright yellow color as shown in Fig. 2.4, which was synthesized by sodium citrate at the water/cyclohexane interface. The color of AuAg colloid steadily changed to orange with the bimetallic polyhedral nanoparticles.

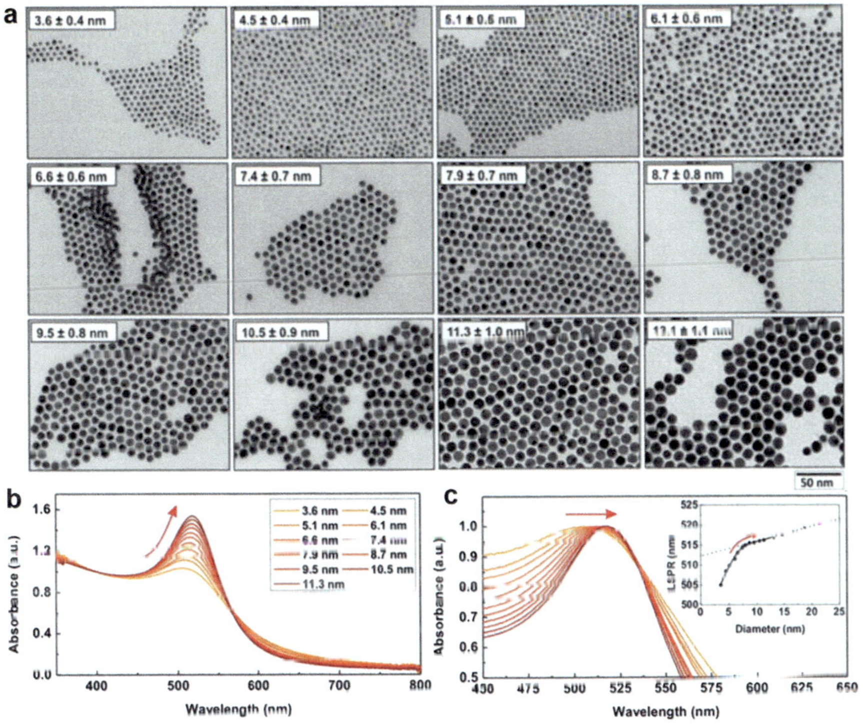

Fig. 2.3 **a** Transmission electron microscopy images of an improved citrate-based Au nanocrystal synthesis using seed particles (3.6 nm) and further growing them in solution. The particle size can be tuned from 3.6 ± 0.4 nm of the seeds to 13.1 ± 1.1 nm, with narrow particles size distributions of 10% dispersion or less. Optical properties of selected Au colloids obtained after different growth steps **b** normalized at 400 nm and **c** normalized at the LSPR peak maximum [5]. Copyright 2016, American Chemical Society

Table 2.1 Recent metal colloids designed by research fellows during the latest five years in publications

Metal	Denotation	Metal source	Reducer	Protector	Solvent	References
Au	3.0 nm	$HAuCl_4$	$NaBH_4$	PVA	Water	[12]
Au	6–15 nm	$AuCl_3$ in DDAB	N_2H_2	Dodecanethiol	Decanoic acid in toluene	[13]
Au	3.5 nm	$HAuCl_4$	Sodium citrate, tannic acid		Water	[5]
Au	15, 40, 60 80 nm	$HAuCl_4$, 15 nm Au seed	Trisodium citrate	n.a., boiled, spherical and stable NPs	Water	[14]
AuAg	10–28 nm, orange solution	$AgNO_3$, $HAuCl_4$	Sodium citrate		Water/ cyclohexane	[9]
AuCu	Hybrid nanoparticles	Au NPs in CTAB solution, $CuCl_2$	Ascorbic acid solution, post-laser treatment for nanoalloy	Hexadecylamine		[15]
Ni_xPt_{1-x}	FCC alloy, 280 °C, black solution	$Pt(acac)_2$, $Ni(acac)_2$	1,2-hexadecanediol	Oleylamine (OAm), oleic acid (OAc), argon blanket	Benzyl ether	[16]
FeCoNiPtRu	5 nm, multimetallic nanoalloys, black solution	Acetyl or acetylacetonate metallic precursors	1,2-hexadecanediol	OAm, OAc, argon blanket	Benzyl ether	[17]
Fe	2.9 nm	$FeCl_3{\cdot}6H_2O$ in isopropyl alcohol	UV-irradiated with a xenon pulsed lamp	PVP		[18]
Ni	3–8 nm, black solution	Seed-mediated growth, $Ni(acac)_2$	Borane *tert*-butyl amine complex	OAm, OAc, Ar		[19]

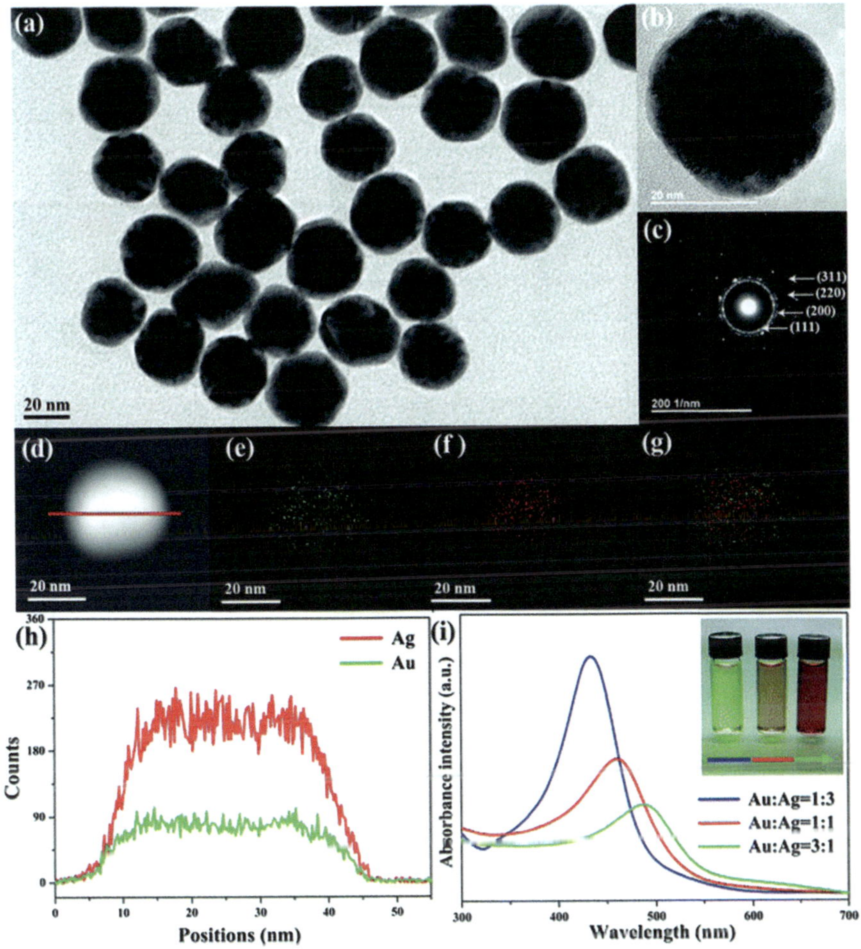

Fig. 2.4 Structure, composition, and plasmonic feature study of AuAg ANPs. **a, b** Low and high-magnification TEM images of AuAg ANPs. **c** SAED image of the ANPs as shown in (**b**). **d** STEM image of the single ANP. **e–g** EDX elemental mapping images of Au and Ag. **h** EDX line profile of the single ANP as shown in (**d**). **i** Absorption spectra of AuAg ANPs (Au/Ag = 1:3, 1:1, and 3:1). The insets denote their corresponding alloy colloid solution photographs [9]. Copyright 2023, Royal Society of Chemistry

Bimetallic metal colloids such as NiPt and AuCu particles can also be prepared. If the transition metal was present in the bimetallic colloids, it would be very difficult to obtain a stable colloid in metallic states. As illustrated by Kranz et al. [15], it was impossible to form uniform AuCu alloy particles unless specific post-treatment by laser was in use. The importance of the post-laser treatment can be revealed by the TEM images and elemental mappings in Fig. 2.5. No matter how, the synthesis methods with elaborate design determines the growth orientation of

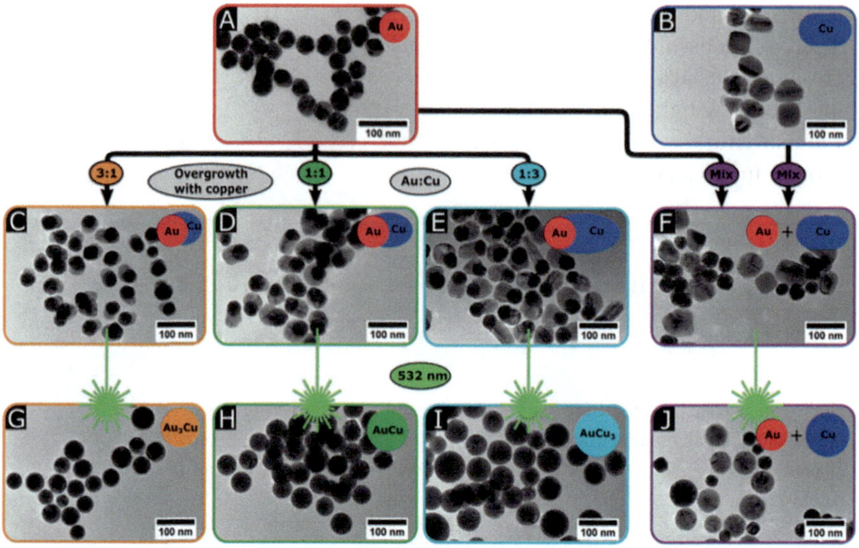

Fig. 2.5 **a–f** Laser-induced composition-controlled alloying of Au–Cu hybrid nanoparticles with different ratios of Au and Cu and comparison with a mixture of Au and Cu nanoparticles. **g–j** After the laser treatment, the hetero particles were transformed into spherical alloy nanoparticles [15]. Copyright 2023, with permission from Elsevier

different bimetal crystals. Silva and colleagues displayed mature technology for monodispersed bimetallic Ni_xPt_{1-x} colloids using 1,2-hexadecanediol in benzyl ether as the reducer in argon blanket [17]. The synthesis process was very sensitive on synthesis temperature, which determined the final morphology of NiPt particles with the core–shell structure, Janus structure, or atom-uniform alloy structure. They also confirmed the successful formation of Fe–Co–Ni–Pt–Ru multimetallic nanoalloys with high entropy using a similar reaction system. Recently, transition metal-based colloids have also been successfully synthesized including the Ni and Fe nanoparticles.

2.1.2 Metal Colloid as Catalyst Precursor for Heterogeneous Reactions

No matter how, the long-term stability of the metal colloid and the leaching of metal during a reaction may still cause a severe problem if used as a catalyst for the organic reaction. Herein, the metal-based catalyst using ex-situ synthesis by anchoring metal nanoparticles on selected solid supports makes the metal colloid a preferential precursor for the heterogeneous catalyst with easily controlled shapes and size distribution. Herein, one of the superiority using ex-situ colloid mobilization is the facility to study the size effect and supporting roles by controlling the size distribution of the particles at the target level [20]. In many works concerning supported precious metals, colloid synthesis is pre-synthesized as the catalyst precursors. After anchoring on the surface or pores of the supporting materials, the initial size distribution and shapes of colloidal particles can be maintained. These ideal model catalysts can be used for many organic reactions and even oxygen reduction reactions [21–23]. Supporting materials including TiO_2, SiO_2, carbon, Al_2O_3, and other reducible metal oxides (e.g., CeO_2, FeO_x, CuO) can be used to disperse the colloidal metal particles. Colloidal deposition on supports can be fulfilled by direct deposition, surface modification promoted deposition, ion exchange, flame spraying techniques, and Langmuir–Blodgett techniques [20, 24].

It is worthy to note that the usage of a protecting agent, e.g., organic compounds and polymers, may cause the coverage or deactivation of active metal sites during the reactions. Herein, surfactant removal to obtain a clean surface for catalysis through traditional approaches (such as solvent extraction and thermal decomposition) is necessary in most cases, and the increase of particle sizes during these post-treatments are frequently reported in literature. It even easily induces the sintering of metal nanoparticles, greatly hampering their use for novel catalysts. However, the supported metal particles become less stable if the protecting agent was entirely removed after thermal treatment. Research follows also engaged to understand and avoid the passive roles of these protecting agents. Zhan and coworkers announced the pre-annealing of the Au/TiO_2 by colloid immobilization in N_2 instead of direct calcination in air [25]. Octadece-9-nyl amine existed in the gold colloid as the organic capping agent/protecting agent, which easily blocked the activity sites and limited the mass diffusion of reactants. After annealing in N_2 (500–700 °C), the protecting agent can be carbonized by partially encapsulating the particles. The carbonaceous coatings had the great role to enhance the synergy between metal particles and the support, making the metal particles stably dispersing on the surface as illustrated in Fig. 2.6. The Ostwald ripening process can be thus suppressed in thermal treatment and reaction processes.

2.2 In Situ Self-Growth of Supported Metal Particles

For the in situ synthesis method of nanoparticles, inorganic metal salts or organometallic salts are used as the metal precursors, which will be anchored on the surface or into the pores of the supporting materials, either by physical/chemical adsorption, reduction of metal cations, or chemical reaction with supports. Metal nanoparticles are in situ deposited/grown on the support after the synthesis or a specific post-thermal treatment. The selected solvent, synthesis temperature, gas atmosphere (in air or inert gas), the metal precursor and the supports, are essential factors influencing the synthesis efficiency of the supported metal particles. Traditional synthesis method include the impregnation method (IMP), deposition–precipitation method (DP), sol–gel method (SG), chemical vapor deposition (CVD) method, co-precipitation method (CP), and other promoted methods.

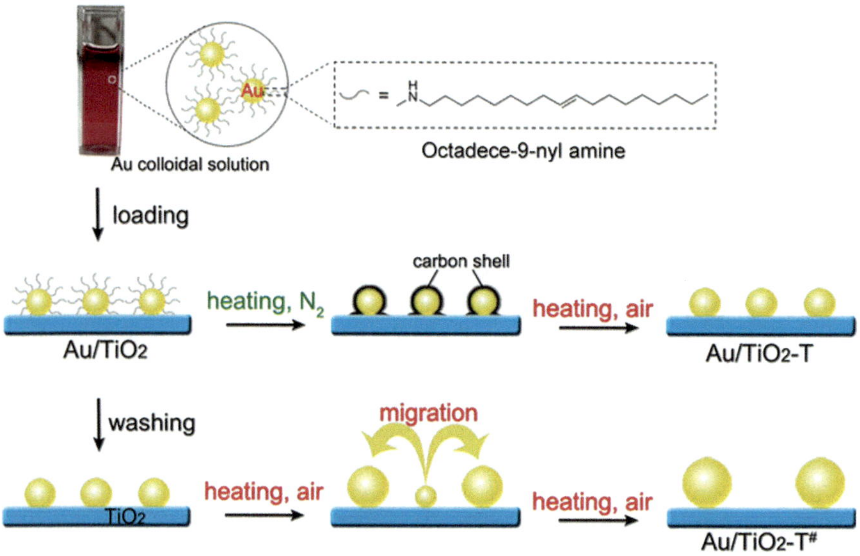

Fig. 2.6 Depiction of the in-situ construction of carbonaceous architectures from surface-bound surfactants to stabilize Au colloids on solids [25]. Copyright 2017, Wiley–VCH Verlag GmbH & Co. KGaA

2.2.1 *Impregnation Method*

The impregnation method can be divided into wet impregnation synthesis and incipient-wetness impregnation depending on the solution volume compared to the pore volume of the solid support. The incipient-wetness impregnation is carried out using the mother liquid with the similar volume to the pore volume of the support. The mother liquid can be thoroughly adsorbed by the pores of the support. In an ideal synthesis, the metal ions in the mother solution could be fully anchored in the porous structure of the support, and metal particles could be formed and crystallized after a subsequent calcination process. On the contrary, the wet impregnation does not restrict the volume of the mother solution of metal precursors. The volume of thw solution is obviously more than the pore volume of support, and the concentration of the solution can be controlled at a desired value. One of the superiority by using the impregnation method is the much-simplified synthesis steps compared to other preparation methods. It is also not necessary to consider the post-shaping of the catalyst, since a modeled solid support and an even shaped monolith can be directly used during the impregnation, which is a preferential strategy of the heterogeneous catalyst for many industrial production lines. It is also necessary to mention that the metal nanoparticles acting as the active sites for many reactions can be finely dispersed on the surface of the supports by impregnation. Considering the significant phase interface between the solid catalyst and the reactants, the utilization rate of the metal obviously increases under lowered raw cost.

No matter how, the efficiency of the impregnation depends on the property of the support and the interaction/attraction between the metal ions in solution. Taking the famous gold catalysts for example, gold nanoparticles can be dispersed very well on supports such as TiO_2, CuO, CeO_2, etc., if the impregnation method was used. The particles can be kept at small sizes and nice catalytic performances could be obtained for many oxidation reactions. However, if supporting materials such as silica and carbon materials were applied, it is impossible to get small gold particles with fine stability using the impregnation method. It is mainly caused by the different isoelectric point of the supports. Oxides including TiO_2, CeO_2 and MgO with higher isoelectric point could be positively charged in aqueous solution and easily adsorb the negatively charged $AuCl_4^-$. However, supports such as carbon and silica with a quite low isoelectric point display a negative-charged surface, which are repelled to the $AuCl_4^-$ ions during the synthesis and caused the failure loading of gold species. Herein, researchers try to replace the $HAuCl_4$ with Au precursors stabilized by the organic compound (e.g., $Au(en)_2Cl_3$, en denotes ethylenediamine) and successful reports of Au loaded on silica could be obtained [26].

Another problem using the impregnation method is the risks of the maintenance of the counterpart anions during further thermal treatment. Waste gas could be produced including NO_x. And the presence of typical anions, e.g., Cl^-, easily induced the variation and pollution of the active metal centers. The presence of Cl^- in the mother liquid become the main reason for the sintering of gold particles during calcination in air. Zhao et al. [27] suggested the variation of zeta potential of gold particles with a similar size controlled by the concentration of chloride ions that the negative charge of gold particles changed from −36 mV to −19 mV with the Cl^- concentration increasing from 1 to 20 mM.

2.2.2 Deposition–Precipitation and Co-precipitation

Deposition–precipitation and co-precipitation are two different strategies, both needing the presence of the precipitator during metal loading. These methods are appropriate strategies to synthesis of metal-based catalysts with multiple metal compositions. Solid powder of the supporting material is mixed with the metal salts in solution. Precipitators such as NaOH, $NaCO_3$, $NaHCO_3$, urea, and their mixture will be using to modify the pH value of the solution for the successful metal loading under suitable condition. The deposition–precipitation method is also modified with promotion for better loading efficiency and metal dispersion. In the previous work, research fellows changed the addition sequence of the support and the precipitators during the synthesis of TiO_2, Al_2O_3, ZrO_2 supported Au nanoparticles [28]. The gold precursors were changed according to the detailed pH value, from $[AuCl_x(OH)_{4-x}]^-$ (x = 0–4) to $[Au(OH)_4]^-$, which directly influenced the anchorage of the fine gold particles on the surface of the support [29].

Of cause, the selection of the precipitator also greatly modulated the deposition and growth of the metal particles [30]. It is suggested that fine particles were formed in Ni/Al_2O_3–OH if NaOH was used as the precipitator due to the very fast and uniform growth of particles in the solution. On the contrary, metal particles growing/depositing using Na_2CO_3 or NH_3 (Ni/Al_2O_3–CO_3 or Ni/Al_2O_3–NH) went through a much slow process with bigger particles due to uneven growth of metal seeds. Consequently, the surface states of these metal particles were also varied as a direct result of the metal-support coordination and the electronic structure according to the particle sizes, including the formation of surface $Ni^0/Ni^{\delta+}$ species, the weak acid sites, and the H_2 dissociation ability. All the above factors become the essential factors to regulate the kinetics of a selective hydrogenation process of nitriles, obeying the second-order reaction kinetics as displayed in Fig. 2.7. Deposition–precipitation can be applied for many precious metal-based catalysts and transition metal catalysts with clear function of each metal composition.

Co-precipitation method is also viewed as one of the most efficient synthesis methods for metal particles with high-stability, since strong metal-support interaction is easily formed by the uniform element distribution. Different from the deposition–precipitation method, active metal composites and the precursors/metal salts of supports are simultaneously added in the mother solution. Precipitators are subsequently added into the mother solution, the active metal particles, and the support (mostly metal oxide) can be in situ formed at the same time. In this case, the metal active phase can be very well formed with or capped by the surrounding support. Herein, the greatly enhanced metal dispersion and strong metal-support interaction are the advantages of using the co-precipitation method. No matter how, the catalytic activity of the metal particles by the co-precipitation may be decreased due to their insufficient exposure on the surface, although their stability can be very-well maintained.

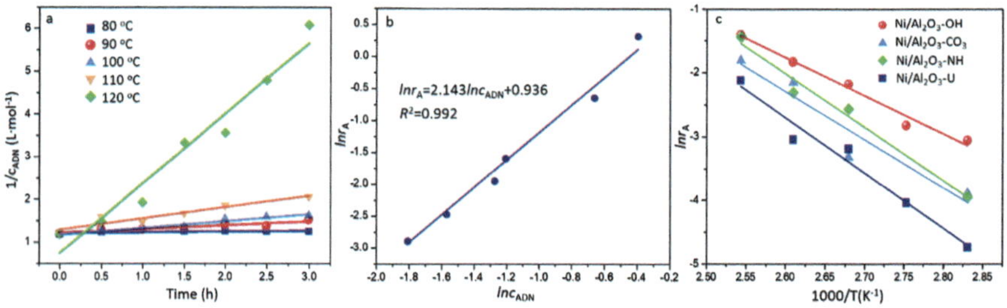

Fig. 2.7 **a** The linear relationship between $1/c_{ADN}$ and reaction time, according to second-order reaction kinetics. **b** The linear relationship between $\ln r_A$ and $\ln c_{ADN}$ at 90 °C of the adiponitrile (ADN) hydrogenation by Ni/Al_2O_3-OH. **c** The Arrhenius plot for the hydrogenation of ADN by different Ni/Al_2O_3 catalysts over a temperature range of 80–120 °C, 7.0 MPa H_2, 800 rpm for 1.5 h, molar ratio of ADN/Ni = 200, if not specified [30]. Copyright 2021, Royal Society of Chemistry

2.2.3 So-Gel Method

During the sol–gel method, compounds containing active metal species can be used as precursors by uniformly mixing the raw materials in the liquid phase. Hydrolysis and condensation chemical reactions happen during the synthesis to form a stable transparent sol system in the solution. The sol system slowly polymerizes during aging to form a three-dimensional network structure of the gel. The gel network is filled with the solvent that has lost liquidity. After aging for a certain time, the gel is dried, sintered and solidified to get the catalyst with metal particles dispersing in the sample.

Compared with other methods, the sol–gel method has many unique advantages. Since the raw materials used in the sol–gel method are first dispersed in the solvent to form a solution with low viscosity, the molecular level uniformity can be obtained in a very short time. It is also a facile method to get the even composition with precise quantity even if trace amounts of a specific element was added, achieving uniform doping at the molecular level. Compared with the solid phase reaction, the chemical reaction in the liquid phase will be easy to carry out, and only requires a lower synthesis temperature. It is generally believed that the diffusion of components in the sol–gel system is in the nano-meter range, while the diffusion of components in the solid phase reaction is in the micrometer range. Researchers suggested that the catalyst sample, e.g., Cu-based catalysts with binary and ternary compositions by the sol–gel method displayed much better catalytic conversion for biomass transformation [31, 32], probably due to the greatly improved uniform distribution of active metal particles as displayed in Fig. 2.8. If other methods such as deposition–precipitation or impregnation method was used, very clear phase segregation would be formed due to the incompatibility of different metal atoms.

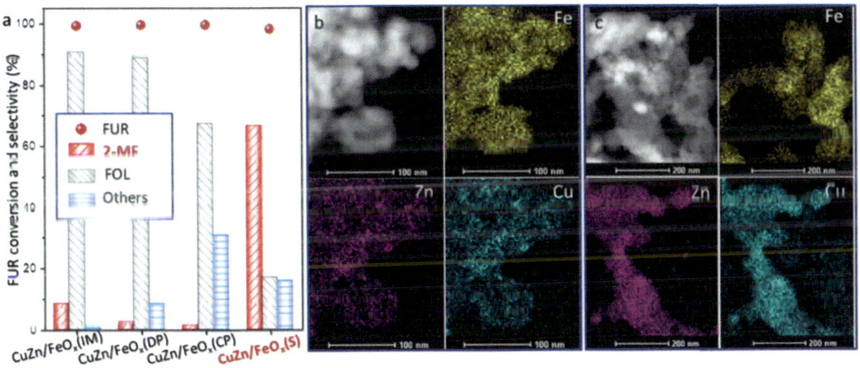

Fig. 2.8 **a** Catalytic performance of $CuZn/FeO_x$(S) by sol–gel method to transform furfural to 2-methylfuran. Comparison of the STEM elemental mappings by **b** $CuZn/FeO_x$(S) via sol–gel method with uniform ternary composition and **c** $CuZn/FeO_x$(DP) by deposition–precipitation method with clear CuZu-Fe phase segregation [31]. Copyright 2023, Royal Society of Chemistry

No matter how, the sol gel method also has some problems. The raw materials are usually organic compounds, which are expensive and harmful to human's health. Most of the sol–gel processes take a long time, several days or even weeks. It further increases the time cost for an industrial production. Besides, during the drying process, many gases and organic substances will escape and shrink. And micropores are not generated in the gel, which may be harmful for the transformation and diffusion of reactants during industrial applications.

2.2.4 Atomic Layer Deposition

Atomic layer deposition (ALD) technology is a promoted chemical vapor deposition (CVD) technology based on surface ordered self-saturation reactions. Atomic layer deposition was put forward in the 1960s and 1970s, first reported by former Soviet scientists Aleskovskii and Koltsov. However, due to the complex surface chemical processes, atomic layer deposition technology did not achieve significant development at the beginning. It was until the 1990s with the rise of the semiconductor industry that requires new development of component sizes, integration, and other aspects, the ALD technology became greatly developed.

With the progress of ALD, it is now an advanced strategy in the heterogeneous catalysis field to produce supported metal particles with high metal dispersion and controllable sizes. The principle is based on the alternating injection of two precursor vapors and the occurrence of "self-limiting" reactions under molecular level on the surface of the carrier. The precise deposition of the target metal particles on the carrier can be achieved [33].

Atomic-dispersed metal catalysts receive increasing attention due to their unique catalytic performances and maximum atomic utilization efficiency. ALD technology obtaining single atom catalysts from the gas phase poses challenges, as ALD growth generally occurs at high temperatures and metal aggregation significantly intensifies [34, 35]. But by precisely changing the number, order, and type of deposition cycles, control of the active sites on nanoparticles can also be obtained, providing researchers a new strategy for "bottom-up" precise and controllable synthesis [33, 36]. In recent years, the research group of Lu developed a new strategy for the synthesis of metal nanoparticles using ALD technology [37, 38], which involves one to two metal components with controllable sizes and metal sites in one particle as shown in Fig. 2.9. Research suggested that metal substrates promoted the nucleation and ALD growth of metal precursors under lower temperature, which would be another case and cannot be homogeneously grown on a metal oxide substance.

The precise synthesis of catalysts at the atomic level is crucial for elucidating the mechanisms of catalytic reaction and for designing advanced catalysts. The unique self-limiting characteristics of ALD can achieve uniform and controllable deposition of catalytic materials on high specific surface materials, achieving a step-by-step and "bottom-up" approach to constructing heterogeneous catalyst with complex structure at the atomic level. These ALD catalysts have high uniformity, which enables them

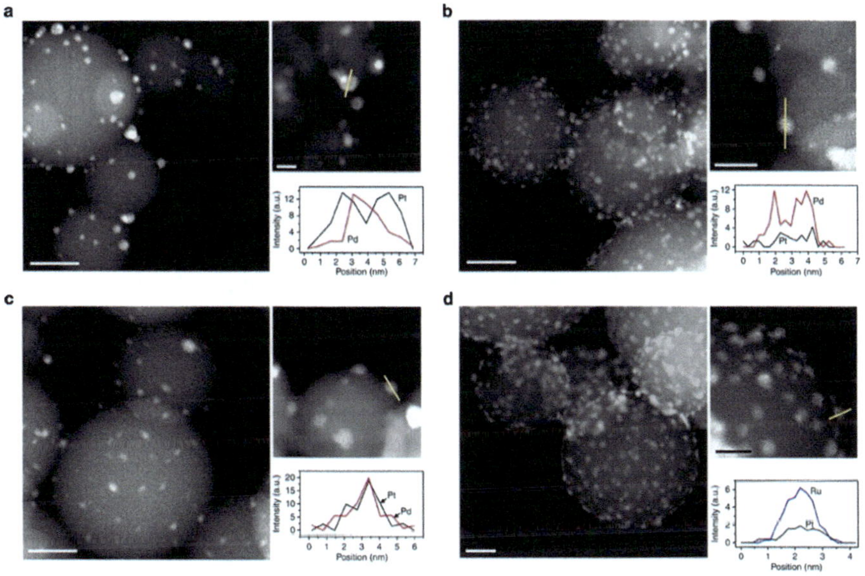

Fig. 2.9 Representative aberration-corrected HAADF-STEM images and corresponding EDS line profiles of **a** 5Pd-core 15Pt-shell, **b** 12Pt-core 20Pd-rich-shell, **c** 12Pt10Pd alloy and **d** 1Pt-core 35Ru-rich-shell bimetallic nanoparticles on spherical alumina support [38]. Copyright 2014, Nature Publishing Group

to have better or considerable catalytic performance compared to the ones prepared by traditional methods. They can also serve as model catalysts to help elucidate the structure-performance relationship of catalysts.

2.2.5 *Thermal Treatment After Synthesis*

Thermal treatments, including the calcination, thermal reduction, pyrolysis under different gas atmosphere, are essential for the growth of stable catalysts and are necessary in most synthesis. During such process, the metal active sites can be formed with specific surface compositions/chemical states. The texture structure of the supporting materials will be consolidated and become stabilized after the thermal treatment. If organic compounds were used during the synthesis, thermal treatment by e.g., air, oxidative gas or inert gas, could be helpful to eliminate the organic compounds or even surface carbonaceous layers, the presence of which would be harmful for the activity of metal nanoparticles. The temperature for a thermal treatment can be selected in a wide range of 250–800 °C and even higher temperature (e.g., for pyrolysis), and the time varies depending on the specific situation.

For most reaction systems using metal nanoparticles by traditional methods such as impregnation, deposition–precipitation, co-precipitation, calcination in gas atmosphere (air, O_2, N_2, Ar, etc. or their mixture) was indispensable to form the active metal particles with surface exposure. During the calcination process, the catalyst undergoes oxidation–reduction reactions with changes in physical and chemical states, which can remove residues, increase the surface area of the catalyst, and change the polarity or acid–base sites of the catalyst. In the oxidative conversion of many organic compounds, bulk metal oxides can be reactive. For the supported nanoparticles with small sizes, the maintenance of positively charged species is considered the key point in many catalyst systems. Supported precious metal catalyst can be viewed as a very interesting example with high efficiency and properties towards outstanding conversion of organic compounds. Supported gold nanoparticles are one of the most typical samples. Most research study suggested the necessary thermal calcination of gold catalyst to remove the Cl^-, NO_3^-, and organic coordinated species. The temperature is generally kept at 300 °C to avoid further aggregation of gold nanoparticles. Metallic Au^0 species is major with small amount of $Au^{\delta+}$ ($0 < \delta < 1$) species formed after calcination due to the specific reduction potential of gold. No matter how, the presence of the later species ($Au^{\delta+}$) facilitated the activation of oxygen species during oxidative conversion of the reactants.

For the hydrogenation reactions of organic compounds, metallic metal species become the essential active sites. Thermal-reduction will be helpful to reduce the metal ions and metal oxides in the catalyst precursors to their metallic states. During the thermal treatment, reduction under lower temperature may be not enough to get a satisfied reduction degree of metal particles, and deactivation of the catalyst easily happens during reaction. But if the reduction temperature was too strict, the phase segregation may simultaneously happen due to the different reduction potential of the existing metal species.

2.3 Essential Roles for Catalytic Performances

There are plenty of essential factors influencing the catalytic performances for the catalytic conversion of organic compounds, which can be divided into selective hydrogen and oxidation, polymerization, isomerization and so on. Herein, cases become complicated depending on the types of the reactions, the active metal centers, the reaction operating parameters, etc. But several significant factors can be viewed as the universal problems for a controllable high-efficient performance. In this part, we will discuss several key factors for the catalytic conversion of organic compounds, which include the size effect, metal-support interaction/synergy, the consolidated spatial and surface structure of active sites.

Size Effect

Size effect is a direct influencing factor for many heterogeneous reactions. The term is mainly used for the precious metal nanoparticles. Precious metals such as gold and platinum lose their metallic properties as their sizes decrease to the nano-meter scale. On the contrary, the catalytic performances of metal nanoparticles can be greatly regulated by the number of low-coordinated metal atoms as the active centers.

The size effect of metal nanoparticles has a significant impact on the catalytic activity and selectivity for different reactions. From a geometric perspective, low coordination atoms are gradually exposed on the surface as the size of metal particles decreases. The structure and proportion of active centers can be promoted in return. From the perspective of electronic structure, the electronic energy levels of metal particles also undergo significant changes due to quantum size effects, greatly affecting the orbital hybridization and charge transfer between the metal center and reactants. Due to the synchronous changes in the geometric and electronic structure by size variation of metal particles, it is very difficult to effectively distinguish the contribution of these structural effects on catalytic reaction activity, selectivity, and size dependence. Research fellows are also engaged into the understanding of the inherent nature of the size effect of metal nanoparticles during the recent years. Experimental and theoretical collaborations were carried out to reveal the modulating roles of geometric and electronic effects on catalytic reactions with size changes. A strategy to separate the geometric and electronic effects was creatively proposed using an "oxide selective encapsulation" strategy [39]. Atomic layer deposition (ALD) technology was used to synthesize the low and high coordination atoms of Pd particles selectively encapsulated by Al_2O_3 and FeO_x. The exposed ratio of the low/high coordinated Pd atoms can be divided for further investigation. It was suggested that on larger (>4 nm) and smaller (<4 nm) particles, geometric and electronic effects respectively controlled the catalytic performances, resulting in a "volcanic" trend in the catalytic selectivity and activity of reactions.

Size effect is another amazing study topic, especially for the precious metal particles. To reveal the essence of dynamics, research fellows used environmental electron transmission microscopy and computational chemistry simulations to reveal the dynamic interaction mechanism between the reactant and metal nanoparticles [40]. In a gold catalyst, large particles (>4 nm) only underwent surface reconstruction when interacting with carbon monoxide molecules, without damaging the overall structure of the particles. But the structure of particles was destroyed and became an amorphous dynamic structure, if smaller particles (~2 nm) was exposed in carbon monoxide.

Researchers studied the possible reconstruction of Pt on carbon nanofiber (CNF) as a function of the particle size as displayed in Fig. 2.10 [41]. The chemical attraction by the carbon edge atoms on surface causes the Pt atoms to detach from the Pt particle/cluster. Some of the Pt atoms migrate into the carbon interspace with the simultaneous formation of Pt-C bond. By comparing the situation of Pt atom distribution between small (Pt_{100}) and big (Pt_{400}) size model, the migration and dispersion of Pt atoms is extremely severe on smaller Pt. It subsequently resulted in the modulation of the geometric structure and electronic/coordination environment of Pt as a function of sizes.

Even transition metal particles have size effect although it may be a complicated subject. The decreasing of particle sizes may not be simply positive-correlated with

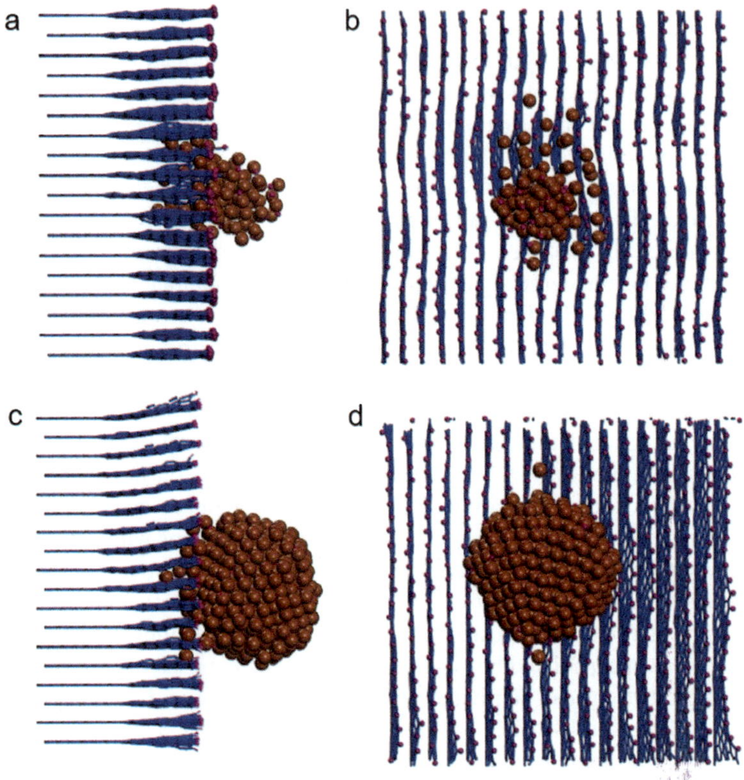

Fig. 2.10 Snapshots of the geometries of the Pt cluster interacting with CNF in equilibration state from lateral (**a, c**) and top (**b, d**) views. **a, b** is Pt_{100} cluster, and **c, d** is Pt_{500} cluster. Red and brown spheres represent H and Pt atoms, respectively, and the blue part represents graphene layers of CNF [41]. Copyright 2018, with permission from Elsevier

their catalytic performances for organic reactions. The famous Fishcher-Tropsch reactions are usually considered as structural sensitivity reactions [42]. Jong et al. [43] found that when the Co particle size increased from 2.6 nm to 16.0 nm, the turnover frequency (TOF) of the catalyst increased from 1.4×10^{-3} s^{-1} to 23.0×10^{-3} s^{-1}, and the C_{5+} selectivity increased from 51 to 85%. When the active phase Co particle size was greater than 8 nm, there was no significant change in the catalytic activity and product selectivity, indicating the size effect between the conversion rate, product selectivity and the Co particle. As the Co grain size is approximately 8–10 nm, the catalyst generally exhibits the highest intrinsic activity. Other research work based on Ni catalysts also claimed the size effect of Ni nanoparticles on the selective hydrogenation reactions of nitriles displaying a volcanic relationship between the Ni sizes and diamine yields [30].

In general, the smaller the particle size of a catalyst, the higher its catalytic activity. This is because small-sized catalysts have more active sites and larger surface areas, which can adsorb reactant molecules and promote the catalytic reaction more effectively. In addition, small-sized particles also have higher surface energy and greater curvature, which improves the reaction rate and selectivity. However, a catalyst particle size that is too small can cause the sites to be too dense, preventing reactants from entering the sites, thereby affecting the reaction rate and the specific selectivity. Therefore, in a catalyst design, it is necessary to select appropriate particle sizes based on the needs of a reaction.

Metal-Support Interaction/Synergy

The interaction between the metal and the carrier is another key factor for the charge transfer between the metal and the carrier. The supporting materials in catalysts not only play a role in dispersing and stabilizing metal nanoparticles, but also interact with the metal particles, which often leads to interfacial charge transfer, metal structure changes, molecular adsorption modulation, and other phenomena, thereby affecting the activity, selectivity, and stability of the catalyst. There are usually three types of interactions between metals and carriers: (1) Metal particles dispersed neighboring to the support with a well-defined metal-support interface; (2) Dispersed metal atoms on particles melting into the lattices of support or partially forming mixed oxides with the carrier; (3) the surface of metal particles entirely or partially coated by the support forming partially encapsulated layer or even the core–shell structure [44]. It is precisely due to the interaction between the metal and support that the electronic fluid transfers between different composites resulting in charge displacement. The formation of strong metal-support interaction regulates the catalytic performances during an organic reaction process.

The definition of metal-support interaction (MSI) requires a significant charge transfer between the metal and carrier, leading to reversible structural changes at nanoscale. In the MSI phenomenon, both metals and carriers are very important. In recent years, researchers have conducted in-depth research on metal carrier interactions based on different types of metals and carriers, which directly encourage the development of outstanding metal catalysts.

A prominent example of metal carrier interaction is the loss of chemically adsorbed hydrogen on titanium supported platinum (Pt/TiO_2) during high-temperature reduction treatment (HTR). Tauster et al. [45] first discovered that Pt could be reduced through a series of high-temperature oxidation and subsequent low-temperature reduction steps. Initially, the inhibition of chemical adsorption was attributed to electronic perturbations in the system, namely the bonding between Pt and Ti in a reducing atmosphere. Subsequent studies have shown that in situ reduction activation leads to Pt NPs being encapsulated in a thin, partially reduced TiO_2 layer [46, 47]. The driving force of this so-called strong metal-support interaction (SMSI) state is attributed to the minimization of surface energy.

The formation of SMSI can also be modified via the activation of support by introducing hetero-atoms, e.g., P species (Fig. 2.11), which leads to the simultaneous formation of metal phosphide nanoparticles (NPs) and the core–shell nanostructures by migration of support onto particle surface [48]. The encapsulation state and charge transfer of metal phosphides are similar to classical SMSIs, and can be directional regulated via synthesis and post-treatment. The $Co_2P_L/LaPO_4$ catalyst displayed excellent catalytic activity and stability even after prolonged exposure to air or in high concentrations of acidic substrates. Multiple platform compounds such as maleic anhydride, maleic acid, fumaric acid, citric acid, levulinic acid, vanillin, cinnamaldehyde, glucose, xylose, and sucrose can be completely hydrogenated into corresponding value-added fine chemicals compared to commercial Pt/C and Pd/C catalysts. This SMPSI structure can be extended to other systems, such as $Ni_2P_L/Mn_3(PO_4)_2$, $Co_2P_L/LaPO_4$, and $CoP_L/CePO_4$ catalyst systems.

Similar encapsulation effects have also been reported for Co, Ni, Au, and Cu NPs. Recent studies have shown that high-temperature oxidation treatment can also lead to the encapsulation of metal nanoparticles. The encapsulation caused by SMSI restrained the nanoparticles from the coalescence and Ostwald maturation. In addition, the selectivity of the catalyst can be adjusted by changing the adsorption strength of molecules on the catalyst surface and blocking specific active sites [49]. Therefore, the development of synergistic effects between metal nanoparticles and their carriers under controllable conditions is of great significance for developing the catalysts and processes.

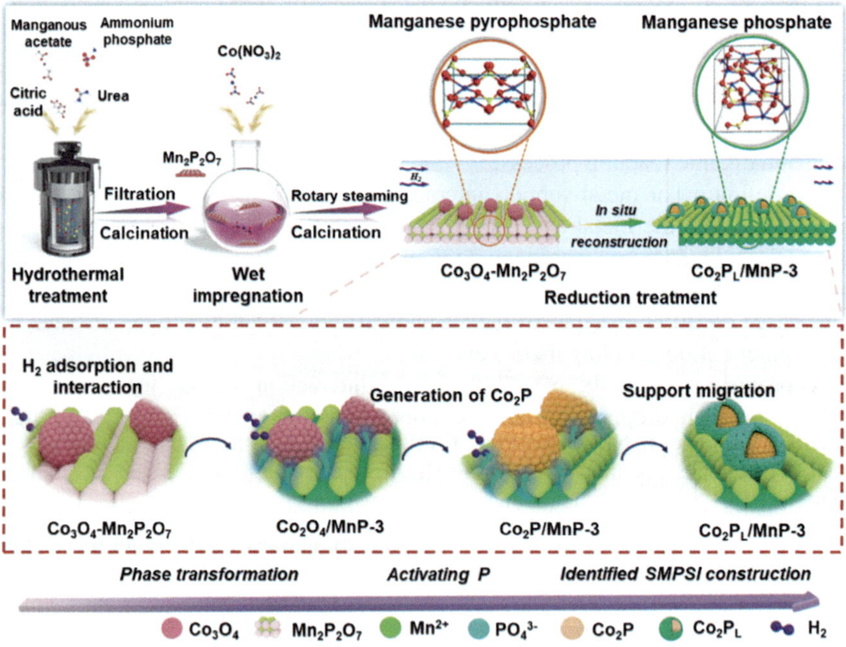

Fig. 2.11 Synthesis route of Co_2P_L/MnP-3 catalyst and formation process of SMPSI construction [48]. Copyright 2021, Wiley–VCH Verlag GmbH & Co. KGaA

Consolidated Spatial and Surface Structure

The transport of reactants on catalysts is one of the most important processes in chemical reactions, which mainly includes the mass transfer and diffusion. Catalysts promote the reactions by modifying the transfer process via different adsorption modes and reaction mechanisms. The major factors correlated with (1) a better transfer process via the advanced porous structure, (2) well-dispersed metal sites in support with designed spatial structure, (3) modified surface with hetero-atoms/functionalities. All the above factors can be ascribed to the consolidation of a well-designed spatial and surface structure of the support and even the entire catalyst.

The porous structure of the support directly affects the adsorption of the reactants and intermediates. The detective surface area of the catalyst is determined by the internal surface area, which mainly depends on the real texture structure of the support. Active centers are mostly distributed on the internal surface that are usually formed on the defects and vacancies of the inner surface. Herein, the organic molecules (reactants and intermediates) diffuse through the pores to reach the active centers, where activation, adsorption/dissociation of organic molecules happen and trigger the reaction.

In general, intermolecular collisions induced great diffusion resistance if the pore size of the catalyst was larger than 10 μm, where the influence of pore the structure can be ignored. But if the pore size was less than 2 nm, the collision of the gas molecules with the pores becomes intensive than that between organic molecules. Diffusion resistance is quite high under such circumstance (D_{pore} < 2 nm), and the collision between molecules can be ignored. If the pore size is between 2 and 10 nm, the collision of gas molecules with either the pore wall or with organic molecules cannot be ignored, since the diffusion flux and pore size exhibit a non-linear relationship. For most high-efficient catalysts during the organic conversion reactions, mesoporous structure is reported to be advantageous for the mass transfer and diffusion. Although the synthesis of the catalyst with rich micropores easily results in a catalyst with huge surface area, the micropores (<2 nm) were too small for the diffusion of many organic molecules.

Due to the presence of certain defects in single type of pores, research fellows are committed to synthesize supports with the combining superiority of different types of pores, namely hierarchical porous materials or multi-level porous materials. It refers to a composite with two or more types of pore structures, or composite including two or more different sizes of the same level of pores. Ordered porous structure with different pore sizes can thereby be constructed in the hierarchical structure (from micropores to macropores, e.g., hierarchical structure consisted by mesopores and macropores). Catalysts with hierarchical porous structure were reported with great application prospects due to their high specific surface area, high pore volume, adjustable pore size distribution, and excellent processing performance [50]. It facilitated to improve the material transfer efficiency during the catalytic process. The presence of micropores provides a large surface area to enhance charge storage capacity, while mesoporous and macroporous structures improve the electrolyte permeability and ion diffusion.

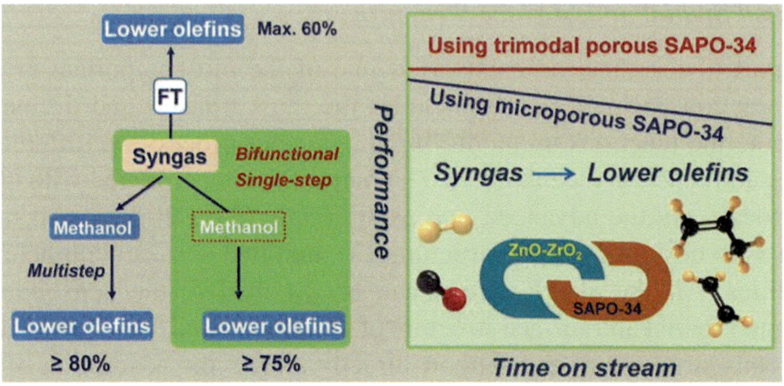

Fig. 2.12 Hierarchical structure in a well-designed ZnO-ZrO_2/SAPO-34 system prevents the formation of polyaromatics, which resulted in the improved catalytic performance for syngas-to-olefins reactions [51]. Copyright 2021, with permission from Elsevier

Taking the Fischer–Tropsch synthesis as an example, the growth of C–C chains on the catalyst surface follows the polymerization mechanism. The selectivity of synthesizing low-carbon olefins from syngas is generally less than 60%. The bifunctional catalyst obtained by integrating the hierarchical ZnO–ZrO_2/SAPO-34 offers a C_2–C_4^- selectivity of 80% at 25% CO conversion in syngas conversion. Figure 2.12 shows the ZnO–ZrO_2/SAPO-34 molecular sieve successfully synthesized by Wang et al. [51], which displayed a multi-level pore structure with micropores, mesopores, and macropores. The density and strength of Brönsted acid of the molecular sieve are similar independent of the multi-level porous structure. Introducing a multi-level porous structure in the Fischer–Tropsch catalyst contributes to the reduction of carbon deposition rate, thereby significantly improving the stability of the catalyst.

The second strategy is to design well-dispersed metal sites under pore confinement by a specific support structure. The definition of the confinement effect of molecular sieves was first discussed by the Belgian scientist with formulaic definition. The van der Waals adsorption energy (W) of spherical molecules with radius of d confined by micropores with pore size of a can be estimated by the following Eq. (2.2):

$$W = \frac{-C}{4d^3\left(1 - \frac{s}{2}\right)^3} \tag{2.2}$$

In the formula, C is a constant, $s = d/a$, two limiting cases are the $s = 0$ (plane) or $s = 1$ (pore diameter is equal to the molecular radius). From the above equation, it is easily obtained that $W_{s=1} = 8W_{s=0}$, meaning that the adsorption energy originating from the van der Waals confinement by pores is 8 times of the adsorption energy on flat plane. Confinement effect caused by the porous and specific spatial structure of support has attracted tremendous attention, especially after the development of highly structured mesoporous molecular sieves such as MCM-41 and SBA-15, as well as carbon nanotubes, metal organic framework materials (MOFs), covalent organic framework materials (COFs), etc. It affects the diffusion/adsorption of reactants, the electronic effect on the metal sites and the entire catalyst, and the catalytic kinetics as well.

Numerous studies have shown that mass transfer happens between spreading molecules and surrounding molecules through random collisions in unrestricted systems. In a confined structure, the spreading molecules also interact with the inner wall of pores, thereby affecting their diffusion behavior. Research fellows studied the diffusion of adsorbate molecules in the molecular sieve and found that the atoms commonly diffuse in a peristaltic form. However, a collective migration of atoms occurs if the large number of atoms suddenly swarm into small pores at the same time.

The adsorption behavior of molecules is also influenced by the confinement effect of support structure, which is a surface phenomenon based on diffusion and a prerequisite for the activation of reactant molecules. As a result, the electronic environment and the reactant concentration inside and outside the structure are totally distinctive. No matter how, a satisfied catalytic activity requires moderate adsorption to achieve faster adsorption and desorption rates. The effectiveness of a catalyst depends on the accessibility of the reactants to the active center, the electron transfer and geometric matching between the catalyst surface and the reaction species. Therefore, factors affecting the accessibility of active centers and mass transfer during catalytic conversion can be used to regulate the reaction performances of catalysts. The confinement effect displays an important role on the diffusion and adsorption in catalytic reactions. The current "*Nano-scale Confined Catalysis*" put forward by Bao et al. successfully modulated the electronic energy states of many catalytic systems with precise regulation of active centers by using the nanoscale space and interface confinement effect [52].

In a recent work, they found an induction period in the catalytic conversion of synthesis gas using oxide-zeolites (OXZEO), where the dynamic confinement of molecular sieves regulates the selectivity of the product [52]. The bi-functional OXZEO can efficiently proceed the conversion of synthetic gas for the selective preparation of high value-added chemicals and fuels such as low-carbon olefins, aromatics, gasoline, etc. The confinement effect of porous structure by molecular sieves plays an important role in modulating the selectivity of products.

The confinement effect can also be used to solve the stability problem of non-noble metal catalysts. Construction of 2D cover encapsulating non-noble metal catalysts (chainmail catalysts) has been proposed as a promising strategy in recent years by Deng et al. [53, 54]. In a typical chainmail catalyst, graphene layer was used to cover the surface of metal particles. It boosts the electron penetration from encapsulated nanometals to graphene as displayed in Fig. 2.13, while the outer cover completely protects the encapsulated metals from damage under harsh reactions. This unique strategy can promote both the catalytic activity and stability of inner metal particles, which was described as "chainmail for catalyst." Encouragingly, previous research has confirmed that electron transfer between the inner metals and graphene layers affected the catalytic reactions. Other 2D materials including the hBN and g-C_3N_4 can also be adopted to encapsulate the metal particles. Benefiting from the advantages

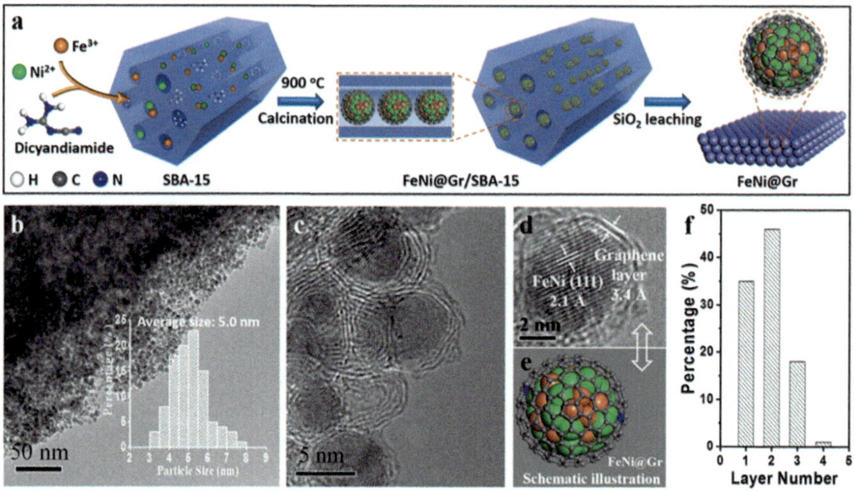

Fig. 2.13 Synthetic route, morphology and structural characterizations of FeNi@Gr. **a** Schematic illustration of the synthetic route of FeNi@Gr. **b** TEM image of FeNi@Gr-900. **c**, **d** HRTEM images of FeNi@Gr-900, showing the encapsulated FeNi nanoparticles within graphene layers. **e** Schematic illustration of the FeNi@Gr structure shown in (**d**). **f** Statistical analysis of the layer number of the graphene layers encapsulating the FeNi nanoparticles [53]. Copyright 2021, with permission from Elsevier

of 2D cover encapsulating metal, the chainmail catalysts have been extended into many important catalytic processes with satisfied thermal stability and reactivity.

Operating factors influencing the catalytic behavior of supported metal nanoparticles may include the initial chemical/property of supporting materials, the synthesis method, the detailed condition during thermal treatment. No matter how, it induces the variation or regulation of the active metal sites via the above-mentioned structure change directly or indirectly.

2.3.1 Characterization

Characterizations are taken to understand the formation of metal particles in supported catalysts, which would mainly include the crystal structure, surface composition, chemical states of different metal species, real percentages of different elements, the porous structure, sample morphology with precise spatial and geometric structure of the metal sites.

2.3.2 Chemical Composition and Surface States

The chemical composition and surface states/valences are the most fundamental information to study the supported metal particles. Characterization such as X-ray Diffraction (XRD) pattern gives the information of crystal phase/structure of the bulk catalyst. In the sample with clear phases, the positions and intensities of diffraction peaks can be calculated if the "crystal structure" (spatial group, atomic position, bond length, bond angle, etc.) was already known and depicted by a software. Surface composition including the metal surface ratio, chemical valances of metal species, and electronic interaction between the metal-support can be reflected by X-ray Photoelectron Spectroscopy (XPS) with deconvolution of peaks. Current analysis software provides semi-quantitative results to reveal the surface and crystal structure of samples in a more precise path using both XRD and XPS. No matter how, a credible conclusion can be made after the combining analysis by many basic and fine characterization techniques. Different from the XPS results displaying the surface information, the X-ray Absorption Fine Structure (XAFS) is used as a very specific characterization to understand the chemical environment and valences of metal species in a sample. In some special cases, the results of XAFS and XPS could be consistently described, e.g., if the fine metal particles/clusters were highly dispersed on the surface of catalyst. For metallic elements, the full spectrum of the absorption spectrum includes XANES (near edge absorption structure) and EXAFS (extended edge absorption structure) which provide the chemical valences and coordination information. Similar as the valence states of XPS determined by the binding energy position, all the analyzed results require comparison by the standard data. But there is no universal source for the analysis of absorption spectra, the detection of a metal foil and crystals of metal salt/oxide under the same operation conditions can be used as the standard data for analyzing the detected samples.

XAFS is one of the most powerful tools to describe the localized structure of the detected samples. Before the operation, the X-ray energy was adjusted to be consistent with the inner electron layer of the studied element. The function relationship between the number of absorbed X-rays and their energy can be monitored. If the accuracy of the instrument is precise enough, the spectrum could exhibit small oscillations due to the change of adsorption behavior influenced by the localized environment. Based on the analysis of the spectrum, precise information can be obtained, which includes the distance between the absorbing atoms and neighboring atoms, the number and type of these atoms, and the oxidation state of the absorbing elements. All the above-mentioned parameters reflected the localized structure of the detected sample. Such information of any target element in a sample or metal particles can be revealed by directly selecting the energy of the X-rays. EXAFS has the following characteristics:

(1) EXAFS depends on the short-range ordering instead of the long-range ordering in a sample, making it suitable to measure a wide range of samples. It can be used for amorphous, liquid, molten, active (metal) centers of catalysts, metalloproteins, and impurity/hetero-atoms in crystals.

(2) The X-ray absorption edge reflects the elemental characteristics. If atoms of different elements Co existed in the sample, the atomic neighboring structure of different elements in the same sample can be studied by adjusting the incident X-ray energy.
(3) The fluorescence method can be used to measure samples consisting by a specific element with concentrations as low as one millionth.
(4) The surface structure and the bond angles of atoms in oriented samples can be measured by polarized X-rays.
(5) Preparation of samples is relatively simple and single crystal is not necessarily synthesized. The operation period for data collection can be short that several minutes would be enough using a synchrotron radiation X-ray source to get a spectral line.

Experienced researchers easily obtain the information on the valence state and coordination structure by analyzing the XAFS profiles, which mainly include the element valence states via X-ray absorption near edge structures (XANES) and the coordination numbers and atomic spacing via extended X-ray absorption fine structures (EXAFS). In recent years, supported metal particles have been widely used in the transformation of organic compounds for energy conversion and storage. It becomes the key factors of the XAFS for exploring a high-efficient catalyst with well-understanding reactive center in terms of the valence states, coordination structures, and electronic structures of the metal atoms on particles. Wavelet transform can also be illustrated to represent three-dimensional information. Different colors on the image represent the height of peaks. The distance (i.e. bond length) of coordination atoms and the types of coordination atoms (qualitatively, the larger the atomic number, the more right the peak position) can be visualized by a two-dimensional image. But this technique is basically consistent with the R-space.

Taking the k-edge XANES spectrum of Ni for example, the white-line peak appears at about 8348 eV, as shown in Fig. 2.14. The EXAFS spectra of Ni after annealing and subsequent acid treatment give clear information on the oxidation state of Ni, which is between 0 and 2 + valence. In the EXAFS spectrum converted by Fourier transform, two protruding peaks at 1.4 Å and 1.8 Å suggest the coordination of Ni-C or Ni–N. The absence of Ni–Ni coordination peak in the first shell layer at 2.2~2.3 Å indicated the formation of small Ni particles, which was eliminated after the sample was leached by acid (denoted as "leached" in Fig. 2.14). Herein, the structure information including the chemical states and coordination states of the Ni active center could be understood by analyzing the XAFS profiles.

No matter how, the XAFS technology may be more favorable to precisely confirm the formation of single atom catalysts, which is currently a hot topic in the field of catalysis and will be discussed in another section of the book. The results of XAFS should also be discussed together with the surface information given by XPS deconvolution results and crystal phases by XRD patterns as well as other fundamental characterizations for a global prove of an assumption.

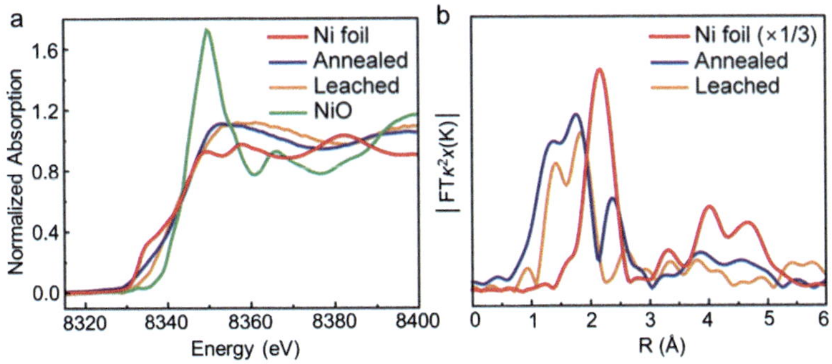

Fig. 2.14 **a** Ni K-edge XANES of the annealed and the leached residue carbon in comparison to Ni metal and NiO references, **b** FT-EXAFS spectra in the R space [55]. Copyright 2023, Royal Society of Chemistry

2.3.3 *Morphology and Spatial Structure of Active Sites*

Scanning electronic microscopy (SEM) and Transmission electronic microscopy (TEM) are practical technology to show the general morphology and surface distribution of metal particles or even element signals directly by a magnified vision. It is commonly interpreted in many foundational books discussing the characterizations and analysis of materials. Herein, the most used electronic microscopy is not the objective topic in this section. We would like to discuss the most essential development of current TEM techniques by either Aberration-corrected high-resolution images or in situ experiments, which provides the intrinsic evidences to disentangle the specific formation of metal sites in a visualized way, including e.g., the arrangement of metal atoms at the metal-support interface, metal nanoparticles with specific spatial/geometric structure, and even the defective particles with rich meta-stable but super active low-coordinated metal sites.

Spherical Aberration Corrected Transmission Electron Microscope (AC-TEM) uses a spherical aberration correction device as a decelerating lens to correct the spherical aberration occurring in an electron beam or an ion beam. It is an analytical instrument for the research fields including the materials science, physics, basic medicine, and chemical engineering. The biggest advantage of AC-TEM is the severely increased resolution after the reduction of aberration. Different from the lower resolution of traditional TEM at only nanoscale or subnano-meter level, the resolution of AC-TEM reaches the angstrom level or even sub angstrom level (The reported maximum resolution of current AC-TEM is even higher than 0.06 nm). The increase in resolution makes the accurate understanding of the sample structure more conceivable.

One of the most practical operation modes of AC-TEM is the TEM mode in high-resolution. The main factor affecting the imaging resolution is the spherical aberration of the objective lens, if the TEM mode was in use. The corrector is generally installed at the position of the objective lens in most cases. Of course, there are also cases that two correctors are installed on one AC-TEM (both objective lens and condenser lens), which is known as double spherical aberration corrected TEM. Qualitative analysis of the chemical composition of both metal nanoparticles and carriers can be carried out through high-resolution AC-TEM images, where clear lattice phases and diffraction images can be observed for a precise crystal phase. The contrast during TEM observation determines the image quality, which depends on the number of electrons participating during imaging. The existence of more electrons results in more intensive scattering, less transmitted electrons, and a darker image. The contrast of the sample depends on the thickness/density of the sample therefore, as well as the atomic number by different element in the sample, with the former more significantly impacting the contrast. An outstanding image with anticipative assumption usually depends on the detailed composition of the tested materials. For example, if precious metal nanoparticles on materials composing of light element (e.g., carbon & SiO_2), perfect images with strong contrast between different composition can be obtained under AC-TEM, as shown in Fig. 2.15a, b. But if metal species such as Au, Pd and Ce co-existed in a catalyst, it brings problems to recognize each component directly by the contrast. The measurement of lattice fringes would be helpful to identify each species, which is however a huge workload considering the objectivity of TEM observation. No matter how, in situ AC-TEM experiments and environmental TEM images are conducted under the TEM mode in many cases.

When we use the STEM mode, the electron beam is converged through the condenser lens to image the sample. The spherical aberration of the condenser lens is the main reason to result in the low resolution. Therefore, the spherical aberration correction device will be installed at the position of the condenser lens if the STEM mode is in use, namely AC-STEM.

The images collected in STEM mode can be proceeded in terms of either light-filed or dark-field probes, which can be divided into High-angle annular dark-filed (HADDF), LADDF (Low-angle), Bright field (BF), Annular bright field (ABF, JEOL). The images display clear atomic structure, where the contrast of the image is directly related to the atomic number. For a simple system with bimetallic components, a STEM image with good contrast can be obtained to show the detailed structure/special location of compositions, if a certain difference in the atomic numbers of the metal and oxide carriers existed. Experienced operators can semi-qualitatively determine the simple components by comparing light and dark contrast, making the data analysis fast and simple. No matter how, if the sample is consisted by multiple metal compositions with thickness, the lattice fringes observed in the STEM mode are not recommended for the qualitative analysis, since multiple layers of different lattice planes can be observed due to the nice contrast under STEM mode and easily caused the overlapping of lattices by different planes. The STEM mode will be a perfect strategy to understand single-atoms/clusters dispersed on the surface of support (Fig. 2.15c, d).

The Energy Dispersive X-ray Spectroscopy (EDS) and Electron Energy Loss Spectroscopy (EELS) can be operated under the STEM mode. The former mainly focuses on semi-quantitative analysis of metal elements (heavy elements), while the latter analyzes light elements. The line scanning or/and surface scanning can be used

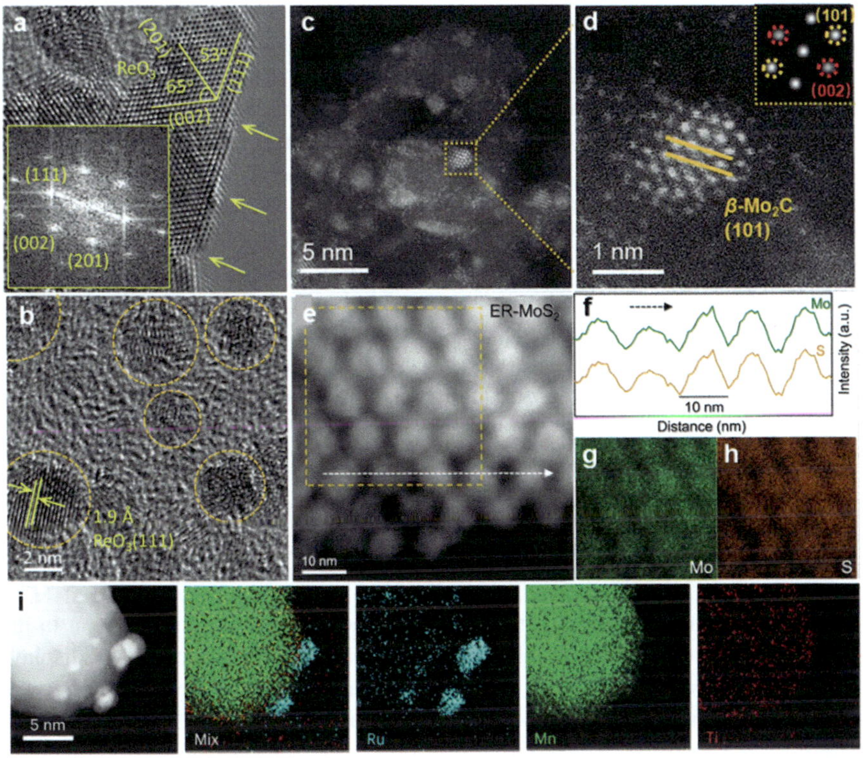

Fig. 2.15 **a, b** AC-TEM images of Re/C with clearly displayed lattice fringes and surface steps [56]. Copy right 2020, with permission from Elsevier; **c, d** AC-STEM of β-Mo_2C with dispersed single atoms on the fine particles [57]. Copy right 2022, with permission from Elsevier; **e–h** HAADF-STEM elemental line scan and mappings of MoS_2 particles [58]. Copy right 2023, Nature; **i** HAADF-STEM elemental line mappings of Ru particles supported on Ti modified MnO_x [59]. Copy right 2023, Nature

to analyze the location of different compositions and the spatial position/structure of metal nanoparticles in samples.

For example, by analyzing the trend of signal lines of elements, the position/composition of different elements can be preliminary inferred. For metal nanoparticles with A@BO_x core–shell structure, the signals of A and B metal elements in the line scan should be differently separated from each other, with the A signal maximized at the center of the particle and the B signals reaching its maximum on particle sides. Only weak B signal can be observed at the center of the spherical particle. If such sample is also scanned to show the EDS elemental mappings, it can be found that the element signals of A and B in the colorful maps are roughly at the same position, and the signal peak area of B will be larger than that of A as an outer layer surrounding A element. However, this situation can only be used as evidence in terms of high-resolution STEM images, with the resolution generally above 1.4 Å. When a similar signals of A and B appear in the line scanning, it indicates that the two elements are mixed in a certain state to form a bi-element particle, either an ordered alloy or a disordered binary element mixture (Fig. 2.15e, i). However, it is impossible to get precise information or to distinguish the real bimetallic structure under common TEM without spherical aberration correction. AC-STEM can give quite accurate information to distinguish the specific morphology of the bimetallic mixture.

No matter how, the structure validation of metal nanoparticles must be based on objective and extensive observation by abundant images from different viewing zones, as both TEM and STEM modes are localized characterization. Therefore, a single TEM/STEM image is not sufficient to provide direct conclusions. It needs to be combining analyzed with multiple characterizations including surface information (XPS, XAFS), spectra (FTRIR, Uv–vis), crystal structure (XRD), and many other required techniques for more accurate information and conclusions.

2.3.4 *In Situ FTIR Spectra for Studying Kinetics and Mechanism*

In situ infrared technology (In situ FTIR) has received increasing attention in the fields of energy, environment, catalysis, and other fields. Such technology displays unparalleled advantages to study the microstructure and reactivity of materials, as well as to explore the reaction mechanisms. Diffuse reflection method is the fundamental of in situ FTIR. Reflection, absorption, scattering, and transmission occur as the infrared light falls on the rough surface of a sample. The diffuse reflection is collected and delivered to the spectrometer detector to generate a diffuse reflectance infrared spectrum.

Diffuse reflectance infrared spectroscopy is a research method based on absorption and scattering, particularly suitable for the determination of surface structure and adsorbed species on solid powder. The experimental system of in situ FTIR generally consists of diffuse reflection attachment, in situ cell, vacuum system, gas source, purification and pressure device, heating and temperature control device, and FTIR spectrometer. The system is an ideal analysis method without destroying the sample structure and tableting is not necessary.

In situ FTIR spectra give rich and precise information on molecules. In situ FTIR has been widely used to study the surface properties, among which the most effective and widely used is the so-called "probe molecules" adsorbed on the catalyst surface, such as NO, CO, CO_2, NH_3, C_3H_5N, etc. Infrared spectroscopy provides information on the "active centers" and adsorbed species on surface, especially under in situ reaction conditions. Therefore, it is very important to reveal the catalytic reaction mechanism. Taking the in situ CO-FTIR for example, the experiment will be carried out in the presence of CO balanced with inert gas (e.g., 5–10% CO/Ar). Due to its excellent coordination properties with metal, CO becomes the commonly used probe molecule in characterizing supported metal particles.

The adsorption behavior of CO on the catalyst depends on the chemical structure of the metal active centers. Electronic environments around the active centers also induces the changes of FTIR spectra including the band position, number, intensity, etc. [61, 62]. Herein, different types of metal active centers can be distinguished, and the reaction mechanism can be inferred according to the detailed results of CO-FTIR spectra. The adsorption peaks generally appear in the 2250–1650 cm^{-1}. The coordination states of precious metals can be observed, e.g., the well-coordinated ($N_{Pt\text{-}Pt}$ ~ 9)/under-coordinated Pt sites (UC, $N_{Pt\text{-}Pt} \geq 7$), and high-under coordinated Pt sites (H-UC, $N_{Pt\text{-}Pt} \leq 6$) at 2080–2060 cm^{-1} and 2055–2000 cm^{-1} (Fig. 2.16a) [60]. The later ones are proved to be easily enriched on the surface of smaller Pt particles. The experimental temperature also influences the in situ CO oxidation behavior by FTIR, reflecting the catalytic performances to activate the C=O bond. As displayed by Fig. 2.16, the preferential activation/consumption of adsorbed CO in the presence of O_2 pulse could be revealed by the peak intensity of the Pt UC sites and Pt H-UC sites as a function of operating temperature. With the increasing temperature, the adsorption peak on reactive Pt centers shifts from the H-UC Pt sites to UC Pt sites, depending on the ignition behavior of the catalyst.

Even in a supported catalyst with bimetallic active centers, e.g., Au-Pt, Au–Pd, Au-Cu systems, the formation and detailed structure of nanoparticles can be revealed by in situ FTIR spectra in the presence of CO adsorption. There would be distinguishable peak shifts of CO adsorption mode on different metal sites. Therefore, the initial reference spectra of the mono-metallic particles should be taken for comparison. Taking Au-Cu bimetallic particles for instance, the CO adsorbed on metal step sites shifts from 2073 to lower than 2058 cm^{-1}, which can be divided into the adsorption of CO molecules on different step sites of Au and Cu (unsaturated metal coordination sites) as displayed in Fig. 2.17 [63]. The formation of either CuO nanoclusters oppositely supported on Au particles (CuO*/Au) or a typical Au@CuO core–shell structure can be analyzed according to the competition adsorption mode of CO molecules on either Au or Cu step sites.

The precise information of the catalytic mechanism can be obtained based on the analysis of in situ FTIR in the presence of reaction atmosphere, either in gas phase or liquid phase depending on the detecting instrument and the reaction conditions. In situ FTIR can be commonly used to identify the adsorbed substances over time for kinetic study and track the states/changes of active sites for mechanism study. In the biodiesel industry, the yield of by-product is one of the most essential parameters to determine the efficiency of an organic reaction. The catalytic conversion of organic compounds is a practical reaction for industry, such as alcohols transforming into high-value added ketones and aldehydes through chemical catalytic oxidation. For example, catalytic oxidation of glycerol to produce C_3 products (such as dihydroxyacetone, glyceraldehyde, and glycerol acid) can be widely used in cosmetics, pharmaceutical intermediates, and chemical industries. However, there are still challenges to improve the catalytic activity and product selectivity, which directly affect the yield of products. Research fellows used the in situ FTIR to understand the tough task on glycerol over supported Pt nanoparticles. FTIR spectra gives the information of adsorption tendency of Pt sites for either the secondary hydroxyl group (C_2=O) or

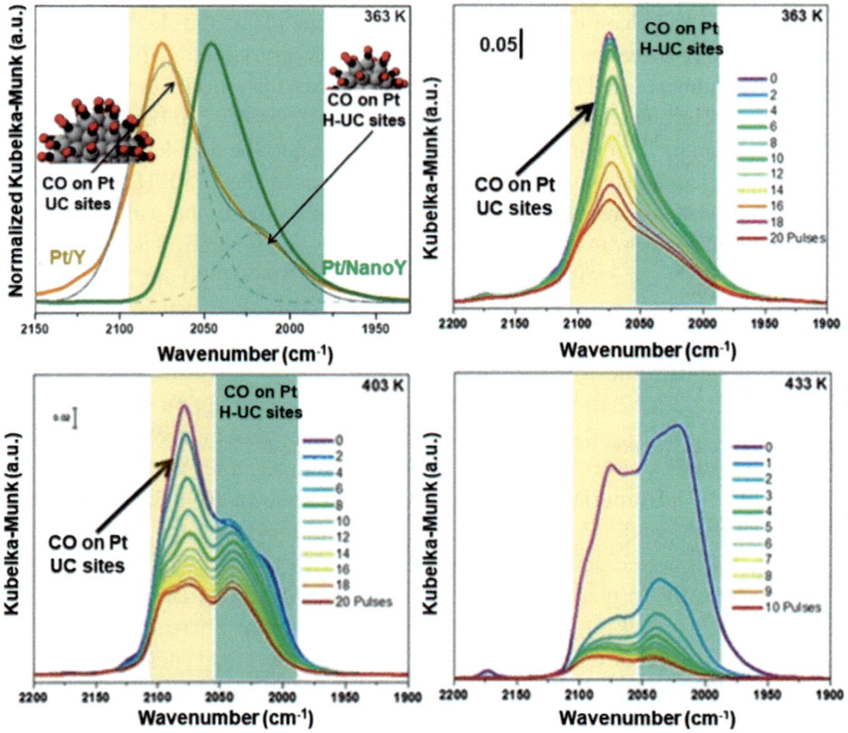

Fig. 2.16 **a** Normalized DRIFT spectra of CO adsorption at 363 K for Pt/Y (orange) and Pt/NanoY (green). IR spectra recorded for CO adsorbed on Pt/Y after every two pulses of O_2 at 363 K (**b**), 403 K (**c**) and 433 (**c**) [60]. Copyright from 2023, American Chemical Society

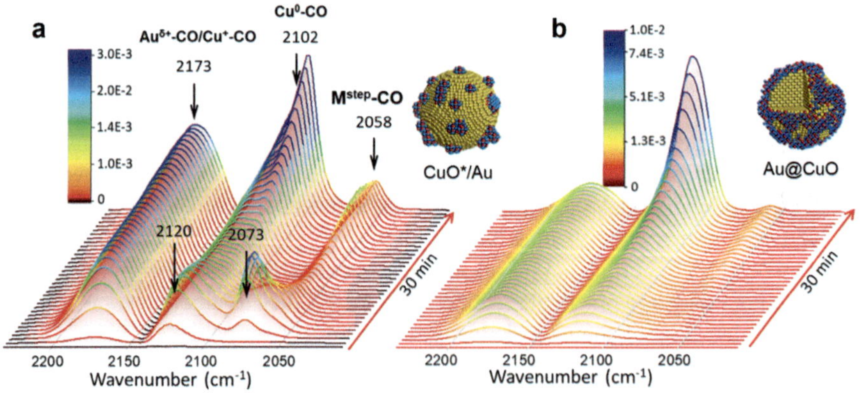

Fig. 2.17 In-situ DRIFT spectra of **a** CuO*/Au and **b** Au@CuO during CO adsorption in He at 25 °C. The inset picture illustrated the structure models of different catalysts in different catalytic systems [63]. Copyright 2019, American Chemical Society

primary hydroxyl group (C_1=O) depending on the supports (CuO or TiO_2), which can be disentangled by the adsorption states of $C_2=O_{DHA}$ at 1731 cm^{-1} and $C_1=O_{GLD}$ at about 1641 cm^{-1} [64]. Characterization of the adsorption, activation, and desorption of reactant/intermediate by using in situ FTIR in the presence of reaction atmosphere provides more information of the competition adsorption of different gas molecules and reaction mechanisms of organic reactions on the active centers of supported metal particles [65, 66].

2.3.5 *Thermoanalysis Technology*

Thermoanalysis technology is a kind of technique carried out to understand the adsorption/activation and transformation of reactants under temperature-programmed operations. It may include the phase transition, pyrolysis, and reaction kinetics of the detected catalyst, as a function of incremental temperature or reaction time. Thermoanalysis technology is based on the dynamic measurement and displays advantages compared with the static methods. It is simpler and faster with continuous experimental operations, which becomes a powerful tool for studying the properties and state changes of the catalyst surface in oxidation/reduction atmosphere and even reaction condition. Thermoanalysis technology can be divided into several common projects including the thermogravimetric analysis (TGA), differential thermal analysis (DTA), differential scanning calorimetry (DSC), the temperature-programmed reduction/ oxidation/ desorption/ surface reaction (TPR/ TPO/ TPD/ TPSR), as well as the pulse titration/chemisorption.

Thermoanalysis can be used to characterize the phase properties of catalysts including the decomposition, oxidation, reduction, phase to phase transformation of the carrier or metal components, active species, etc. The composition of a catalyst can be determined by the information on the valence state of the active metal, the interaction between the active metal and carrier, the dispersion degree/threshold of the active metal.

TGA, DTA and DSC are similar techniques to understand the physical and chemical evolution with temperature, mostly in an inert atmosphere. TGA is an analytical method determining the changes in sample mass as a function of the incremental temperature. The TGA profiles mainly help research fellows to study the physical and chemical changes such as thermal stability, thermal decomposition, and oxidative decomposition of materials in air or inert atmospheres. DTA is a method of analyzing the temperature difference between a test sample and a reference substance over time under temperature-controlled conditions. It is mainly used to detect the melting, crystallization transformation, secondary transformation, oxidation–reduction reaction, cracking reaction, etc. Whilst the DSC is an analytical method that measures the power difference (reaction heat) between the input sample and reference over time/controlled temperature. It can be used to measure the melting point, boiling point, glass transition, heat capacity, crystallization temperature, crystallinity, reaction temperature, purity and reaction heat.

For each test, only very small amount of sample is required. The small quantity of the sample reduces the temperature gradient happening inside the sample under a more realistic temperature. The diffusion of gas products could be promoted with inhabited reverse reaction. The separation ability of adjacent peaks for DSC will be enhanced although the sensitivity is reduced.

In contrast, DSC has higher sensitivity and resolution than the DTA, making the quantification more accurate. The thermal effect can be directly measured by the area enclosed by the exothermic and endothermic peaks of the DSC curve and the baseline. However, the peak area needs to be corrected by an instrument constant to obtain an accurate thermal effect value. The instrument constant can be calculated through a standard sample and determined only by the instrument. Researchers also use the combined TG-DSC to analyze catalysts, which displays obvious advantages of both methods.

The notable features of this technology are the dynamic and in situ test with better reflection of the actual properties of a catalyst under convenient operation. The information obtained by thermoanalysis can be easily integrated with other technologies. The detailed operation conditions can be designed according to the required information on the reactive metal phase of the catalyst.

The design and optimization of efficient catalysts require a deep understanding of the surface structure and chemical properties. Chemical adsorption provides the essential information to evaluate a catalyst and the advanced instrument has multiple functions conducted under controlled temperature, which mainly include the temperature-programmed desorption (TPD)/ reduction (TPR)/ oxidation (TPO)/ surface reaction (TPSR on fresh catalysts. The temperature-programmed oxidation (TPO) can also be carried out on the deactivated catalyst to understand the surface evolution during reaction and the deactivation mechanism. The pulse titration/chemisorption technology in the chemical adsorption instrument can be used to study the acidity of catalysts, surface metal dispersion, and the interaction between the metal and the carrier. These analytical methods play a crucial role in the development of supported metal particles, support and their synergy.

One of the most basic functions of chemical adsorption is to find the real active sites according to different requirements. TPR measures the reducible ability of the surface and bulk phase in the entire substance. The detection generally happened in a reducible atmosphere using e.g., CO/Ar or H_2/Ar (v/v of 1–10%) from room temperature to 800–900 °C under a constant heating rate. TPR profiles are used to study the information such as the interaction between the active metal and the carrier, the variation of metal species during reduction, hydrogen overflow phenomenon and alloying effect. The location and shape of the reduction peaks in TPR would change according to the valence state and the metal-support interaction/synergy. And semi-quantitative calculation of samples can be operated by the consumption of the reducing gas during the reduction process.

The TPD profiles give better information on the type and quantity of adsorption centers on the surface metal particles. The practical gas atmosphere can be used including the H_2, O_2, CO_2, NH_3, and even the liquid reactant carried by inert gas flow from room temperature to e.g., 800 °C. The strength and quantity of different centers can be calculated by the gas/reactant consumption. Alkaline gases (e.g., NH_3) adsorb on acidic centers; those adsorbed on strong acid centers are more stable and difficult to desorb than those adsorbed on weak acid centers. A temperature-programmed desorption induces the desorption of NH_3 desorb on acidic centers at different temperature ranges. The NH_3 adsorbed on weak acid centers desorb at lower temperature, whilst those adsorbed on strong acid sites desorb at high temperature (e.g., 400–650 °C). CO_2-TPD provides the information of surface basic sites of the catalysts. If gas atmosphere such as H_2 (for reduction reactions) and O_2 (for oxidation reactions) or inert gas flow carrying reactant liquid were in use, the active (metal) centers to activate/adsorb the H_2/O_2 and reactant could be reflected to guide the hydrogenation/oxidation mechanism of organic compounds. The results of TPD and TPR can be operated for a contrastive analysis. An typical example is displayed in Fig. 2.18 for selective oxidation of benzyl alcohol [67]. The peak area in H_2-TPR suggested the reducible ability and the metal species in oxidative states. The more intensive reduction peaks correspond to a larger amount of oxidized metal species in catalysts. The huge peaks at 400 °C are ascribed to the bulk lattice O in support and Au exists mainly in metallic states. Whilst the peak location and the peak area in O_2-TPD suggest the ability for O_2 activation/desorption and the number of active metal sites. The larger desorption peaks in Au/Mg_xAl-L (x = 2.5–3.5) under lower temperature are related with the better activation ability of oxygen molecules. The combining analysis of TPR and TPD give the preliminary inference that better catalytic performances would be behaved by these samples for the benzyl alcohol oxidation.

TPO detects the type and quantity of surface attachments (such as carbon deposits) different from the TPD that measures the properties of the surface active (metal) center, which can be used to determine the states, deactivation and regeneration of metal centers after reaction. And O_2 balanced with inert gas will be in use.

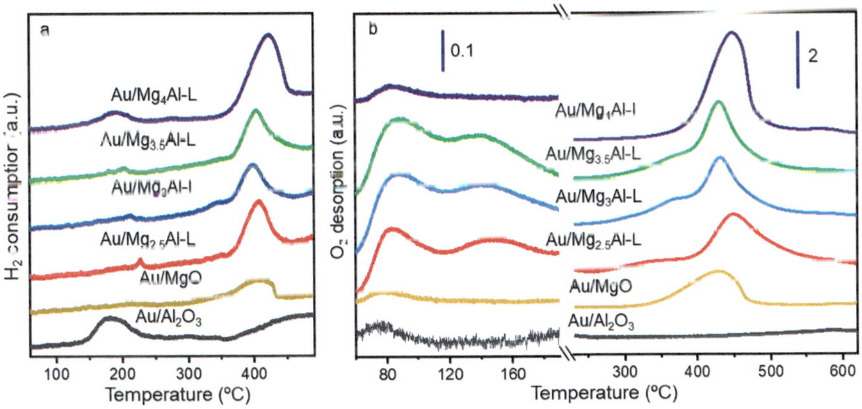

Fig. 2.18 **a** H_2-TPR profiles and **b** O_2-TPD profiles by Layered double hydroxides (LDH) supported Au catalysts with different Mg/Al atomic ratios [67]. Copyright 2023, with permission from Elsevier

The oxidation performance of the supported metal nanoparticles and the surface carbon deposition can be revealed by TPD profiles. If sulfur-containing species existed in the reaction system, the state of the metal species and the poisoning phenomena can also be understood in terms of the TPO profiles especially a GC-mass which was connected to analyze the outlet SO_2, CO and CO_2.

During TPSR, the reactants were pre-adsorbed on the surface of the catalyst at a relative-low temperature (e.g., 25–50 °C) for a certain time and the atmosphere is changed into an inert gas after then. The temperature increases at a constant heating rate. Evolution of the adsorbed reactant on the surface or the reaction between adsorbates happen, with the desorbed products easily detected by the GC-Mass coupling device. The catalytic reaction occurring between reactants on the surface of the metal centers, all belong to the research object of TPSR. During the heating process, the catalyst surface may simultaneously undergo decomposition or surface variation. Temperature-programmed experiments clearly reveal different properties and changes of the active metal centers as well as the preliminary theory of the reaction mechanism.

TPD technology is limited to the desorption investigation of a certain component adsorbate, and therefore cannot obtain important information about the adsorbed species on the catalyst surface under real reaction conditions. Herein, it is not facile to obtain the most important reflections of the reaction mechanism. TPSR compensates the shortcomings of TPD by combining TPD with surface reactions, providing a new means for in-depth research and revealing the essence of the catalytic reaction [68].

Another measurement-Pulse titration/chemisorption is measured on the similar instrument for temperature-programmed experiments. It can be used to calculate the surface dispersion of metal species. During each test, adsorption of gas molecules such as H_2, CO or N_2O is carried out under constant temperature (e.g., 50 °C) without additional desorption procedures. Each pulse is a fixed amount, such as 0.5 mL gas. Under normal temperature and pressure, the physical adsorption of gas molecules on metal catalysts can be ignored. After several pulses, the gas adsorption can be saturated on the exposed metal sites, and the metal dispersion on the surface can be calculated according to the recorded data. Precious metal such as Au and Pt can be detected directly by CO titration/chemisorption, whilst the dispersion of Cu active sites can be more accurately determined if N_2O titration/chemisorption was used.

No matter how, the programmed-heating processes also has some drawbacks, such as the possibility of the side reactions during the heating process, leading to an increase in the purity of the product. The process also requires dedicated device and technical support with high costs. The temperature-increasing rate and design of adsorption/reaction gas atmosphere need to be well-designed and optimized according to the research requirement on the metal sites of the supported particles.

2.3.6 *Evaluation of the Supported Nanoparticles*

The characterization and analysis technologies in the field of catalysis have been extensively applied to understand the essential factors of the catalyst towards an organic reaction. It displays an important role in the exploring, design, and optimization of a high-efficient catalyst. No matter how, the catalyst will be finally evaluated by a target organic reaction to show its efficiency, where structure-performance relationship will be taken into an insight for further development of the catalyst systems. Herein, it comes the most intrinsic procedures for the typical evaluations.

1. Evaluation of the catalytic activity and selectivity: By analyzing the physical and chemical properties of the catalyst, the activity and selectivity characteristics of the catalyst can be understood. For example, by analyzing the surface acidity/alkalinity, the distribution and strength of the acid-base sites of the catalyst can be determined, thereby predicting the activity and selectivity of the catalyst in acid-base catalytic reactions. The activation/adsorption ability of a catalyst can be determined by the temperature-programmed experiments in the presence of reactants, and the structure-performance relationship can be assumed therefore.
2. Research on the catalyst deactivation mechanism: Catalysts may experience deactivation during long-term operation, and the study of the deactivation mechanism is of great significance for the continuous operation and prolongation of catalyst life. The mechanisms of catalyst deactivation can be revealed through analysis of catalyst deformation, which may include characters such as the changes in surface structure, chemical/geometric variation of both metal sites and support, loss of acidity and alkalinity, and accumulation of toxic substances.
3. Design and development of new catalysts: Through catalyst characterization and analysis, the impact of catalyst structure and composition on reaction activity and selectivity can be understood. Based on this information, more efficient catalysts can be designed, for example, by regulating the pore size and metal-support interface of the catalyst to improve its mass transfer and number of reactive centers.

In summary, catalyst characterization and analysis play the important role in chemical technology. With the continuous progress of science and technology, the characterization methods of catalysts also constantly develop. The application of these methods benefits the perceiving of the chemical structure and surface properties of the supported metal particles, which in turn promotes the understanding, design, and optimization of the catalytic systems. In the future, further improvement of the characterization and analysis technology brings great impetus for the development of catalytic science and technology.

2.4 Organic Reactions

Supported metal catalysts can be used for a variety of organic reactions, with the active metal centers contributed by transition metals (e.g., Ni, Cu, Co, etc.) and precious metals (e.g., Pt, Pd, Au, Re, etc.). Satisfied catalytic activity and selectivity towards the target products can be obtained via selecting a suitable metal catalyst system by organic reactions. The selection of supported metal nanoparticles largely depends on the reactants and reaction conditions. In this section, typical catalysts with supported metal nanoparticles will be discussed, which could be widely used for organic reactions.

It has to be admitted that precious metal centers display their specific characters for a variety of organic reactions according to the supporting materials, the porous structure, metal-support interaction and surface formation of different components as discussed above. No matter how, precious metals such as Au, Pt, Pd, and Re have their own superiority towards different types of reactions. Taking the supported gold nanoparticles for instance, it will be a definite shining star for the catalytic oxidation/coupling reactions of many organic compounds including the alcohols, alkynes, C_1 compounds, etc. However, supported gold nanoparticles are hardly reported for hydrogenation or H_2 involved reactions due to the very weak adsorption/activation ability of H_2 molecules on Au centers. Precious metal catalysts such as supported Pd and Pt nanoparticles were reported as outstanding catalysts for selective hydrogenation and other H_2 involved reactions. However, if CO was present in the reaction, precious catalysts such as Pt and Pd would be easily deactivated due to excessive CO poison. Re-based catalysts can be efficient during many organic reactions for petroleum refining processes. Herein, the supported metal nanoparticles can be designed with the specific ability to activate/dissociate the reactant molecules for a target reaction.

2.4.1 Supported Gold Nanoparticles

Gold catalyst was for the first time reported by Haruta et al. in 1990s for CO oxidation reaction. It since then opened the new world of gold catalysis. Efficient catalysts with supported gold nanoparticles are generally synthesized using chloroauric acid as the metal precursor. It can be loaded on the surface of industrial carriers such as carbon, titanium oxide, aluminum oxide, or other individual supports such as metal oxide, hydrotalcite and envn hydroxyapatite, exhibiting excellent catalytic activity under extremely low mass percentage (0.5–5.0%). The reaction activity is closely related to the size and preparation of gold particles. Despite the high cost of pure gold raw materials, the low mass percentage of gold in the catalyst makes it a promising catalyst for industrial applications. Currently, they exhibit unique performance in reactions such as elimination of CO and NO_x by oxidation, water gas transformation, selective oxidation of alcohols and aldehydes, and elimination of nitrogen oxides [69, 70]. At present, the development of highly stable nanoscale gold catalysts is still one of the research focuses in related fields.

Water–gas shift reaction (WGS) Beside the most simplified reaction model of gold catalysts (CO oxidation), low-temperature water–gas shift reaction (WGS) is one of the most typical reactions using gold catalysts. In recent years, research on automotive fuel cell power systems has been continuously innovating with breakthrough, resulting in extensive research progress. This process involves pure hydrogen products, making the WGS transformation in the presence of CO and water to CO_2 and H_2 an applicable process. Traditional Cu-based catalysts are not suitable to be used for fuel cells due to the risks by exposure to air and dew during the start-up and shutdown of automobiles, while Au catalysts can effectively solve this problem. Typical catalysts for the WGS reaction can be Au/Fe_2O_3, Au/TiO_2, Au/α-MoC and Au/CeO_2, with the latter being particularly stable [71–73].

The dissociation of water is the essential procedure in such a reaction since it is thermodynamically and kinetically hard on gold. Mechanisms such as formate involved theory and surface hydroxyl involved theory are discussed in literature [69]. It was suggested that H_2O molecule could be easily dissociated on Au_{15}/α-MoC(111) to form H atoms and OH species via Eq. (2.3) at lower coverage. The subsequent reaction between OH species immediately happens without a barrier by forming a surface O atom as shown by Eq. (2.4). The surface O atoms could further promote water dissociation on the boundary of Au and α-MoC(111) in Eq. (2.5). The water-assisted pathway (WAP) was also discussed by researchers (Fig. 2.19), where one or more water molecules can participate depending on the location of the adsorbed CO in relation to the oxidized-hydroxylated substrate [74]. In a single-water participation event, the substrate anchors water molecules through H-bonds near Au clusters. During the WGS reaction, more than one water molecules take part in the process. The adsorbed CO binds with an OH from water molecule to trigger the WAP pathway. The "linked" water directly supplies an OH to *CO on the metal to form *COOH, while simultaneously depositing an H on the oxidized-hydroxylated substrate.

$$CO + 2H_2O \rightarrow CO + 2OH + 2H \quad (2.3)$$

$$CO + 2OH + 2H \rightarrow CO + H_2O + O + 2H \quad (2.4)$$

$$CO + 3O + H_2O \rightarrow CO + 2O + 2OH \quad (2.5)$$

Acetylene hydrochlorination The industrial production of vinyl chloride from acetylene still relies on $HgCl_2$ catalyst. However, $HgCl_2$ is highly toxic and easily reduced to metallic Hg under reaction conditions. Its volatilization seriously affects the environment and human health. Such problem needs urgent resolution in the industrial facilities. The mercury free catalysts suitable for the process of acetylene hydrochlorination mainly include precious metal catalysts, precious metal catalysts, and non-metal catalysts. Precious metal catalysts mainly use gold as a catalyst, which

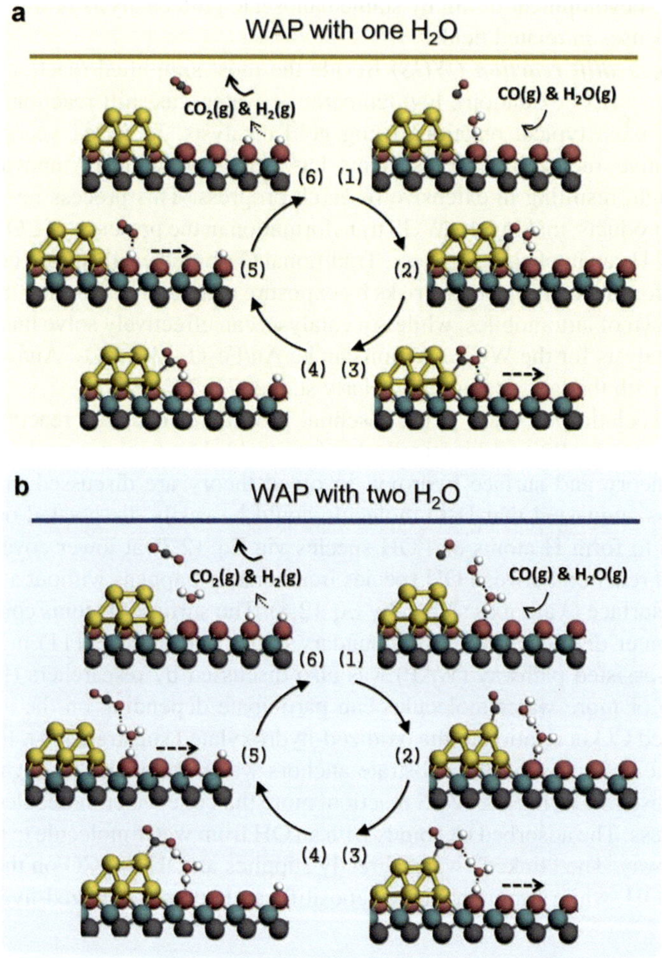

Fig. 2.19 Schematic illustration of the WGS reaction under WAP [74]. Copyright 2021, with permission from Elsevier

has excellent performance and stability. An industrial demonstration project of more than 8000 h has been carried out in enterprises. Gold catalyst is currently the best performing and most promising mercury-free catalyst for acetylene hydrochlorination. Research fellows conducted the studies using modified gold nanoparticles supported on carbon materials in many cases. Oxidized amorphous carbon was reported playing a key role in generating the active sites consisted by isolated Au (e.g., volatile Au_2Cl_6 species by O'Connell et al. [75]).

Kinetic studies of acetylene hydrochlorination over Au catalysts suggested that the reaction is of first order with respect to both acetylene and HCl [77], and formation of a C_2H_2/Au/HCl complex has been suggested. Previous work has also suggested

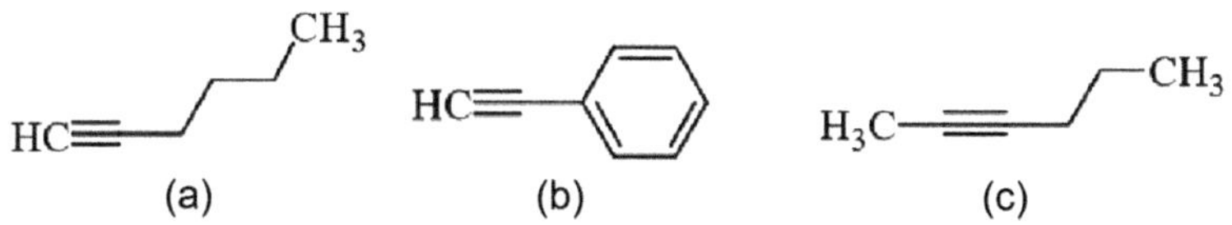

Fig. 2.20 Substituted alkyne substrates used to investigate the mechanism of addition of HCl across the alkyne bond. (1) Hex-1-yne; (2) Phenyla-cetylene; (3) Hex-2-yne [76]. Copyright 2016, with permission from Elsevier

that the reaction may proceed via a redox cycle involving the cationic gold species. Deactivation of gold catalysts happened if the reduction of cationic gold species to Au^0 happened during the reaction. The exposure to acetylene led to the deactivation of gold sites since acetylene is a typical reducing agent. Due to the symmetric nature of acetylene, mechanistic data is very difficult to be operated due to the fact that each possible product is identical [76]. Herein, experiments with substituted alkynes were used as listed in Fig. 2.20. It was found that the activity was affected by steric hindrance of the substrates. No matter how, precious metals are expensive, and the chlor alkali industry itself has low profits, so large-scale industrial applications are limited.

Propylene epoxidation This is also one of the typical reactions proceeded by gold catalysts in the co-presence of H_2 and O_2. Propylene epoxidation with molecular oxygen has been proposed as a green and alternative process to produce propylene oxide (PO). Titanosilicate-1 (Ti-SiO_2, TS-1) with or without modification can be used as very suitable supporting materials for gold nanoparticles [78, 79]. The PO formation requires the presence of H_2 and O_2 via the speculation with a peroxide species formed on gold. It may be subsequently transferred to a neighboring Ti site to epoxidize propylene into PO. A "simultaneous" mechanism requiring the proximity of the Au and Ti active sites were established as shown in Scheme 2.1 [70, 80].

Selective oxidation Gold catalysts are also good candidates for liquid-phase selective oxidation of oxygen-containing compounds such as carbohydrates, aldehydes, and alcohols. Water or other green solvent can be used with oxygen or air as the oxidant, making a typical green chemical process with the mild reaction conditions. Typical oxidation reactions include the selective oxidation of glucose to gluconic acid and derivatives, oxidation of alcohols (e.g., benzyl alcohol and glycerol) to aldehydes, aerobic coupling of amine for imine, oxidation of aldehydes to acids.

CH_4 hydrogenolysis The current well-designed gold catalysts can also be used for the transformation of CH_4 to CH_3OH and even hydrogenolysis [81, 82]. Comparing with the platinum group metals, the outstanding advantages of gold catalysts for liquid-phase oxidation reactions include the relative-low cost, milder reaction conditions, high selectivity of target products without peroxidation or deactivation due to product adsorption (e.g., CO), and better anti-poisoning performances. The excellent performances of gold catalysts in liquid-phase selective oxidation of oxygen-containing organic compounds make the potential application of supported gold nanoparticles more available.

$$2H^* \rightleftharpoons H_2 + 2^* \quad (2.6)$$

$$O_2 + {}^* \rightleftharpoons O_2^* \quad (2.7)$$

$$H_2 + O_2^* + {}^* \rightleftharpoons HOO^* + H^* \quad (2.8)$$

$$O_2 + 2H^* \rightleftharpoons HOO^* + H^* \quad (2.9)$$

$$H_2 + HOO^* + {}^* \rightleftharpoons HOOH^* + H^* \quad (2.10)$$

$$C_3H_6 + \square \rightleftharpoons C_3H_6\square \quad (2.11)$$

$$C_3H_6\square + HOOH^* + H^* \rightarrow C_3H_6O\square + H_2O^* + H^* \quad (2.12)$$

$$C_3H_6O + {}^* \rightleftharpoons C_3H_6O^* \quad (2.13)$$

$$C_3H_6O + \square \rightleftharpoons C_3H_6O\square \quad (2.14)$$

$$H_2O + {}^* \rightleftharpoons H_2O^* \quad (2.15)$$

Scheme 2.1 Proposed elementary steps for propylene epoxidation on Au/TS-1 catalysts via the "Simultaneous" mechanism [70]. * is an Au site and □ is a Ti site, and they are assumed to be in proximity, such that a gold-adsorbed HOOH and titanium-adsorbed propylene can react to form propylene oxide. Copyright 2022, American Chemical Society

2.4.2 Platinum Group Metal

Platinum group metals, including the ruthenium (Ru), rhodium (Rh), palladium (Pd), osmium (Os), iridium (Ir), and platinum (Pt), are typical precious metals with high chemical stability and excellent corrosion/oxidation resistance. They have extremely strong atomic bonds, resulting in significant interatomic bonding and maximum packing density. The atomic coordination number of all platinum group metal elements is 6, which determines their unique physical and chemical properties. Platinum group metals and alloys have many important industrial applications. The main use of platinum group metals is in the catalyst field. The catalytic activity, thermal stability, and selectivity can be satisfied after fine promotion, which can be competent for the chemical industry such as reforming processes in the refining industry.

In the field of chemical synthesis, supported platinum group metal nanocatalysts are widely used in organic synthesis reactions. They can catalyze carbonization reactions, hydrogenation reactions, and oxidation reactions. Supported platinum group metal nanocatalysts not only have efficient catalytic activity, but also have the characteristics of reusability and good catalytic selectivity, which greatly improve the efficiency and yield of synthesis reactions.

Carbon–carbon bond reduction The reduction of unsaturated C–C bonds by transition-metal-catalyzed hydrogenation is one of the most relevant reactions in organic chemistry. For reduction of the C=C bond, Pd is the most prevalent active site, and the CTH process could occur via both the gas–solid and liquid–solid modes at relatively mild conditions. Using Pd/C catalyst and formic acid hydrogen donor, carbon–carbon double bonds in carboxylic acids or ketones could be reduced to saturated fatty acids or deeply decarbonylated to hydrocarbons [83]. Partial reduction of the C≡C bond to C=C is highly desirable because of its wide applications in the industry such as monomer manufacture.

Hydrogenation of nitroarenes Hydrogenation of nitroarenes is a catalytic reaction of high interests catalyzed by platinum-group metals, owing to the importance of the aniline products for the chemical industry. The chemoselective reduction of the nitro group in the presence of other reducible functionalities in the nitroaromatic substrates remains a challenging goal. Among the studies of catalyst systems with metal nanoparticles, platinum-group metals, especially the Pt nanoparticles appear as a most promising one with outstanding nitro reduction ability and product selectivity. The general mechanism accepted by the scientific community for the hydrogenation of nitrobenzene was proposed according to the early work of Haber et al. [84] and is depicted in Scheme 2.2. The accumulation of the reaction intermediates such as hydroxylamines, azo and azoxy derivatives by the condensation of hydroxylamines and nitroso species, makes the successful operation with high product yield very difficult. Another difficulty to be faced concerns the case of halonitroarenes, where the dehalogenation reaction may simultaneously happen to give a mixture of the byproducts haloaniline and dehalogenation products.

Electrocatalysis Platinum-group metals as a representative class of electrocatalysts play a vital role in contemporary energy conversion and storage, including but not limited to fuel cells, metal-air batteries, and precise electrosynthesis. In the challenging twenty-first century, the development of platinum group metal materials directly restricts the speed of development in various high-tech fields and affects the international position in the world economy. For example, the study of platinum nanocatalysts on the electrocatalytic oxidation behavior of organic small molecules such as methanol, formaldehyde, and formic acid, which can be used as fuel for fuel cells, has both fundamental theoretical research significance and broad application prospects [85]. Research has shown that the main catalysts with certain electrocatalytic activity towards organic small molecules are mostly platinum group precious metals.

2.4.3 Rhenium Nanoparticles for Petroleum Refining

To solve the urgent demands required by the chemical industry, the development of efficient catalysts for hydro-converting the renewable resources to fuels and high value-added chemicals has spurred continuous interests [86, 87]. For many years, the platinum-group metal and typical transition metal (e.g., Ni and Cu) are the metals

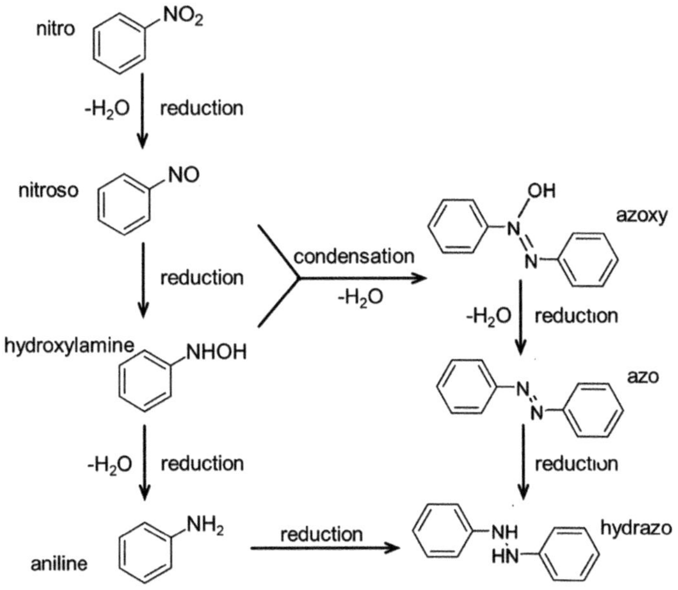

Scheme 2.2 Accepted mechanism for the hydrogenation of nitroarenes [84]. Copyright 2009, European Chemical Society publishing

of choice for designing an efficient hydrogenation catalyst [32, 88, 89]. However, the limited availability of metals prompts researchers to explore alternative metals in the periodic table. Among the platinum group metals and transition metals, rhenium (Re) is a rare precious metal and plays an important role in many fields including the petrochemical industry and aerospace field. Different from other precious metals, rhenium species is very oxophile with poor oxidation resistance to give multiple chemical valences from +1 to +7, presenting variable opportunities for the $Re^{\delta+}$ species used in chemistry and industry applications.

Selective hydrogenation of acid, ester, C=O, N/S-containing ccompounds One of the typical reactions is the above-mentioned hydrogenation of acid and even diacid with one to two carboxylic acid groups [90], as displayed in Fig. 2.21. The carboxylic acid group connects with either a long chain hydrocarbon or a phenyl group. The hydrogenation of carboxylic acids involves the activation of the C=O bond and the cleavage of the C–OH bond, which is still a tough industrially subject. The high-value added γ-butyrolactone (GBL) and 1,4-butanediol (BDO) can be produced by cyclodehydration if succinic acid was used as substrate [91, 92]. As a counterpart, esters can also be activated and hydrogenated on the surface of the monometallic Re catalyst. The hydrogenation process comprises of a series of C=O and C–O scission steps, due to the presence of multiple functionalities in esters. The selective hydrogenation of esters is more difficult than the hydrogenation of counterpart acids, which may be a preliminary step to produce acid as an intermediate under special conditions.

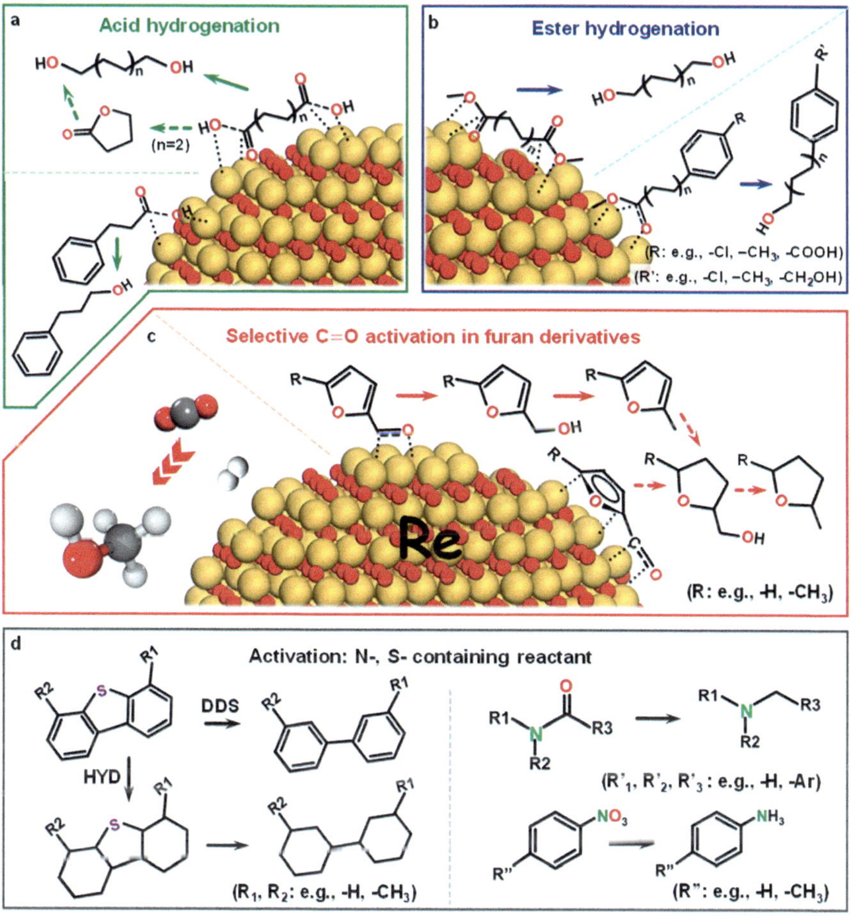

Fig. 2.21 Schematic illustration of monometallic Re catalysts for selective hydrogenation reaction [96]. Copyright 2024, American Chemical Society

Another very important reactions involved the activation of C = O bond are the selective hydrogenation of furan and its derivatives (e.g., furfural, 5-hydroxymethylfurfural (5-HMF), N-methylpyrrolidone). The oxophilicity of rhenium can effectively activate polar bonds and improve the reaction rate. However, the adsorption model of furan derivatives, e.g., furfural, was still not clear. Different from the metals (Pt, Pt, Ni, etc.) gave a η^2-(C,O)-aldehyde adsorption mode to produce 2-mythyl furan (2-MF) and tetrahydro-products [93], where hydrogenation of furan directives on Re seemed different. It was also not a simple tilted η^1-(O) configuration as reported by Cu and Ru catalysts to produce mainly furfural alcohol [94]. The main hydrogenation products for furfural by monometallic Re were furfuryl alcohol (FFA) and 2-MF [95]. Rhenium was also competent for hydrogenation of some N- and S-containing reactants, including the selective hydrodesulfurization of thiophene and its derivatives, as well as hydrogenation of amides or nitryl groups into amines.

2.4.4 Non-precious Transition Metal Nanoparticles

Transition metal catalysts display their own characteristics which are different from the precious metal nanoparticles supported on carriers, They can be used for the reforming reaction, hydrogenation reaction, oxidation reaction with high catalytic performances. The most amazing advances by using transition metal nanoparticles are the low-price and easy availability of the metal sources.

Ethanol steam reforming Ethanol steam reforming (ESR) is a promising reaction as a sustainable, carbon–neutral hydrogen production process. Direct hydrogen production by ethanol steam reforming is a highly efficient energy production process that avoids the high energy loss resulting from the external heat of alcohol combustion enthalpy (Eqs. 2.16–2.17) [97, 98]. Among non-noble metal catalysts, the highly active Co, Ni, and Cu catalysts are used widely for ESR [99]. Catalyst deactivation by coke formation and methane formation as a by-product pose severe difficulties [9, 108]. To suppress those outcomes in ESR, solutions have been proposed. The use of basic support or promoter is essential. Ethylene is usually formed by ethanol dehydration over an acidic site. Basic support decreases the selectivity toward ethylene and the coke formation can be largely suppressed in return. Using redox metal oxides with high oxygen mobility and oxygen storage capacity (e.g., CeO_2), they can contribute directly to the elimination/oxidation of coke. The particle size of active metal nanoparticles can be efficient for the coke resistance, and coke deposition occurs frequently on the terrace site with high coordination number on big particles.

Ethanol combustion

$$C_2H_5OH\ (g) + 3O_2(g) \rightarrow 2CO_2(g) + 3H_2O(g)\ \Delta H_r^{\ominus}(298) = -1277.4\text{kJ}/mol \tag{2.16}$$

Ethanol steam reforming

$$C_2H_5OH\ (g) + 3H_2O(g) \rightarrow 2CO_2(g) + 6H_2(g)\ \Delta H_r^{\ominus}(298) = 173.5\text{kJ/mol} \tag{2.17}$$

Co-based and Ni-based catalysts are suitable for industrial ESR processes due to their nice activity and lower price [99]. Highly active and stable Co-based catalysts usually contain both metallic Co^0 and Co^{2+} species, in which Co^{2+} species favor water activation as well as the oxidation of the deposited coke or coke precursor. Controlling the atomic ratio between metallic $M^0/M^{\delta+}$ on metal nanoparticles is one means of avoiding deactivation by carbon deposition and increasing the catalytic performances.

C–O/C=O Selective hydrogenation Selective hydrogenation is one of the most crucial transformations of organic compounds for synthesis of fine chemicals or pharmaceuticals and is applied both in the laboratory and industry. Transition metal nanoparticles supported on different carriers can also be used as promising candidates for the selective hydrogenation reactions including the aldehydes/ketones, acids, cellulose, and other biomass-derived oxygenates [100]. Transition metal-based catalysts, especially Cu and Ni nanoparticles display the outstanding catalytic performance for the selective hydrogenation of organic compounds. Both the hydrogenation in the presence of gaseous H_2 molecules or via a catalytic hydrogen transfer reaction using H-donor (e.g., formic acid and isopropanol) can be catalyzed under appropriate modification of the reactive metal centers. Most of these hydrogenation reactions provide high-value added downstream chemicals, which could be used as platform chemicals or the monomers for polyesters, which are the significant building blocks of the economy and industry development. There are several typical hydrogenation reactions reported by efficient transition metal nanoparticles: (i) Selective hydrogenation of oxygenated furan derivatives, e.g., furfural hydrogenation to produce e.g., 2-methyl furan, γ-valerolactone (GVL), cyclopentanone, tetrahydrofurfuryl alcohol, depending on the detailed reaction conditions. The product distributions can be directly regulated with 1–2 target chemicals by turning the active metal particles according to the adsorption modes (e.g., η^1-(O)-furfural, η^1-(C)-furfural, η^2-(C,O)-furfural, Fig. 2.22); (ii) Selective hydrogenation of diacids to diols (e.g., succinic acid & adipic acid to 1,4-butanediol (BDO)/butyrolactone (GBL) & 1,6-hexanediol); (iii) Hydrogenation of N-containing organic compounds, e.g., adiponitrile to hexamethylenediamine [101], which is the pivotal monomer to produce nylon-66 with adipic acid; (iv) Hydrogenation of the biomass including hemicellulose, cellulose, lignin, and triglycerides, which transfers the renewable resources/wastes into fine chemicals for a sustainable economy.

Reforming of CO_2 with CH_4 Syngas (mixture of CO_2 and H_2) generation from the reforming of CO_2 with CH_4, commonly known as dry reforming of methane (DRM), is a viable method to effectively combine these two global warming gases to generate high energy chemicals and fuels. One of the major problems found in DRM is the inevitable catalytic deactivation due to coking owing to low (H/C = 2 and O/C = 1) ratios in the feed gas [103]. Non-precious transition metal, particularly the Ni catalysts, with high reactant conversion comparable to the noble metals (Rh, Ru and Pt) at similar CH_4/CO_2 feed ratios is more desirable in DRM from a practical standpoint [104]. The coke formation can be successfully suppressed on efficient Ni nanoparticles if a secondary metal was introduced as the doping element. On Ni–Cu, the dissociation of CH^* requires higher energy barrier. The facilitated activation of CO_2 simultaneously contributes to the oxidization of amorphous carbon formed on Ni–Cu, and the coke cannot be easily formed on such promoted bimetallic metal nanoparticles. On Ni–Fe, Fe can partially oxidize to $Fe^{\delta+}$ by the activated CO_2. The FeO follows a subsequent redox cycle to gasify the surface coke on Ni sites with the FeO reduced to re-form the original Ni–Fe structure. If Ni–Co bimetallic particles were in use, high carbon resistance is mainly attributed to high surface

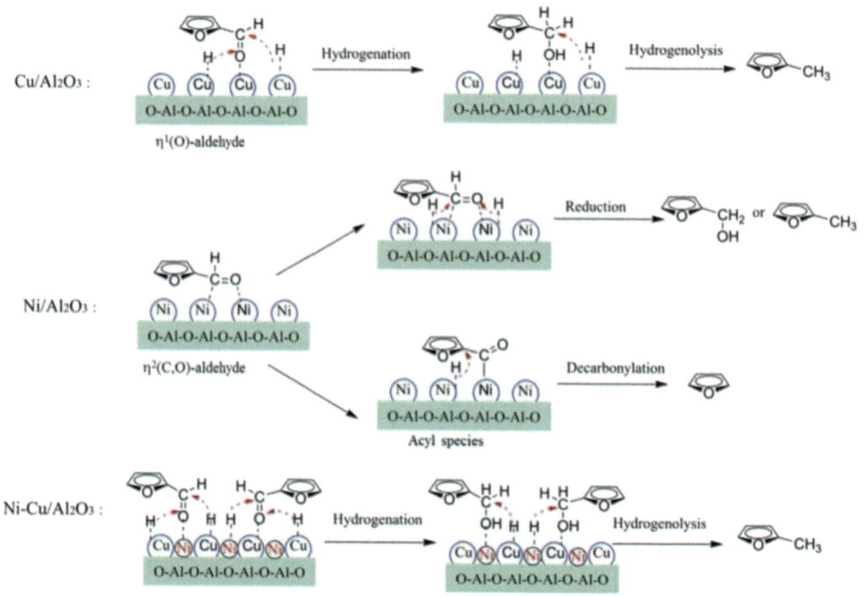

Fig. 2.22 Possible reaction pathways of furfural over Cu/Al_2O_3, Ni/Al_2O_3, and $Ni–Cu/Al_2O_3$ catalysts [102]. Copyright 2017, with permission from Elsevier

carbon mobility, where the interplay between forward and backward C* diffusion rates plays a big role in negating carbon coupling on Ni–Co [104].

Fischer–Tropsch (FT) process Fischer–Tropsch (FT) process can also be catalyzed by the transition metal catalysts, which is a promising method for producing liquid fuels and other valuable chemicals through CO hydrogenation. Iron can be used as one of the inexpensive supported catalysts with high availability. But it easily leads to the water–gas shift reaction (WGS, $CO + H_2O \rightarrow CO_2 + H_2$) and will perform better under low H_2/CO ratio syngas [105]. Cobalt shows high activity, especially to produce heavy hydrocarbons, including diesel fuel and wax, while iron-based catalysts are suitable for gasoline production. Co catalysts possess a very low intrinsic WGS activity, and it is more suitable for syngas with a high H_2/CO ratio [106]. Nickel catalysts are useful for the selective methane production. Despite its higher activity than cobalt and iron, due to the high price (~50,000 times higher than iron), ruthenium is not economically preferred for commercial FT production, and it is usually used as a reduction promoter, mostly for cobalt-based FT catalysts.

2.5 Summary and Outlooks

Based on the above discussion, it can be seen that supported metal particles can be used as the active sites for a variety of organic reactions. The surface dispersion degree, supporting materials, chemical states, acid/basic sites as well as the geometry and electronic environment of the active metal nanoparticles all greatly influence the catalytic performances and selectivity towards target products. In the economy and industry world, these catalytic systems greatly lower the energy consumption and the raw cost of a chemical technology. The number of publications related to the use of supported metal nanoparticles for organic reactions has grown exponentially in recent years, which has now become a gradually mature catalytic system compared to other metal-based heterogeneous systems such as the single atom metal catalyst or metal clusters. No matter how, promotions and optimized operations should be taken to avoid the coke deposition or other deactivation of the active metal centers on nanoparticles.

References

1. Turkevich J, Stevenson PC, Hillier J (1951) A study of the nucleation and growth processes in the synthesis of colloidal gold. Discuss Faraday Soc 11:55–75. https://doi.org/10.1039/DF9511100055
2. Leheny AR, Rossetti R, Brus LE (1985) Molecular resonance Raman observation of tetrathiafulvalene oxidation by colloidal platinum crystallites. J Phys Chem 89(2):211–213. https://doi.org/10.1021/j100248a005
3. Mourdikoudis S, Menelaou M, Fiuza-Maneiro N, Zheng G, Wei S, Pérez-Juste J, Polavarapu L, Sofer Z (2022) Oleic acid/oleylamine ligand pair: a versatile combination in the synthesis of colloidal nanoparticles. Nanoscale Horiz 7(9):941–1015. https://doi.org/10.1039/D2NH00111J
4. Wang M, Wang L, Li H, Du W, Khan MU, Zhao S, Ma C, Li Z, Zeng J (2015) Ratio-controlled synthesis of CuNi Octahedra and Nanocubes with enhanced catalytic activity. J Am Chem Soc 137(44):14027–14030. https://doi.org/10.1021/jacs.5b08289
5. Piella J, Bastús NG, Puntes V (2016) Size-Controlled synthesis of sub-10-nanometer citrate-stabilized gold nanoparticles and related optical properties. Chem Mater 28(4):1066–1075. https://doi.org/10.1021/acs.chemmater.5b04406
6. Losch P, Huang W, Goodman ED, Wrasman CJ, Holm A, Riscoe AR, Schwalbe JA, Cargnello M (2019) Colloidal nanocrystals for heterogeneous catalysis. Nano Today 24:15–47. https://doi.org/10.1016/j.nantod.2018.12.002
7. Ji X, Song X, Li J, Bai Y, Yang W, Peng X (2007) Size control of gold nanocrystals in citrate reduction: the third role of citrate. J Am Chem Soc 129(45):13939–13948. https://doi.org/10.1021/ja074447k
8. Bastús NG, Comenge J, Puntes V (2011) Kinetically controlled seeded growth synthesis of citrate-stabilized gold nanoparticles of up to 200 nm: size focusing versus Ostwald ripening. Langmuir 27(17):11098–11105. https://doi.org/10.1021/la201938u
9. Han Q, Lu Z, Gao W, Wu M, Wang Y, Wang Z, Qi J, Dong J (2020) Three-dimensional AuAg alloy NPs/graphene/AuAg alloy NP sandwiched hybrid nanostructure for surface enhanced Raman scattering properties. J Mater Chem C 8(36):12599–12606. https://doi.org/10.1039/D0TC02752A

10. Malik AS, Bali H, Czirok F, Szamosvölgyi Á, Halasi G, Efremova A, Šmíd B, Sápi A, Kukovecz Á, Kónya Z (2022) Turning CO_2 to CH_4 and CO over CeO_2 and MCF-17 supported Pt, Ru and Rh nanoclusters-Influence of nanostructure morphology, supporting materials and operating conditions. Fuel 326:124994. https://doi.org/10.1016/j.fuel.2022.124994
11. Alzoubi FY, Al-zou'by JY, Theban SK, Alqadi MK, Al-khateeb HM, AlSharo ES (2021) Size, stability, and aggregation of citrates-coated silver nanoparticles: contribution of background electrolytes. Nanotech Environ Engin 6(3):69.https://doi.org/10.1007/s41204-021-00165-6
12. Comotti M, Li W-C, Spliethoff B, Schüth F (2006) Support effect in high activity gold catalysts for CO oxidation. J Am Chem Soc 128(3):917–924. https://doi.org/10.1021/ja0561441
13. Jana NR, Peng X (2003) Single-phase and gram-scale routes toward nearly monodisperse au and other noble metal nanocrystals. J Am Chem Soc 125(47):14280–14281. https://doi.org/10.1021/ja038219b
14. Yahaya ML, Zakaria ND, Noordin R, Abdul Razak K (2022) Synthesis of large and stable colloidal gold nanoparticles (AuNPs) by seeding-growth method. Mater Today: Proc 66:2943–2947. https://doi.org/10.1016/j.matpr.2022.06.563
15. Kranz D, Bessel P, Rosebrock M, Niemeyer M, Dorfs D (2023) Composition-controlled laser-induced alloying of colloidal Au–Cu Hetero nanoparticles. Part Part Syst Char 40(8):2300021. https://doi.org/10.1002/ppsc.202300021
16. Moreira Da Silva C, Girard A, Dufond M, Fossard F, Andrieux-Ledier A, Huc V, Loiseau A (2020) Nickel platinum (Ni_xPt_{1-x}) nanoalloy monodisperse particles without the core–shell structure by colloidal synthesis. Nanoscale Adv 2(9):3882–3889. https://doi.org/10.1039/D0NA00450B
17. Moreira Da Silva C, Amara H, Fossard F, Girard A, Loiseau A, Huc V (2022) Colloidal synthesis of nanoparticles: from bimetallic to high entropy alloys. Nanoscale 14(27):9832–9841. https://doi.org/10.1039/D2NR02478K
18. Solovov R, Orlichenia V, Ershov B (2022) Iron nanoparticles in isopropyl alcohol: preparation and properties. J Mol Liq 368:120743. https://doi.org/10.1016/j.molliq.2022.120743
19. Vrijburg WL, van Helden JWA, van Hoof AJF, Friedrich H, Groeneveld E, Pidko EA, Hensen EJM (2019) Tunable colloidal Ni nanoparticles confined and redistributed in mesoporous silica for CO_2 methanation. Catal Sci Technol 9(10):2578–2591. https://doi.org/10.1039/C9CY00532C
20. Dimitratos N, Lopez-Sanchez JA, Anthonykutty JM, Brett G, Carley AF, Tiruvalam RC, Herzing AA, Kiely CJ, Knight DW, Hutchings GJ (2009) Oxidation of glycerol using gold–palladium alloy-supported nanocrystals. Phys Chem Chem Phys 11(25):4952–4961. https://doi.org/10.1039/B904317A
21. Inaba M, Zana A, Quinson J, Bizzotto F, Dosche C, Dworzak A, Oezaslan M, Simonsen SB, Kuhn LT, Arenz M (2021) The oxygen reduction reaction on Pt: why particle size and interparticle distance matter. ACS Catal 11(12):7144–7153. https://doi.org/10.1021/acscatal.1c00652
22. Jaydev SD, Martín AJ, Pérez-Ramírez J (2021) Direct conversion of polypropylene into liquid hydrocarbons on carbon-supported platinum catalysts. Chemsuschem 14(23):5179–5185. https://doi.org/10.1002/cssc.202101999
23. Hao B, Xiao M, Wang Y, Shang H, Ma J, Liao Y, Mao H (2018) Recyclable amphiphilic metal nanoparticle colloid enabled atmospheric oxidation of alcohols. ACS Appl Mater Interf 10(40):34332–34339. https://doi.org/10.1021/acsami.8b12989
24. Zhang Y, Grass ME, Kuhn JN, Tao F, Habas SE, Huang W, Yang P, Somorjai GA (2008) Highly selective synthesis of catalytically active monodisperse rhodium nanocubes. J Am Chem Soc 130(18):5868–5869. https://doi.org/10.1021/ja801210s
25. Zhan W, Shu Y, Sheng Y, Zhu H, Guo Y, Wang L, Guo Y, Zhang J, Lu G, Dai S (2017) Surfactant-assisted stabilization of Au colloids on solids for heterogeneous catalysis. Angew Chem Int Edit 56(16):4494–4498. https://doi.org/10.1002/anie.201701191
26. Zhu H, Ma Z, Clark JC, Pan Z, Overbury SH, Dai S (2007) Low-temperature CO oxidation on Au/fumed SiO_2-based catalysts prepared from $Au(en)_2Cl_3$ precursor. Appl Catal A-Gen 326(1):89–99. https://doi.org/10.1016/j.apcata.2007.04.004

27. Zhao L, Jiang D, Cai Y, Ji X, Xie R, Yang W (2012) Tuning the size of gold nanoparticles in the citrate reduction by chloride ions. Nanoscale 4(16):5071–5076. https://doi.org/10.1039/C2NR30957B
28. Azizi Y, Petit C, Pitchon V (2010) Role of support in the oxidation of acetylene over gold catalysts. J Catal 269(1):26–32. https://doi.org/10.1016/j.jcat.2009.10.011
29. Ivanova S, Pitchon V, Petit C (2006) Application of the direct exchange method in the preparation of gold catalysts supported on different oxide materials. J Mol Catal A: Chem 256(1):278–283. https://doi.org/10.1016/j.molcata.2006.05.006
30. Zhao Q, Yang W, Luo J, Liang C (2022) Regulation of Ni/Al_2O_3 catalysts by metal deposition procedures for selective hydrogenation of adiponitrile. New J Chem 46(4):1498–1506. https://doi.org/10.1039/D1NJ04776K
31. Wang T, Cheng Y, Niu H, Li S, Luo J, Liang C (2023) Construction of a sol–gel derived ternary $CuZn/FeO_x$ nanostructure for catalytic transfer hydrogenation of furfural. Sustain Energ Fuel 7(5):1187–1195. https://doi.org/10.1039/D2SE01638A
32. Luo J, Cheng Y, Niu H, Wang T, Liang C (2022) Efficient Cu/FeO_x catalyst with developed structure for catalytic transfer hydrogenation of furfural. J Catal 413:575–587. https://doi.org/10.1016/j.jcat.2022.07.013
33. Setthapun W, Williams WD, Kim SM, Feng H, Elam JW, Rabuffetti FA, Poeppelmeier KR, Stair PC, Stach EA, Ribeiro FH, Miller JT, Marshall CL (2010) Genesis and evolution of surface species during Pt atomic layer deposition on oxide supports characterized by in situ XAFS analysis and water−gas shift reaction. J Phys Chem C 114(21):9758–9771. https://doi.org/10.1021/jp911178m
34. Lu J, Low K-B, Lei Y, Libera JA, Nicholls A, Stair PC, Elam JW (2014) Toward atomically-precise synthesis of supported bimetallic nanoparticles using atomic layer deposition. Nat Commun 5(1):3264. https://doi.org/10.1038/ncomms4264
35. Gu J, Xu Y, Lu J (2023) Atom-precise low-nuclearity cluster catalysis: opportunities and challenges. ACS Catal 13(8):5609–5634. https://doi.org/10.1021/acscatal.3c01449
36. Fonseca J, Lu J (2021) Single-atom catalysts designed and prepared by the atomic layer deposition technique. ACS Catal 11(12):7018–7059. https://doi.org/10.1021/acscatal.1c01200
37. Lei Y, Lee S, Low K-B, Marshall CL, Elam JW (2016) Combining electronic and geometric effects of ZnO-promoted Pt nanocatalysts for aqueous phase reforming of 1-propanol. ACS Catal 6(6):3457–3460. https://doi.org/10.1021/acscatal.6b00963
38. Yi H, Du H, Hu Y, Yan H, Jiang H-L, Lu J (2015) Precisely controlled porous alumina overcoating on Pd catalyst by atomic layer deposition: enhanced selectivity and durability in hydrogenation of 1,3-Butadiene. ACS Catal 5(5):2735–2739. https://doi.org/10.1021/acscatal.5b00129
39. Wang H, Gu X-K, Zheng X, Pan H, Zhu J, Chen S, Cao L, Li W-X, Lu J (2019) Disentangling the size-dependent geometric and electronic effects of palladium nanocatalysts beyond selectivity. Science 5(1):eaat6413. https://doi.org/10.1126/sciadv.aat6413
40. He Y, Liu J-C, Luo L, Wang Y-G, Zhu J, Du Y, Li J, Mao SX, Wang C (2018) Size dependent dynamic structures of supported gold nanoparticles in CO oxidation reaction condition. PNAS 115(30):7700–7705. https://doi.org/10.1073/pnas.1800262115
41. Tuo Y-X, Shi L-J, Cheng H-Y, Zhu Y-A, Yang M-L, Xu J, Han Y-F, Li P, Yuan W-K (2018) Insight into the support effect on the particle size effect of Pt/C catalysts in dehydrogenation. J Catal 360:175–186. https://doi.org/10.1016/j.jcat.2018.02.001
42. Chen Y, Wei J, Duyar MS, Ordomsky VV, Khodakov AY, Liu J (2021) Carbon-based catalysts for Fischer-Tropsch synthesis. Chem Soc Rev 50(4):2337–2366. https://doi.org/10.1039/D0CS00905A
43. Bezemer GL, Bitter JH, Kuipers HPCE, Oosterbeek H, Holewijn JE, Xu X, Kapteijn F, van Dillen AJ, de Jong KP (2006) Cobalt particle size effects in the fischer-tropsch reaction studied with carbon nanofiber supported catalysts. J Am Chem Soc 128(12):3956–3964. https://doi.org/10.1021/ja058282w

44. Parastaev A, Muravev V, Huertas Osta E, van Hoof AJF, Kimpel TF, Kosinov N, Hensen EJM (2020) Boosting CO_2 hydrogenation via size-dependent metal-support interactions in cobalt/ceria-based catalysts. Nat Catal 3(6):526–533. https://doi.org/10.1038/s41929-020-0459-4
45. Tauster SJ, Fung SC, Garten RL (1978) Strong metal-support interactions. Group 8 noble metals supported on titanium dioxide. J Am Chem Soc 100(1):170–175. https://doi.org/10.1021/ja00469a029
46. Frey H, Beck A, Huang X, van Bokhoven JA, Willinger MG (2022) Dynamic interplay between metal nanoparticles and oxide support under redox conditions. Science 376(6596):982–987. https://doi.org/10.1126/science.abm3371
47. Tang M, Li S, Chen S, Ou Y, Hiroaki M, Yuan W, Zhu B, Yang H, Gao Y, Zhang Z, Wang Y (2021) Facet-dependent oxidative strong metal-support interactions of palladium-TiO_2 determined by in situ transmission electron microscopy. Angew Chem Int Edit 60(41):22339–22344. https://doi.org/10.1002/anie.202106805
48. Chen Z, Zeng X, Li X, Lv Z, Li J, Zhang Y (2022) Strong metal phosphide-phosphate support interaction for enhanced non-noble metal catalysis. Adv Mater 34(5):2106724. https://doi.org/10.1002/adma.202106724
49. Liu X, Wang X, Zhen S, Sun G, Pei C, Zhao Z-J, Gong J (2022) Support stabilized PtCu single-atom alloys for propane dehydrogenation. Chem Sci 13(33):9537–9543. https://doi.org/10.1039/D2SC03723H
50. Wang F, Zhang L, Zhang Q, Yang J, Duan G, Xu W, Yang F, Jiang S (2021) Electrode thickness design toward bulk energy storage devices with high areal/volumetric energy density. Appl Energ 289:116734. https://doi.org/10.1016/j.apenergy.2021.116734
51. Wang M, Wang Z, Liu S, Gao R, Cheng K, Zhang L, Zhang G, Min X, Kang J, Zhang Q, Wang Y (2021) Synthesis of hierarchical SAPO-34 to improve the catalytic performance of bifunctional catalysts for syngas-to-olefins reactions. J Catal 394:181–192. https://doi.org/10.1016/j.jcat.2020.08.020
52. Jiao F, Li J, Pan X, Xiao J, Li H, Ma H, Wei M, Pan Y, Zhou Z, Li M, Miao S, Li J, Zhu Y, Xiao D, He T, Yang J, Qi F, Fu Q, Bao X (2016) Selective conversion of syngas to light olefins. Science 351(6277):1065–1068. https://doi.org/10.1126/science.aaf1835
53. Cui X, Ren P, Deng D, Deng J, Bao X (2016) Single layer graphene encapsulating non-precious metals as high-performance electrocatalysts for water oxidation. Energ Environ Sci 9(1):123–129. https://doi.org/10.1039/C5EE03316K
54. Tu Y, Ren P, Deng D, Bao X (2018) Structural and electronic optimization of graphene encapsulating binary metal for highly efficient water oxidation. Nano Energ 52:494–500. https://doi.org/10.1016/j.nanoen.2018.07.062
55. Liu J, Zhang G, Ye K, Xu K, Sheng Y, Yu C, Zhang H, Li Q, Liang Z, Jiang K (2023) Top-down manufacturing of efficient CO_2 reduction catalysts from the gasification residue carbon. Chem Commun 59(5):611–614. https://doi.org/10.1039/D2CC05081A
56. Luo J, Qu E, Zhou Y, Dong Y, Liang C (2020) Re/AC catalysts for selective hydrogenation of dimethyl 1, 4-cyclohexanedicarboxylate to 1, 4-cyclohexanedimethanol: Essential roles of metal dispersion and chemical environment. Appl Catal A-Gen 602:117669. https://doi.org/10.1016/j.apcata.2020.117669
57. Cao S, Ma Y, Chu W, Liu Y (2022) High-density MoC_x nanoclusters anchored on nanodiamond-derived nanocarbon as a robust CO_2 reduction catalyst for syngas production. Fuel 323:124347. https://doi.org/10.1016/j.fuel.2022.124347
58. Hu J, Wei Z, Zhang Y, Huang R, Zhang M, Cheng K, Zhang Q, Qi Y, Li Y, Mao J, Zhu J, Wu L, Wen W, Yu S, Pan Y, Yang J, Wei X, Jiang L, Si R, Yu L, Wang Y, Deng D (2023) Edge-rich molybdenum disulfide tailors carbon-chain growth for selective hydrogenation of carbon monoxide to higher alcohols. Nat Commun 14(1):6808. https://doi.org/10.1038/s41467-023-42325-z
59. Kang H, Zhu L, Li S, Yu S, Niu Y, Zhang B, Chu W, Liu X, Perathoner S, Centi G, Liu Y (2023) Generation of oxide surface patches promoting H-spillover in Ru/(TiO_x)MnO catalysts enables CO_2 reduction to CO. Nat Catal 6:1062–1072. https://doi.org/10.1038/s41929-023-01040-0

60. Liu A, Liu X, Liu L, Pu Y, Guo K, Tan W, Gao S, Luo Y, Yu S, Si R, Shan B, Gao F, Dong L (2019) Getting insights into the temperature-specific active sites on platinum nanoparticles for CO oxidation: a combined in situ spectroscopic and ab initio density functional theory study. ACS Catal 9(9):7759–7768. https://doi.org/10.1021/acscatal.9b02552
61. Ding K, Gulec A, Johnson AM, Schweitzer NM, Stucky GD, Marks LD, Stair PC (2015) Identification of active sites in CO oxidation and water-gas shift over supported Pt catalysts. Science 350(6257):189–192. https://doi.org/10.1126/science.aac6368
62. Kale MJ, Christopher P (2016) Utilizing quantitative in situ FTIR spectroscopy to identify well-coordinated Pt atoms as the active site for CO oxidation on Al_2O_3-supported Pt catalysts. ACS Catal 6(8):5599–5609. https://doi.org/10.1021/acscatal.6b01128
63. Luo J, Liu Y, Zhang L, Ren Y, Miao S, Zhang B, Su DS, Liang C (2019) Atomic-scale observation of bimetallic Au–CuO_x nanoparticles and their interfaces for activation of CO molecules. ACS Appl Mater Inter 11(38):35468–35478. https://doi.org/10.1021/acsami.9b12017
64. Wu G, Liu Y, He Y, Feng J, Li D (2021) Reaction pathway investigation using in situ Fourier transform infrared technique over Pt/CuO and Pt/TiO_2 for selective glycerol oxidation. Appl Catal B 291:120061. https://doi.org/10.1016/j.apcatb.2021.120061
65. Wang Y, Hu X, Guo T, Tian W, Hao J, Guo Q (2021) The competitive adsorption mechanism of CO_2, H_2O and O_2 on a solid amine adsorbent. Chem Eng J 416:129007. https://doi.org/10.1016/j.cej.2021.129007
66. Sun J, Li X, Zhao Q, Liu B (2021) Ultrathin nanoflake-assembled hierarchical BiOBr microflower with highly exposed (001) facets for efficient photocatalytic degradation of gaseous ortho-dichlorobenzene. Appl Catal B: Environ 281:119478. https://doi.org/10.1016/j.apcatb.2020.119478
67. Luo J, Yang S, Ling Y, Yang W, Niu H, Li W, Liu H, Liang C (2023) Defective Au nanoparticles manipulated by Au-Mg_xAl-LDH interplay for alkali-free oxidation of benzyl alcohol. Chem Eng J 473:145171. https://doi.org/10.1016/j.cej.2023.145171
68. Lin L, Zhou W, Gao R, Yao S, Zhang X, Xu W, Zheng S, Jiang Z, Yu Q, Li Y-W, Shi C, Wen X-D, Ma D (2017) Low-temperature hydrogen production from water and methanol using Pt/α-MoC catalysts. Nature 544(7648):80–83. https://doi.org/10.1038/nature21672
69. Zhang Y, Zhang J, Zhang B, Si R, Han B, Hong F, Niu Y, Sun L, Li L, Qiao B, Sun K, Huang J, Haruta M (2020) Boosting the catalysis of gold by O_2 activation at Au–SiO_2 interface. Nat Commun 11(1):558. https://doi.org/10.1038/s41467-019-14241-8
70. Arvay JW, Hong W, Li C, Delgass WN, Ribeiro FH, Harris JW (2022) Kinetics of propylene epoxidation over Extracrystalline gold active sites on Au/TS-1 catalysts. ACS Catal 12(16):10147–10160. https://doi.org/10.1021/acscatal.2c02213
71. Yao S, Zhang X, Zhou W, Gao R, Xu W, Ye Y, Lin L, Wen X, Liu P, Chen B, Crumlin E, Guo J, Zuo Z, Li W, Xie J, Lu L, Kiely CJ, Gu L, Shi C, Rodriguez JA, Ma D (2017) Atomic-layered Au clusters on α-MoC as catalysts for the low-temperature water gas shift reaction. Science 357(6349):389–393. https://doi.org/10.1126/science.aah4321
72. Yang M, Allard LF, Flytzani-Stephanopoulos M (2013) Atomically dispersed Au-$(OH)_x$ species bound on Titania Catalyze the low-temperature water-gas shift reaction. J Am Chem Soc 135(10):3768–3771. https://doi.org/10.1021/ja312646d
73. Rodriguez JA, Ramírez PJ, Asara GG, Viñes F, Evans J, Liu P, Ricart JM, Illas F (2014) Charge Polarization at a Au-TiC interface and the generation of highly active and selective catalysts for the low-temperature water-gas shift reaction. Angew Chem Int Edit 53(42):11270–11274. https://doi.org/10.1002/anie.201407208
74. Tian H, He Y, Zhao Q, Li J, Shao X, Zhang Z, Huang X, Lu C, Wang K, Jiang Q, Ng AMC, Xu H, Tong SY (2021) Avoiding Sabatier's conflict in bifunctional heterogeneous catalysts for the WGS reaction. Chem 7(5):1271–1283. https://doi.org/10.1016/j.chempr.2021.01.018
75. O'Connell KC, Monnier JR, Regalbuto JR (2018) The curious relationship of sintering to activity in supported gold catalysts for the hydrochlorination of acetylene. Appl Catal B: Environ 225:264–272. https://doi.org/10.1016/j.apcatb.2017.12.001

76. Davies CJ, Miedziak PJ, Brett GL, Hutchings GJ (2016) Vinyl chloride monomer production catalysed by gold: a review. Chin J Catal 37(10):1600–1607. https://doi.org/10.1016/S1872-2067(16)62482-8
77. Nkosi B, Adams MD, Coville NJ, Hutchings GJ (1991) Hydrochlorination of acetylene using carbon-supported gold catalysts: a study of catalyst reactivation. J Catal 128(2):378–386. https://doi.org/10.1016/0021-9517(91)90296-G
78. Lu Z, Kunisch J, Gan Z, Bunian M, Wu T, Lei Y (2020) Gold catalysts synthesized using a modified incipient wetness impregnation method for propylene epoxidation. ChemCatChem 12(23):5993–5999. https://doi.org/10.1002/cctc.202001053
79. Lu Z, Piernavieja-Hermida M, Turner CH, Wu Z, Lei Y (2018) Effects of TiO_2 in low temperature propylene epoxidation using gold catalysts. J Phys Chem C 122(3):1688–1698. https://doi.org/10.1021/acs.jpcc.7b10902
80. Harris JW, Arvay J, Mitchell G, Delgass WN, Ribeiro FH (2018) Propylene oxide inhibits propylene epoxidation over Au/TS-1. J Catal 365:105–114. https://doi.org/10.1016/j.jcat.2018.06.015
81. Agarwal N, Freakley SJ, McVicker RU, Althahban SM, Dimitratos N, He Q, Morgan DJ, Jenkins RL, Willock DJ, Taylor SH, Kiely CJ, Hutchings GJ (2017) Aqueous Au–Pd colloids catalyze selective CH_4 oxidation to CH_3OH with O_2 under mild conditions. Science 358(6360):223–227. https://doi.org/10.1126/science.aan6515
82. Hong F, Wang S, Zhang J, Fu J, Jiang Q, Sun K, Huang J (2021) Strong metal-support interaction boosting the catalytic activity of Au/TiO_2 in chemoselective hydrogenation. Chin J Catal 42(9):1530–1537. https://doi.org/10.1016/S1872-2067(20)63763-9
83. López Granados M, Moreno J, Alba-Rubio AC, Iglesias J, Martín Alonso D, Mariscal R (2020) Catalytic transfer hydrogenation of maleic acid with stoichiometric amounts of formic acid in aqueous phase: paving the way for more sustainable succinic acid production. Green Chem 22(6):1859–1872. https://doi.org/10.1039/C9GC04221K
84. Blaser H-U, Steiner H, Studer M (2009) Selective catalytic hydrogenation of functionalized Nitroarenes: an update. ChemCatChem 1(2):210–221. https://doi.org/10.1002/cctc.200900129
85. Li H, Qin X, Zhang X-G, Jiang K, Cai W-B (2022) Boron-doped platinum-group metals in Electrocatalysis: a perspective. ACS Catal 12(20):12750–12764. https://doi.org/10.1021/acscatal.2c04358
86. Sudarsanam P, Peeters E, Makshina EV, Parvulescu VI, Sels BF (2019) Advances in porous and nanoscale catalysts for viable biomass conversion. Chem Soc Rev 48(8):2366–2421. https://doi.org/10.1039/C8CS00452H
87. Besson M, Gallezot P, Pinel C (2014) Conversion of biomass into chemicals over metal catalysts. Chem Rev 114(3):1827–1870. https://doi.org/10.1021/cr4002269
88. Zu Y, Guo ZS, Zheng J, Hui Y, Wang SH, Qin YC, Zhang L, Liu HH, Gao XH, Song LJ (2020) Investigation of Cu(I)-Y zeolites with different Cu/Al ratios towards the ultra-deep adsorption desulfurization: discrimination and role of the specific adsorption active sites. Chem Eng J 380:122319. https://doi.org/10.1016/j.cej.2019.122319
89. Meemken F, Baiker A (2017) Recent Progress in heterogeneous asymmetric hydrogenation of C=O and C=C bonds on supported noble metal catalysts. Chem Rev 117(17):11522–11569. https://doi.org/10.1021/acs.chemrev.7b00272
90. Rozmysłowicz B, Kirilin A, Aho A, Manyar H, Hardacre C, Wärnå J, Salmi T, Murzin DY (2015) Selective hydrogenation of fatty acids to alcohols over highly dispersed ReO_x/TiO_2 catalyst. J Catal 328:197–207. https://doi.org/10.1016/j.jcat.2015.01.003
91. Di X, Shao ZF, Li C, Li WZ, Liang CH (2015) Hydrogenation of succinic acid over supported rhenium catalysts prepared by the microwave-assisted thermolytic method. Catal Sci Technol 5(4):2441–2448. https://doi.org/10.1039/c5cy00004a
92. Hong UG, Park HW, Lee J, Hwang S, Yi J, Song IK (2012) Hydrogenation of succinic acid to tetrahydrofuran (THF) over rhenium catalyst supported on H_2SO_4-treated mesoporous carbon. Appl Catal A-Gen 415–416:141–148. https://doi.org/10.1016/j.apcata.2011.12.022

93. Medlin JW (2011) Understanding and controlling reactivity of unsaturated oxygenates and polyols on metal catalysts. ACS Catal 1(10):1284–1297. https://doi.org/10.1021/cs200336r
94. Shi Y, Zhu Y, Yang Y, Li Y-W, Jiao H (2015) Exploring furfural catalytic conversion on Cu(111) from computation. ACS Catal 5(7):4020–4032. https://doi.org/10.1021/acscatal.5b00303
95. Toledo F, Ghampson IT, Sepúlveda C, García R, Fierro JLG, Videla A, Serpell R, Escalona N (2019) Effect of Re content and support in the liquid phase conversion of furfural to furfuryl alcohol and 2-methyl furan over ReO_x catalysts. Fuel 242:532–544. https://doi.org/10.1016/j.fuel.2019.01.090
96. Luo J, Liang C (2024) Rhenium in heterogeneous catalysis: a rising star for hydrogenation reactions. ACS Catal 14(9):7032–7049. https://doi.org/10.1021/acscatal.4c00298
97. Nanda S, Rana R, Zheng Y, Kozinski JA, Dalai AK (2017) Insights on pathways for hydrogen generation from ethanol. Sustain Energ Fuel 1(6):1232–1245. https://doi.org/10.1039/C7SE00212B
98. Zanchet D, Santos JBO, Damyanova S, Gallo JMR, Bueno JMC (2015) Toward understanding metal-catalyzed ethanol reforming. ACS Catal 5(6):3841–3863. https://doi.org/10.1021/cs5020755
99. Ogo S, Sekine Y (2020) Recent progress in ethanol steam reforming using non-noble transition metal catalysts: a review. Fuel Process Technol 199:106238. https://doi.org/10.1016/j.fuproc.2019.106238
100. Nie R, Tao Y, Nie Y, Lu T, Wang J, Zhang Y, Lu X, Xu CC (2021) Recent advances in catalytic transfer hydrogenation with formic acid over heterogeneous transition metal catalysts. ACS Catal 11(3):1071–1095. https://doi.org/10.1021/acscatal.0c04939
101. Zhang C, Luo J, Zhou Y, Jiang Q, Liang C (2020) Metal oxide sub-nanoclusters decorated Ni catalyst for selective hydrogenation of adiponitrile to hexamethylenediamine. J Catal 381:14–25. https://doi.org/10.1016/j.jcat.2019.10.024
102. Fu Z, Wang Z, Lin W, Song W, Li S (2017) High efficient conversion of furfural to 2-methylfuran over Ni–Cu/Al_2O_3 catalyst with formic acid as a hydrogen donor. Appl Catal A-Gen 547:248–255. https://doi.org/10.1016/j.apcata.2017.09.011
103. Usman M, Wan Daud WMA, Abbas HF (2015) Dry reforming of methane: influence of process parameters a review. Renew Sustain Energ Rev 45:710–744. https://doi.org/10.1016/j.rser.2015.02.026
104. Guharoy U, Reina TR, Liu J, Sun Q, Gu S, Cai Q (2021) A theoretical overview on the prevention of coking in dry reforming of methane using non-precious transition metal catalysts. J CO_2 Util 53:101728. https://doi.org/10.1016/j.jcou.2021.101728
105. Gholami Z, Tisler Z, Rubás V (2021) Recent advances in Fischer-Tropsch synthesis using cobalt-based catalysts: a review on supports, promoters, and reactors. Catal Rev 63(3):512–595. https://doi.org/10.1080/01614940.2020.1762367
106. Ail SS, Dasappa S (2016) Investigations into enhanced wax production with combustion synthesized Fischer-Tropsch catalysts. Energ Convers Manage 116:80–90. https://doi.org/10.1016/j.enconman.2016.02.075

Chapter 3
Organic Synthesis Catalyzed by Metal Nanoclusters

Abstract Metal nanoclusters (MNCs) usually refer to a type of metal nanoparticles with an ultra-small size, i.e., usually <2 nm. In general, metal nanoclusters are generated in situ in the presence of bulky ligands (stabilizers), which prevent agglomeration of the nanoclusters. Atomically precise metal nanoclusters can be considered quasi-homogeneous catalysts because their well-defined molecular-like structures are similar to those of homogeneous catalysts, the metal centers of which are often coordinated with organic ligands (stabilizers). Metal nanoparticles with organic stabilizers exhibit good catalytic efficiency and selectivity in various reactions on account of their well-defined size, composition, surface, and structure. This chapter mainly discusses the preparation, characterization, and application of metal nanoclusters in organic transformation, including hydrogenation, dehydrogenation, oxidation, allylic substitution, polymerization, coupling reactions, and so on.

Keywords Metal nanocluster · Stabilizer · Preparation and characterization · Application · Organic synthesis

Because of the unique physical and chemical properties, such as surface-to-volume effect, quantum-size effect of materials on the nanoscale, nanocatalysts display better catalytic activity and selectivity compared with traditional catalysts, playing important roles in catalysis science.

Metal nanoclusters (MNCs) are particles <2 nm in size or around a few hundred atoms are compared with their larger nanoparticles or nanocrystals counterparts. The study of metal clusters originated in 1958 when trying to repeat a patent which claimed that a mixture of iron and cobalt carbonyls gave improved selectivity in hydroformylation. In 1980s, Haruta et al. discovered metal-oxide-supported nanosized gold particles that showed good catalytic activity in carbon monoxide oxidation at very low temperatures, setting off a wave of research on the catalysis of MNCs. With the advances of technology, in recent years, more attention has been paid to analyze the individual effects of the metal particle's size, composition, support and modifier on the catalytic properties.

M. Bao et al., *Nanometal Catalysis in Organic Synthesis*, Molecular Catalysis 3,
https://doi.org/10.1007/978-981-97-4573-9_3

This chapter mainly introduces the preparation and characterization of MNCs, then discusses the catalytic properties of MNCs (single metals and alloys) in organic transformation.

3.1 Synthesis Methods of MNCs

Due to the high surface free energy of metal nanoclusters, they tend to agglomerate, and it is particularly important to obtain stable metal nanoclusters. Metal nanoclusters can generally be stabilized in three ways: (1) carrier fixation. Fixation of metal atoms or metal clusters by coordination atoms (such as N, P, S, O, etc.) on the carrier and carrier defects; (2) the material structure is fixed. The thermal movement of metal atoms and metal clusters is restricted by the material's own pore or surface covering; (3) Strong metal-carrier interaction. The strong metal carrier interaction was established between metal single atoms and metal clusters and coordination atoms to resist oxidation of metal nanoclusters and improve their stability. Based on the above three points, the preparation methods of metal nanoclusters can be summarized into the following categories.

3.1.1 Direct Synthesis Method

This is an easy method for obtaining single metal nanoclusters, which are protected by the sterically demanding group containing thio-group (–SR). There are usually two steps: first, a rapid initial formation of polymeric, then decomposition of the polymer is stimulated almost instantaneously at ambient temperature by addition of a strong reducing agent in great excess. For example, Schaaff et al. used glutathione (GSH) as the thiol ligand precursor, Au(III) is converted to the Au(I)−SR complex and then directly reduced to Au NCs by a reducing agent (generally $NaBH_4$) [1]. In early research, a series of Au:SG clusters were synthesized by Tsukuda et al. using this method, such as $Au_{10}(SG)_{10}$, $Au_{15}(SG)_{13}$, $Au_{18}(SG)_{14}$, $Au_{22}(SG)_{16}$, $Au_{22}(SG)_{17}$, $Au_{25}(SG)_{18}$, $Au_{29}(SG)_{20}$, $Au_{33}(SG)_{22}$, and $Au_{39}(SG)_{24}$, and pointed out the evolution process of gold-thiolate binary systems (Scheme 3.1).

If the nanoclusters obtained by this method are polydispersed, further processing is required, such as etching by excess ligand, high temperature, and prolonged aging, to make size-focusing of the MNCs. The team of Jin et al. used a similar method with different thiolate ligands, reported many families of Au NCs, such as the family of MBT (methylbenzenethiol): Au_{130}(*p*-MBT)$_{50}$, Au_{104}(*m*-MBT)$_{41}$, and Au_{40}(*o*-MBT)$_{24}$, and another family of TBBT (4-*tert*-butylbenzenethiolate): $Au_{44}(TBBT)_{28}$, $Au_{28}(TBBT)_{20}$, $Au_{36}(TBBT)_{24}$, $Au_{52}(TBBT)_{32}$, et al. [2, 3] (Schemes 3.2 and 3.3).

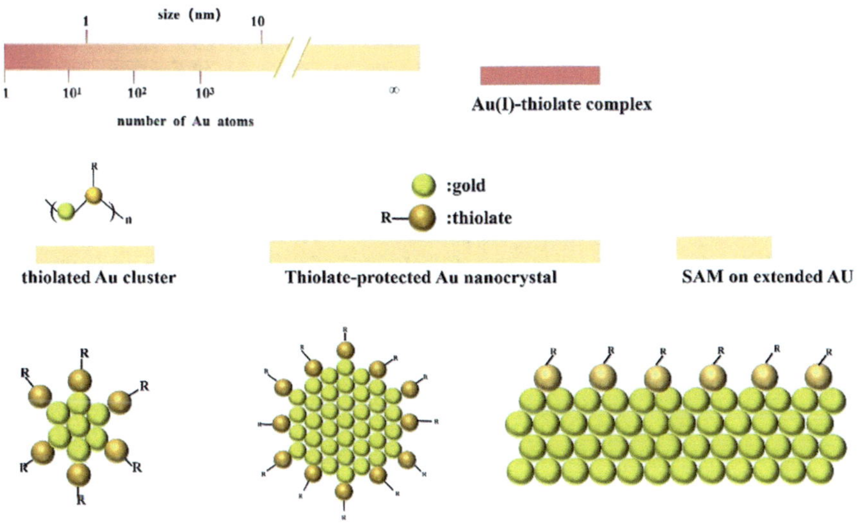

Scheme 3.1 Evolution of gold-thiolate binary systems as a function of the number of gold atoms. Reprinted with permission from Ref. [1]. Copyright © 2005, American Chemical Society

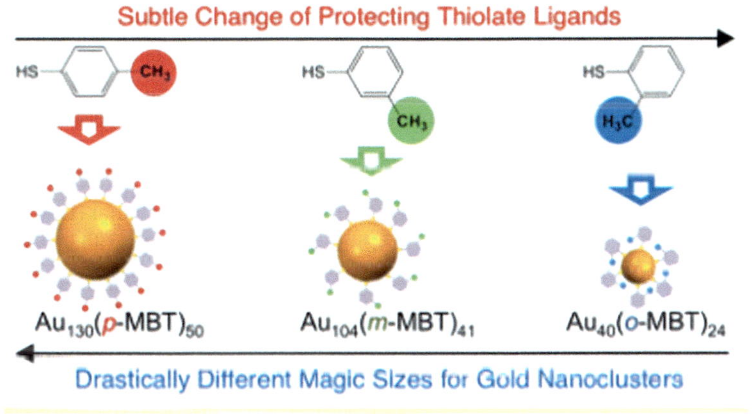

Scheme 3.2 Effect of MBT ligands with different structures on the size of Au NCs. Reprinted with permission from Ref. [2]. Copyright © 2015, American Chemical Society

3.1.2 Ligand-Exchange Method

This method generally requires prepared nanoclusters as the initial raw material, then etched by adding an additional excess of thiol. It should be noted that, the ligands in presynthesized NCs could be phosphor or thiolate ligand which not the same as the thiol added later, this method different from the direct synthesis method.

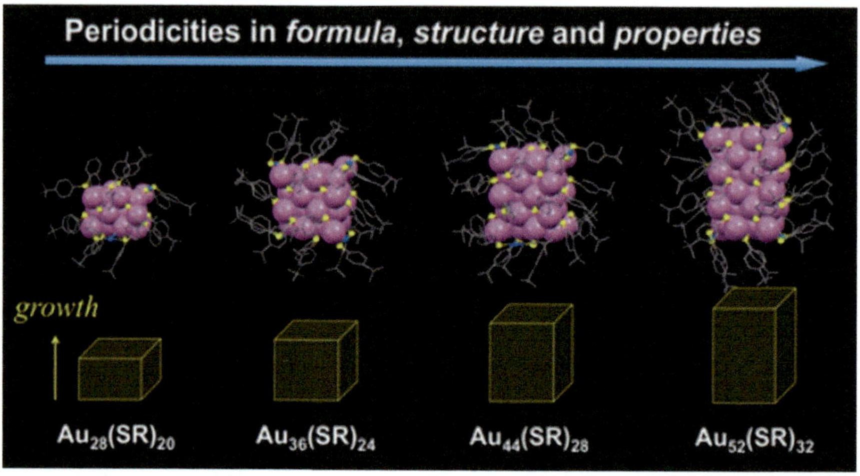

Scheme 3.3 Effect of TBBT ligand on the size of Au NCs. Reprinted with permission from Ref. [3]. Copyright © 2016 American Chemical Society

The MNCs stabilized by phosphine have been studied extensively, these clusters are not stable and tend to decompose even at ambient conditions, which greatly limit their usage in practical applications. But the strong interaction between metal atoms and –SH groups makes the phosphine-stabilized metal clusters be transferred into thiol-stabilized metal clusters by the ligand-exchange reaction. Hutchison and co-workers used $Au_{11}(PPh_3)_8Cl_3$ as the phosphine-stabilized precursor, a diverse family of previously inaccessible thiol-stabilized Au particles were synthesized through the ligand exchange strategy: (a) ligand exchange of 1 in $CHCl_3$ at 55 °C with excess octadecanethiol produces the thiol-stabilized particle 2; (b) interfacial ($CHCl_3/H_2O$) ligand exchange of 1 at 55 °C with (*N*, *N*-dimethylamino)ethanethiol hydrochloride produces the cationic thiol-stabilized particle 3 in a single step; (c) interfacial ($CHCl_3/H_2O$) ligand exchange of 1 at 55 °C with sodium 2-mercaptoethanesulfonate (MESA) produces the anionic particle 4, which contains a mixed ligand shell; (d) treatment of 4 with excess MESA in 1:1 THF/H_2O yields particle 5 containing only MESA in the ligand shell. These Au NCs have the subnanometer core diameters ($d_{CORE} = 0.8 \pm 0.2$ nm) [4] (Scheme 3.4).

Shichibu et al. reported the synthetic method of $Au_{25}(SG)_{18}$ by ligand exchange strategy, the cluster size increases during the ligand exchange reaction of $Au_{11}(PPh_3)_8Cl_3$ by GSH and that thermodynamically stable $Au_{25}(SG)_{18}$ can be selectively obtained on the sub-100 mg scale under similar conditions [5] (Scheme 3.5).

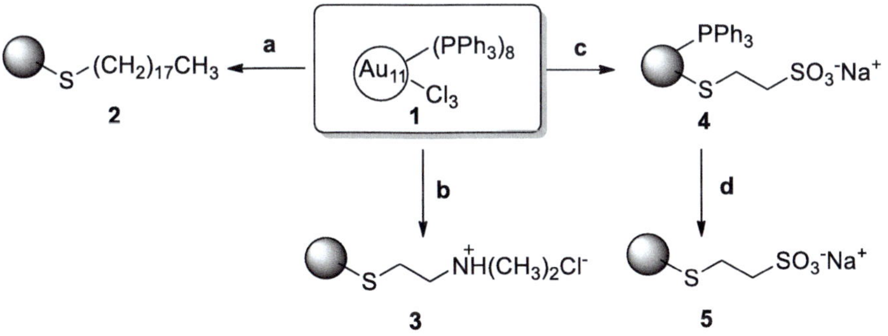

Scheme 3.4 The ligand exchange strategy provides convenient access to a diverse family of previously inaccessible thiol-stabilized Au particles. Reprinted with permission from Ref. [4]. Copyright © 2002 American Chemical Society

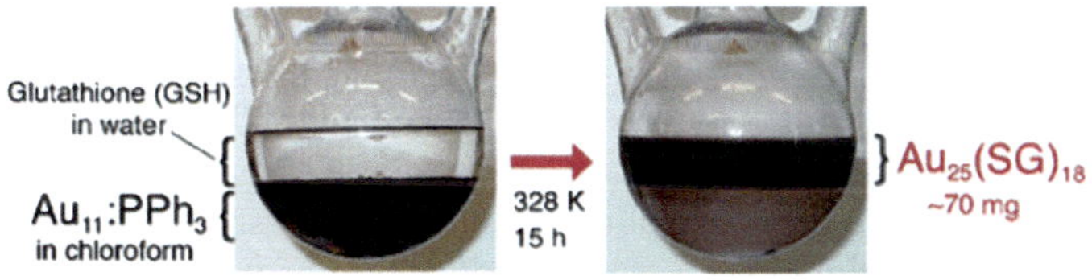

Scheme 3.5 Phase transfer of gold clusters $Au_{11}(PPh_3)_8Cl_3$ from chloroform to water by forming $Au_{25}(SG)_{18}$ clusters upon ligand-exchange reaction. Reprinted with permission from Ref. [5]. Copyright © 2005, American Chemical Society

3.1.3 Chemical Etching Method

Direct etching of bulk metals and template-assisted etching are two effective methods to synthesize MNCs with high metal loads. Due to the strong interaction between metal atoms and protecting ligands, the metal core can be etched by excessive ligands, this strategy is a typical top-down method for the synthesis of MNCs.

Erkang Wang's group has developed an etching method to synthesize smaller Cu NCs, the large-sized Cu nanocrystals (4.4 ± 1.1 nm) which prepared from traditional methods are converted to Cu NCs (1.3 ± 0.4 nm) by etching in the presence of excess ligands [6]. Cu nanocrystals were readily prepared by reduction of Cu^{2+} ions using ascorbic acid (AA) as both the reducing agent and capping agent at room temperature. Typically, $Cu(NO_3)_2$ aqueous solution (0.1 M, 1 mL) was added drop wise into the AA solution (0.1 M, 10 mL) under vigorous stirring. The reaction mixture was left to stir for 1 h at room temperature. Then a natural tripeptide GSH was adopted as a model etching agent for a ligand exchange reaction with the colloidal Cu nanocrystals. The as-prepared Cu nanocrystal solution (0.3 mL) was added to the GSH aqueous solution (0.05 M, 3 mL) and kept stirring for 5 h at room temperature. Large nanocrystals decomposed and clusters with average diameters could be observed by Transmission electron microscopy (TEM) (Scheme 3.6).

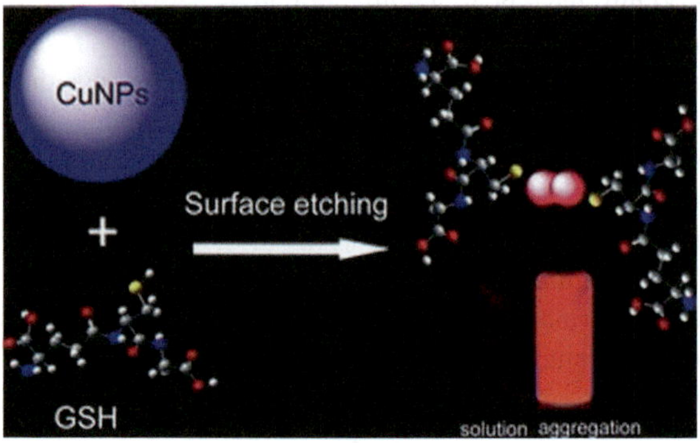

Scheme 3.6 Synthesis of CuNCs by etching method. Reprinted with permission from Ref. [6]. Copyright © 2013 Wiley–VCH Verlag GmbH & Co. KGaA, Weinheim

Han et al. reported an etching method by NaOH to prepare a hollow N-doped carbon sphere with isolated single Co atomic sites (ISAS-Co/HNCS) [7]. The SiO_2 template was first prepared, which was dispersed in Co-complex solution as template before addition of another monomer. The obtained mixed solution performed quaterization. The collected powder was thermally treated under flowing hydrogen/argon and then etched with sodium hydroxide to remove SiO_2 template. Because of the features of the single sites and the hollow substrate, it shows an excellent activity for oxygen reduction reaction (ORR), a good methanol tolerance as well as excellent stability (Scheme 3.7).

Chen et al. chose porous nickel foam as a metal carrier and then used chemical vapor deposition to deposit an atom-thick layer of graphene on the surface of the nickel foam. Ni was completely dissolved by acid etching, and finally an independent porous graphene-supported unit point Ni catalyst was obtained [8]. The author pointed out that because of the *sp–d* orbital charge transfer between the Ni dopants and the surrounding carbon atoms, the resultant local structure with empty C–Ni hybrid orbitals is catalytically active and electrochemically stable.

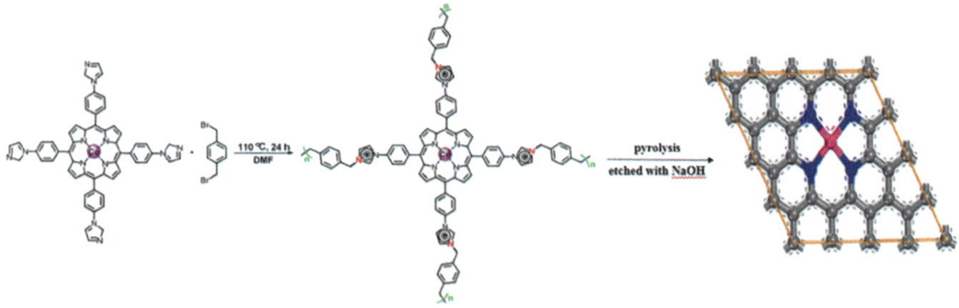

Scheme 3.7 Schematic roadmap of ISAS-Co/HNCS. Reprinted with permission from Ref. [7]. Copyright © 2017, American Chemical Society

3.1.4 Metal-Exchange Method

The method is similar to an atomic "surgery", which leads to the "resection" of some metal atoms in the initial cluster and the formation of a new cluster, without changing the other parts of the starting cluster structure, it's an effective approach for the synthesis of alloy NCs.

Li et al. synthesized $[Au_{23-x}Ag_x(SR)_{16}]^-$ ($x = 1$ to 2), the method is as follows: 20 mg of molecularly pure $[Au_{23}(SR)_{16}]^-$ was first dissolved in 3 mL of DCM, and 1.2 mg of $Ag^I(SR)$ was added, with a molar ratio of elemental Au/Ag in reactants of 1:0.07. The reaction was allowed to proceed for ~1 h. After $[Au_{23}(SR)_{16}]^-$ was converted to $[Au_{23-x}Ag_x(SR)_{16}]^-$, the DCM solution was directly transferred into a glass tube, and pentane was diffused into the solution at room temperature for crystallization of clusters. After ~2 days, crystals of $[Au_{23-x}Ag_x(SR)_{16}]^-$ were obtained on the inner wall and at the bottom of the glass tube [9]. To obtain high-quality single crystals, the crystals were redissolved and then recrystallized for several times. This method offers a promising strategy for atomic surgery on nanoclusters to tailor their structure and functionality (Scheme 3.8).

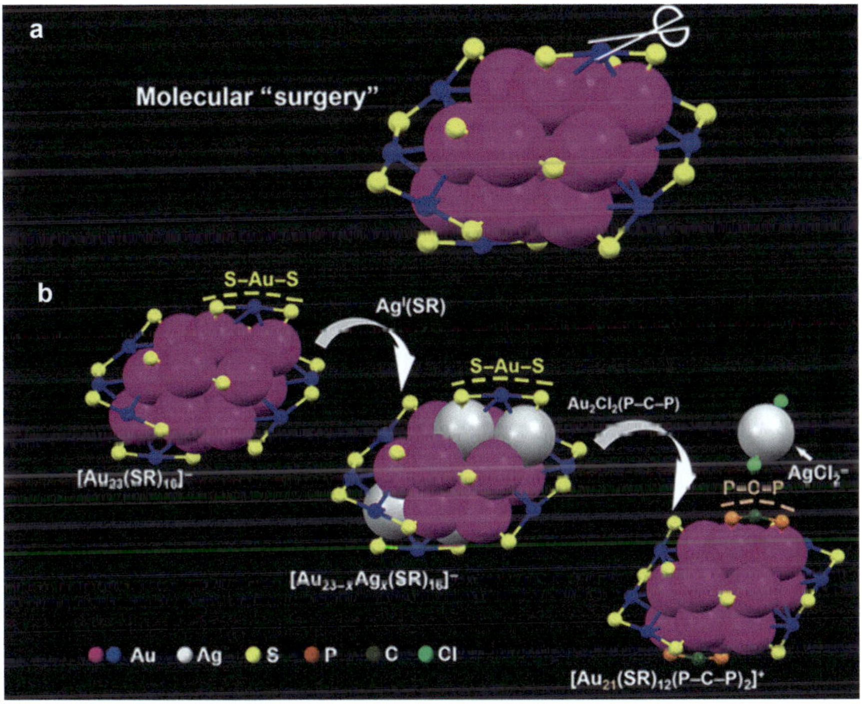

Scheme 3.8 Molecular surgery on the atomically precise 23-gold-atom nanocluster by a two-step metal-exchange method. Reprinted with permission from Ref. [9]. Copyright © 2017, The Authors

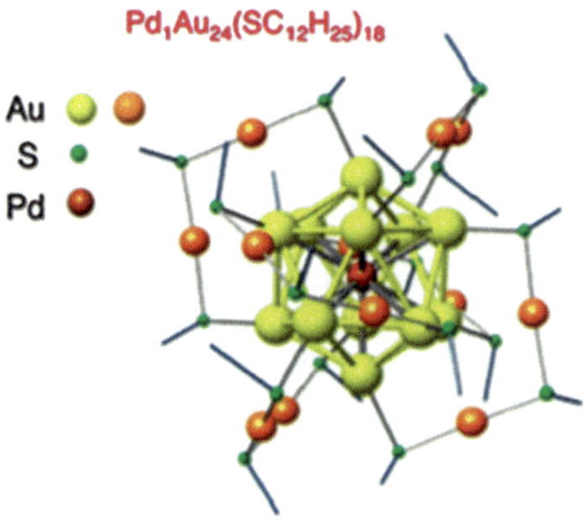

Scheme 3.9 The structure of $Pd_1Au_{24}(SC_{12}H_{25})_{18}$. Reprinted with permission from Ref. [10]. Copyright © 2010 Royal Society of Chemistry

Negishi et al. reported a method to prepare dodecanethiolate-protected $Pd_1Au_{24}(SC_{12}H_{25})_{18}$ cluster [10]. In this method, two metal salts $HAuCl_4$ and Na_2PdCl_4 replaced the single metal salt $HAuCl_4$ in synthesis of Au_{25} nanoclusters, which led to an Au atom in the Au_{25} nanocluster was replaced by a Pd atom. The experimental results show that the doping of a central atom is a powerful method to increase the stability beyond the $Au_{25}(SR)_{18}$ cluster (Scheme 3.9).

3.1.5 Separation Method

The synthesis of metal nanoclusters is usually accompanied a large number of by-products, and the crude products need to be separated and purified to obtain monodisperse atomic-level precise nanoclusters. Therefore, separation technique is a very important method to synthesize high-purity NCs, such as chromatography [high-performance liquid chromatography (HPLC), size-exclusion chromatography (SEC), sub-exchange chromatography (IEC), and so on.], polyacrylamide gel electrophoresis (PAGE), thin-layer chromatography (TLC), etc.

Negishi et al. successfully isolated $PdAu_{24}$ nanoclusters protected by two ligands through HPLC [11]. As the $PdAu_{24}$ nanoclusters are insoluble in methanol but easily soluble in tetrahydrofuran, the nanoclusters are gradually eluded with the increase of the proportion of tetrahydrofuran in the elution phase.

Bigioni et al. synthesized glutathione-protected silver nanoclusters (Ag:SG) with the improved Brust Schiffrin method, and separated the silver nanoclusters by PAGE [12]. Because there are multiple bands of different colors in the mixture, it indicates the presence of multiple components. When less glutathione was added to the synthesis process, other additional bands of higher quality were identified. The authors also separated the gold and silver nanoclusters in the same polyacrylamide gel and found that although both gold and silver provide an electron to the clusters, their banding is significantly different, the most stable clusters in gold and silver nanoclusters may be different in size and structure, possibly due to differences in Au–S and Ag–S bonds.

The team of Pradeep was the first to use TLC for the separation of metal nanoclusters [13]. The method has been successfully applied to the separation of $Au_{25}SR_8$ binary mixed nanoclusters with different ligand protection, and Au_{25} and Au_{144} binary mixed nanoclusters with different number of metal atoms protected by the same ligand. Different clusters are separated by different bands. The separation results fully show that TLC is a simple and effective cluster separation method, can be used to separate nanoclusters of the same metal nucleus protected by different ligands, and can also be used to separate nanoclusters of different metal nuclei protected by the same ligand.

3.1.6 Electrochemical Synthesis Method

In 1994, Reetz and co-workers demonstrated that electrochemical methods could be used to produce certain transition metal colloids in the nanometer region, and that the particle size could be controlled in a simple manner by adjustment of the current density for the first time [14]. This method used an inexpensive two-electrode setup, the sacrificial anode would be transformed into metal clusters which consisted of the bulk metal. Potentiostatic control is not necessary. The electrolyte consisted of tetraalkylammonium salts, which also served as stabilizers for the metal clusters, which hindered the formation of undesired metal powders. This method exhibits a lot of advantages in preparation of metal nanoclusters, such as the low reaction temperature, large scale yield, low cost of the initial materials, and easy manipulation of cluster size by tuning current, voltage, electrolyte, concentration of stabilizers etc.

Based on this method, López-Quintela's group prepared the Ag NCs in acetonitrile containing tetrabutylammonium [15]. The study found that current density plays an important role, not only on particle size but also on the efficiency of the process, at the same time, only particles have been obtained when platinum instead of aluminum was employed, which indicated that the cathode nature maybe also a decisive parameter. In recent years, they synthesized Au NCs and Cu NCs by the electrochemical method, which makes them useful as very small building blocks, bringing new possibilities to construct novel nano/microstructures, with potential applications in fields such as biosensors, biomedicine, etc. [16, 17] (Scheme 3.10).

3.1.7 Support Method

Subject to application conditions, sometimes, MNCs usually need the help of "assistants", such as a support of the catalyst. The interaction between the carrier and the atom of MNCs is favorable for the reaction, and the size of MNCs controls difficultly

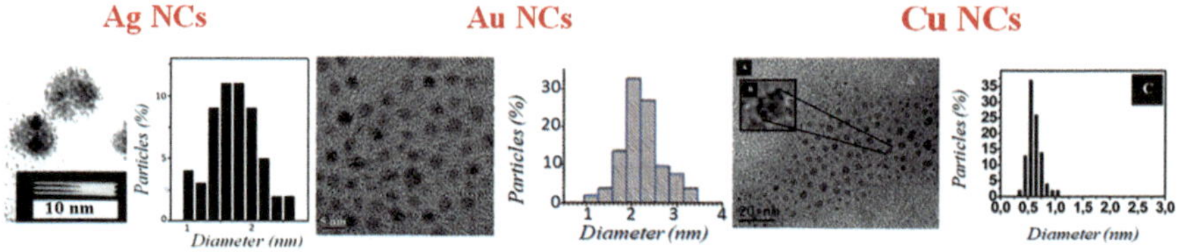

Scheme 3.10 TEM images and size distribution of the electrochemically synthesized Ag NCs, Au NCs and Cu NCs. Reprinted with permission from Ref. [15–17]. [15] Copyright © 2000 American Chemical Society. [16] Copyright © 2008 American Chemical Society. [17] Copyright © 2010 American Chemical Society

in heterocatalysis, that is still a great challenge. In the support method, MNCs are usually prepared by direct synthesis method, then loaded on the support.

Yang's group reported the fabrication of $Au_{25}(Cys)_{18}$/g-C_3N_4 (Cys: L-Cysteine) hetero-structures by a facile wet-impregnation method. The Au_{25}NCs were highly dispersed on the surface of g-C_3N_4 nanosheets. This structure enhances the photocatalytic hydrogen evolution performance of $Au_{25}(Cys)_{18}$/g-C_3N_4 under visible light irradiation [18]. They also reported the synthesis of ligand-free Au nanoclusters (NCs)/g-C_3N_4 ultra-thin nanosheets composite via a facile wet-impregnation method with post-annealing, which led to the loaded Au NCs which were aggregating resistant by post-annealing [19]. Constructing heterojunctions with appropriate inter-band structures between the ligand-free Au NCs and the ultra-thin g-C_3N_4 nanosheets, that ensured the smooth interfacial charge transfer, greatly improves its photocatalytic performance (Scheme 3.11).

Yang et al. synthesized Pt_1Ag_{24}/g-C_3N_4 by the support method. Interfacial electron transfer of the catalyst can be tuned primarily by the incorporation of a single Pt atom into the Ag_{25} structure [20]. This is the first time to use atomically controlled ultra-small silver-based cocatalysts in combination with graphitic carbon nitride (g-C_3N_4) as a model semiconductor, and Pt_1Ag_{24}/g-C_3N_4 has the highest photocatalytic

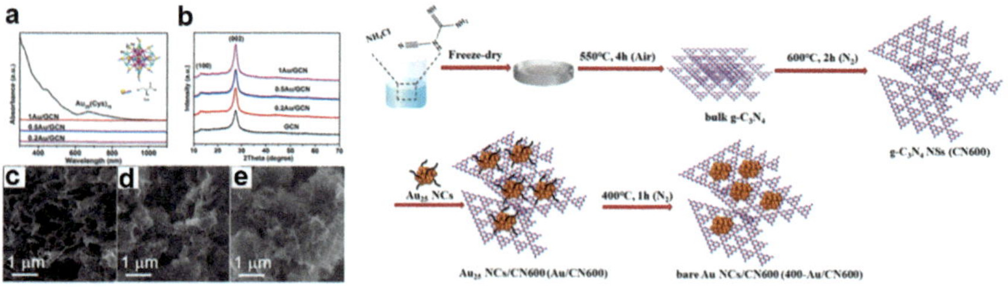

Scheme 3.11 **a** UV–vis absorption spectra of the as-prepared $Au_{25}(Cys)_{18}$ NC solution before (black line) and after (color lines) the addition of GCN. Inset image is the ball–stick model of a $Au_{25}(Cys)_{18}$ NC; **b** XRD patterns of various Au/GCN photocatalysts; and SEM images for **c** 0.2 Au/GCN, **d** 0.5 Au/GCN, **e** 1 Au/GCN and **f** the preparation of ligand-free Au NCs modified g-C_3N_4 NSs. Reprinted with permission from Ref. [18, 19]. [18] Copyright © 2018, American Chemical Society. [19] Copyright © 2021, The Author(s)

activity, which is 330 times and 4 times higher than those of bare g-C_3N_4 and Ag_{25}/g-C_3N_4, respectively.

3.1.8 Atomic Layer Deposition Method

This method is different from the top-down chemical etching method, by selecting suitable nucleation sites, metal atoms are deposited onto a support from the bottom up to form monodisperse MNCs [21]. The synthetic process of atomic layer deposition (ALD) technique usually includes two main steps. The metal precursors initially react with the adsorbed oxygen on the substrate surface, then the introduction of oxygen pulse oxidizes the metal precursors and converts their ligands to metal–oxygen (M–O) species, therefore forming metal single atoms uniformly dispersed on the substrate.

Sun et al. reported a systematic study in which ALD was used to synthesize a class of novel catalysts that consists of platinum single atoms, sub-nanoclusters, and nanoparticles uniformly deposit on the surface of graphene nanosheets [22]. The Pt-SASC exhibited higher methanol oxidation hydrogen evolution reaction activity than that of commercial Pt/C [23] (Scheme 3.12).

In 2017, Lu and co-worker have successfully demonstrated that Pt_2 dimers can be bottom–up constructed on graphene with a high-surface area [24]. They found that the type of surface nucleation sites, selective deposition, the self-limiting nature of ALD, and the high stabilities of Pt_1 single atoms and Pt_2 dimers are the keys factors for the Pt_2 dimers synthesis. It's worth noting that, to avoid any metal aggregation, the temperature of the second ALD cycle was lower than the first ALD cycle (Scheme 3.13).

3.1.9 Photo-Induction Reduction Method

Adsorption of photons and electronically excited states are two key steps in the photo-induction reduction process. This is to ensure the atomic diffusion of the metal ions in the solution is necessary during the process.

Zheng et al. prepared a single-atom palladium catalyst (Pd_1/TiO_2) at room temperature by the photo-induction reduction method using ultra-thin titanium dioxide nanosheets as substrate [25]. The formation of ethylene glycolate free radicals is an important step in the preparation of the Pd_1/TiO_2 catalyst by ultraviolet light, which reduces and stabilizes the Pd atoms on the TiO_2 nanosheets. In the styrene hydrogenation reaction, Pd_1/TiO_2 achieved complete catalytic conversion of styrene and showed good stability.

Wang et al. report a photo-induction method to construct isolated Cu single-atom catalysts (as active species) anchored on UiO-66-NH_2 [26]. They used A 300 W Xearc lamp as a cut-off filter ($\lambda > 400$ nm) as a visible light source to treat the dispersed

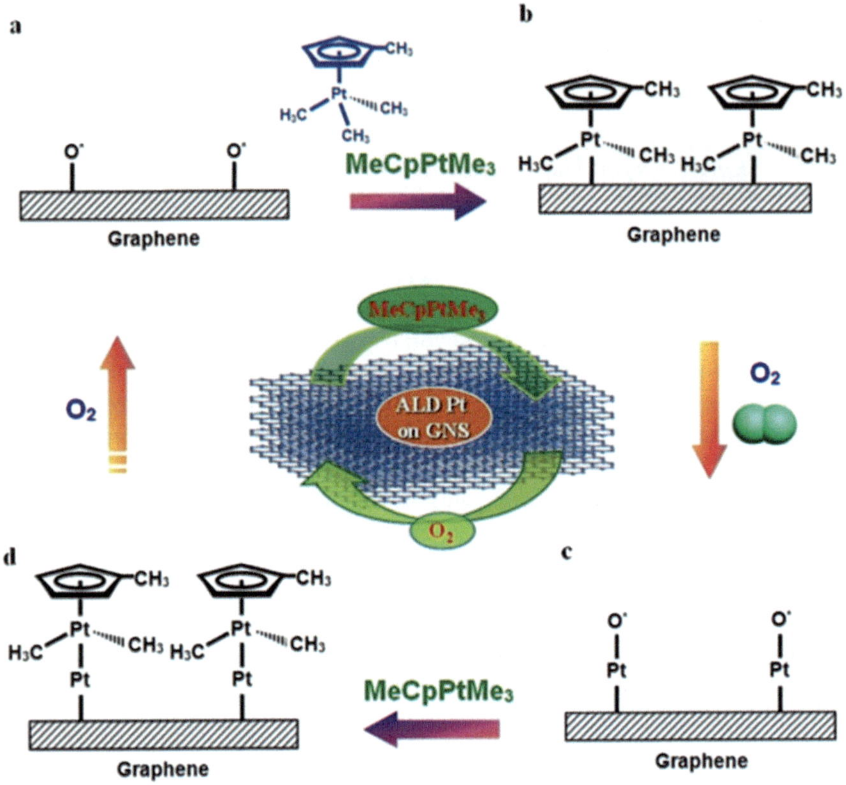

Scheme 3.12 Schematic illustrations of Pt ALD mechanism on graphene nanosheets. Reprinted with permission from Ref. [22, 23]. [22] Copyright © 2013, The Author(s). [23] Copyright © 2016, The Author(s)

mixture. This type of catalyst has excellent catalytic activity, the conversion rates of CO_2 to CH_3OH and C_2H_5OH were 5.33 and 4.22 μmol $h^{-1}\cdot g^{-1}$, respectively (Scheme 3.14).

3.1.9.1 Heat-Induction Synthesis Method

It is extremely difficult to create a high density of thermally stable support defects, and hence the fabrication of high metal loading single atom catalysts (SACs) remains currently a formidable challenge [27]. Heat-induction synthesis method of MNCs can solve this problem opportunely. This method usually requires the preparation of metal nanoparticles, then the precursor is vaporized at high temperatures, finally, the metal atoms are deposited on the support [28] (Schemes 3.15 and 3.16).

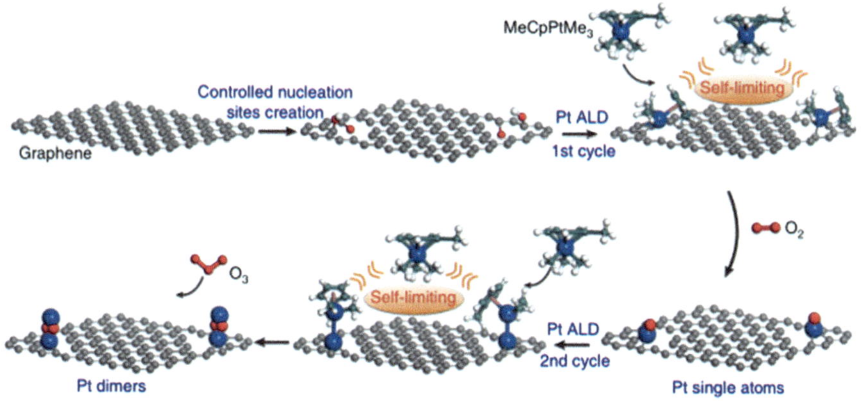

Scheme 3.13 Schematic illustration of bottom–up synthesis of dimeric Pt_2/graphene catalysts, the balls in cyan, white, red, and blue represent carbon, hydrogen, oxygen, and platinum while the ball in gray represents carbon atoms in the graphene support. Reprinted with permission from Ref. [24]. Copyright © 2017, The Authors

3.2 Characterizations of MNCs

The size, composition, morphology and surface properties of MNCs affect their optical, electronic and catalytic activities. As research progresses, characterization techniques for nanoclusters are advancing. The detailed structural characterizations allow people to deeply understand the structure–property relationship, it is beneficial to the cluster structural optimization for different applications. In this part, the commonly used techniques for nanocluster characterizations, such as ultraviolet and visible spectroscopy, infrared spectroscopy, X-ray (XRD, XPS, EDX) spectroscopy, electron microscopy (TEM, HRTEM, SEM), mass spectrometry, nuclear magnetic resonance spectroscopy, and thermogravimetric analysis, etc. are summarized.

3.2.1 UV–Vis Spectroscopy

Metal nanoparticles have unique optical properties which are consistent with Mie scattering, mean that an exponential decay of the absorption profile with increasing wavelength, the absorption wavelength decreases with the decrease of MNCs size.

CuNCs were prepared by a one-pot procedure based on wet chemical reduction by Chen's group, and Cu_n ($n < 8$) nanoclusters with three well-defined absorption bands at 285, 364, and 443 nm are clearly different from the characteristic surface plasmon resonance (SPR) band of large Cu nanoparticles around 560–600 nm [29–31]. Similar to the gold nanoclusters [32], the author considers that the multiband

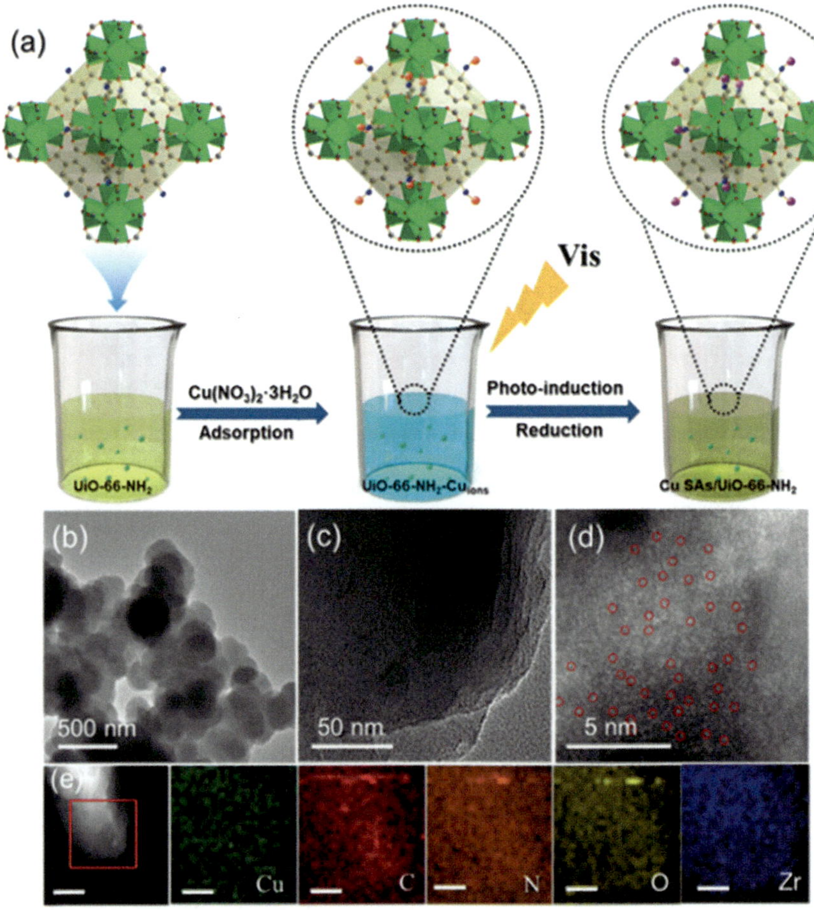

Scheme 3.14 **a** Synthesis process of the Cu SAs/UiO-66-NH_2 photocatalyst. C (gray), O (red), Zr – O clusters (green), N (blue), Cu ions (orange), and Cu SAs (purple). **b** TEM and **c** HRTEM of Cu SAs/UiO-66-NH_2. **d** Aberration-corrected STEM images of Cu SAs/UiO-66-NH_2. Cu SAs were marked with red circles. **e** EDS mapping of Cu SAs/UiO-66-NH_2. Reprinted with permission from Ref. [26]. Copyright © 2020, American Chemical Society

absorption of Cu nanoclusters are also from the interband or intraband electronic transitions due to the discrete energy levels (Scheme 3.17).

3.2.2 *Fourier Transform Infrared (FT-IR) Spectroscopy*

MNCs are usually stabilized by ligands, which influence the surface structure of metal clusters. As a sensitive technique to detect the molecular vibrations, FTIR

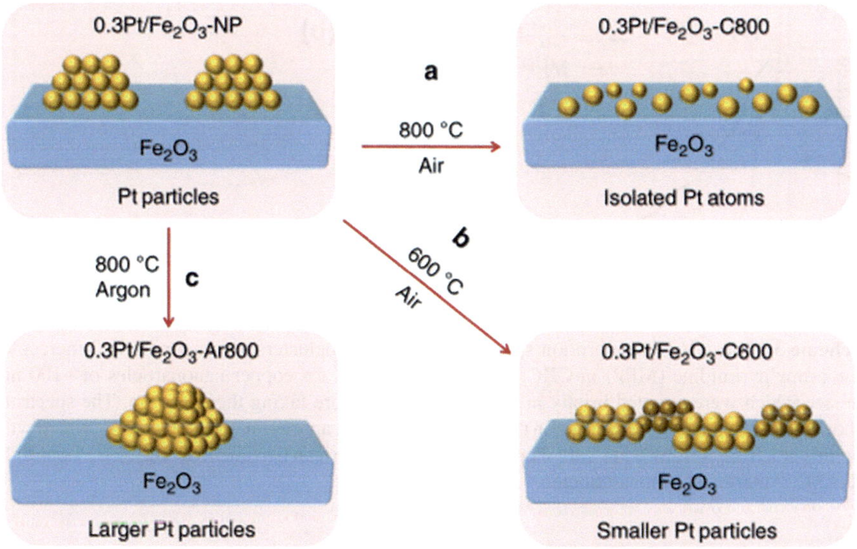

Scheme 3.15 Illustration of thermally induced Pt nanoparticle restructuring. **a**, **b** Calcination under oxygen, or under an inert atmosphere (**c**), resulting in dispersion as single-atoms or particle sintering, respectively. Reprinted with permission from Ref. [27]. Copyright © 2019, The Author(s)

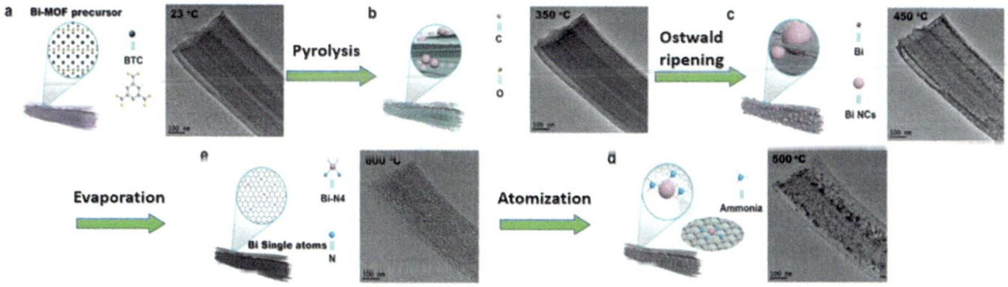

Scheme 3.16 **a–e** Scheme of the transformation from Bi-MOF to single Bi atoms and corresponding representative movie TEM images of Bi-MOF pyrolysed at different temperatures with the assistance of DCD in situ. Reprinted with permission from Ref. [28]. Copyright © 2019, American Chemical Society

spectroscopy can be used to analyze the surface structures of synthesized metal nanoclusters, and the formation mechanism of metal clusters by characterizing the changes of chemical bonds in ligands.

A silver cluster having the composition $Ag_9(H_2MSA)_7$ (H_2MSA = mercaptosuccinic acid) was synthesized by Pradeep et al. [33]. It could be observed by FTIR, the –SH stretching feature at 2552 cm^{-1} in H_2MSA is marked by the dotted-circle in (a) which is absent in other spectra, in agreement with the XPS and mass spectra. H_2MSA features in the region of 2000–500 cm^{-1} confirm the presence of H_2MSA

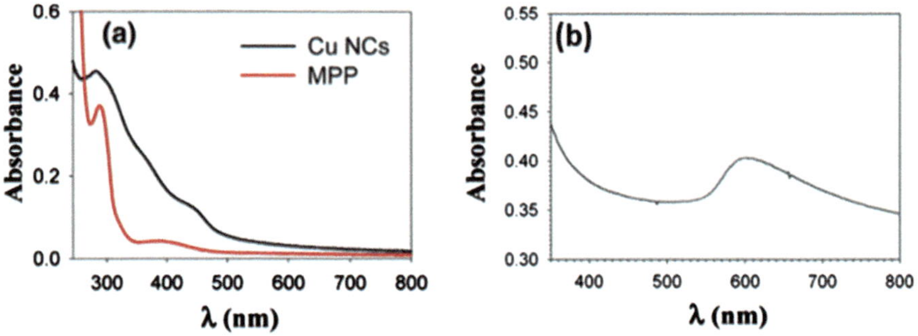

Scheme 3.17 **a** UV–vis absorption spectra of the Cu nanoclusters and monomeric 2-mercapto-5-*n*-propylpyrimidine (MPP) in $CHCl_3$. **b** UV–Vis spectra for copper nanoparticles of ~100 nm in size which were agitated briefly in hexane solution before taking the spectrum. The spectrum shows a well-defined surface plasmon (SP) band centering at approximately 600 nm. Reprinted with permission from Ref. [30, 31]. [30] Copyright © 2011, American Chemical Society. [31] Copyright © 2007, American Chemical Society

in the cluster. The solid FT-IR spectrum of the $Ag_9(MSA)_7$ cluster showed a strong band at 3450 cm^{-1} due to the hydrated water. Similar findings have been reported in other literature [30, 34–36] (Scheme 3.18).

Scheme 3.18 FT-IR spectra of pure **a** H_2MSA, **b** $Ag(I)H_2MSA$, **c** as-prepared $Ag@(H_2MSA)$ nanoparticles and **d** PAGE separated $Ag_9(H_2MSA)_7$ cluster. Reprinted with permission from Ref. [33]. Copyright © 2010, American Chemical Society

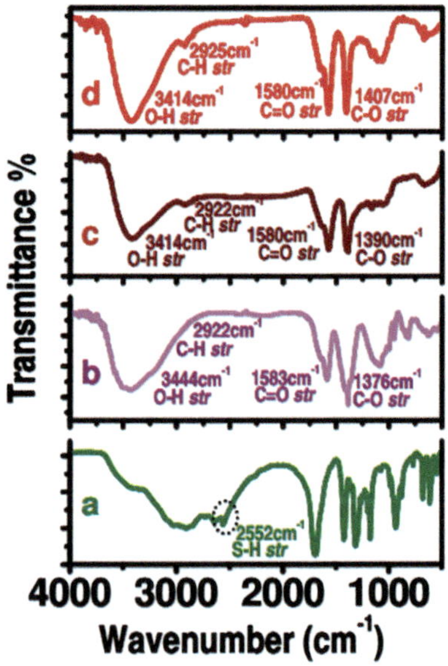

3.2.3 X-ray Diffraction (XRD) Spectroscopy

X-ray diffraction can be divided into powder X-ray diffraction (PXRD) and single crystal X-ray diffraction (SCXRD).

PXRD is usually used to study the crystal structures and evaluate the crystal size of nanomaterials, and reveal whether the clusters are alloy structure or not for bimetallic or multi-metallic clusters. Recently, Wu et al. studied the crystal structures of gold nanoparticles with different core sizes through PXRD measurements [37]. Largest particles (4.0 nm) show a well-defined diffraction pattern with four distinct diffraction peaks at 38.51, 44.61, 64.81, and 77.81, corresponding to (111), (200), (220) and (311) of the Au face-centered cubic (fcc) crystal structure, respectively. When the particle size decreased to 2.0 nm, the weak diffraction peaks indicate that it still adopt the fcc structure, except that the (200) peak is almost merged into the broad (111) peak. However, in the XRD pattern of $Au_{25}(SG)_{18}$ nanoclusters there are two very broad diffraction peaks centered at around 37.51 and 66.51, indicating the non-fcc crystal structure of the Au_{25} clusters. Analysis results indicate that, when the particle size lies in the cluster regime, drastic changes occur to the particle structure (Scheme 3.19).

The object of SCXRD analysis is a single crystal. The atoms or groups of atoms are arranged periodically in a single crystal. Diffraction occurs when X-rays are emitted into a single crystal, and the arrangement of atoms in the crystal can be resolved by analyzing the diffraction lines, that is, the structure of the crystal can be solved. A new method termed "in situ two-phase ligand exchange" was developed to obtain alloy NCs by Zhu et al. and a series of thiolate-stabilized alloy NCs, including $Au_{20}Ag_1(SR)_{15}$(NC-1), $Au_{21-x}Ag_x(SR)_{15}$(x = 4−8, NC-2), $Au_{21-x}Cu_x(SR)15$ (x = 0 or 1, NC-3), and $Au_{21-x}Cu_x(SR)_{15}$(x = 2−5, NC-4) have been successfully synthesized [38]. Interestingly, the SCXRD shows that these alloy nanoclusters have two kinds of configuration (Scheme 3.20).

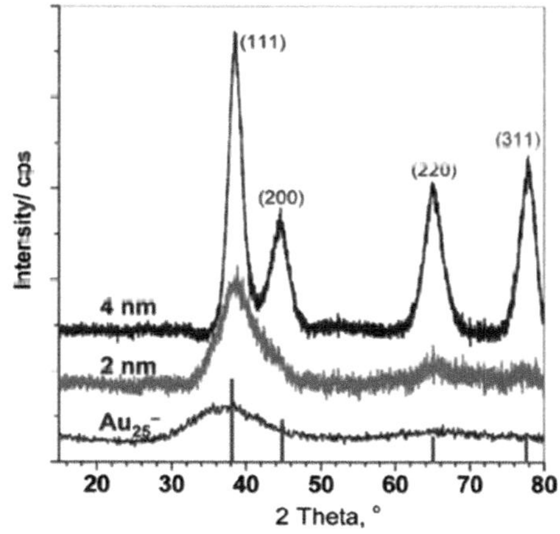

Scheme 3.19 XRD patterns of 4 nm, 2 nm, and $Au_{25}(SG)_{18}$ nanoparticles. The stick pattern corresponds to bulk gold. Reprinted with permission from Ref. [37]. Copyright © 2011 WILEY-VCH Verlag GmbH & Co. KGaA, Weinheim

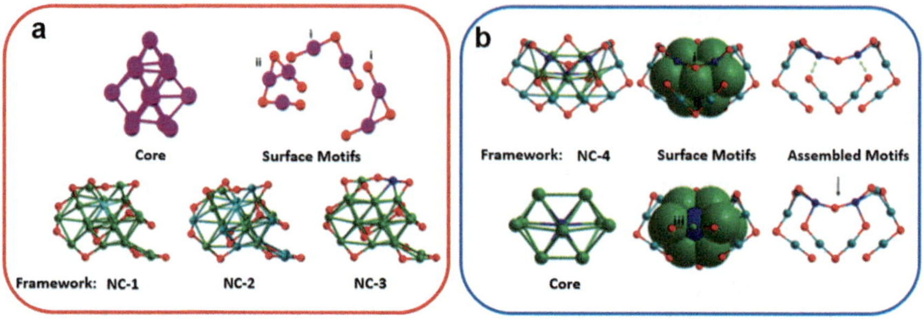

Scheme 3.20 **a** Crystal structure of NC-1–3 (color labels: green = Au, light blue = Ag, blue = Cu/Au, bluish-green = Au/Ag, magenta = Au/Ag/Cu, red = S; no C or H atoms are shown); **b** Crystal structure of NC-4 (color labels: green = Au, bluish-green = Au/Cu, blue = Cu, red = S; no C or H atoms are shown). Reprinted with permission from Ref. [38]. Copyright © 2017, American Chemical Society

As for NC-1–3, they share a common structure, which contains a M_{14} core (M = Au, Ag, and Cu) protected by two M_2S_3 staple motifs (labeled with i), one M_3S_4 staple motif (labeled with ii), and five bridging S atoms. As for NC-4, it also contains a 21-atom core protected by 15 thiolate ligands. However, NC-4 exhibits a similar configuration to that recently reported for Au_{21}(S-Adm)$_{15}$, which is totally different from those of the aforementioned three alloy NCs. The authors argue that these experimental indicating that the metal-doping can remarkably affect the electronic structure of the NCs.

3.2.4 X-Ray Photoelectron Spectroscopy (XPS)

XPS is a surface analysis method, the oxidation states of metal cores in nanoclusters could be investigated, and the stability of metal nanoclusters could be studied by monitoring the change.

Zhu's group examined the TNT support, the Pd_3Cl nanocluster, the Pd_3Cl/TNT catalyst, in-situ Pd_3Cl/TNT catalyst and the recycled Pd_3Cl/TNT catalyst by XPS, in order to identify the presence of metal–support interaction (MSI) effect in Pd_3Cl/TNT catalyst [39]. The MSI formation is usually accompanied by a shift of the peaks of both metals mainly toward opposite direction, as mentioned elsewhere. Test result displayed that the Pd 3d5/2 bond energies of Pd_3Cl/TNT catalyst (335.00 eV) and recycled Pd_3Cl/TNT catalyst (335.00 eV) was significantly lower than that of Pd_3Cl nanocluster (336.57 eV). The substantial negative shift of 1.57 eV evidenced that an enhanced electron density of the Pd_3Cl nanocluster when it was interfaced with TNT. The observed Pd 3d peak shifts could be explained by assuming the presence of MSI between Pd_3Cl nanocluster and TNT support. Just as expected, the Ti 2p2/3 bond energies of Pd_3Cl/TNT catalyst (458.27 eV) and recycled Pd_3Cl/TNT catalyst (458.28 eV) were significantly higher than that of TNT support (457.48 eV).

Therefore, it was proposed that the observed shifts of the Pd 3d peak and the Ti 2p peak were due to the MSI, where electrons were transferred from TNT support to Pd_3Cl.

3.2.5 *Electron Microscopy*

Electron microscopy includes transmission electron microscopy (TEM) and scanning electron microscope (SEM). TEM has traditionally been used to obtain information about the nanoparticle size and shape, and SEM is generally used to observe the appearance of nanoclusters [40]. However, due to the small size of the nanoclusters, the actual size of sub-nanometer sized clusters can't be obtained accurately from TEM measurements. High resolution TEM (HRTEM) is a primary tool in the study of size and crystalline structure of metal nanoclusters.

TEM or SEM are usually equipped with Energy dispersive X-ray spectroscopy (EDS) and EDS can be conveniently obtained during the TEM or SEM measurements. EDS analysis has been proven to be an efficient chemical tool for qualitative and quantitative element analysis [41]. For alloy MNCs, EDS can be used to determine the composition of a single cluster or the overall sample. Sometimes, the function of energy dispersive analysis of X-rays (EDAX) is similar to that of EDS.

3.2.6 *Mass Spectrometry (MS)*

Whetten and co-workers were the first to use laser-desorption/ionization and time-of-flight mass spectrometry (MS) to analyze gold nanoclusters in 1996 [42]. Now, MS is widely used in almost all the structural studies of sub-nanometer sized metal clusters, it gives the molecular weight of the NCs by the ion mass/charge ratio and is a relatively reliable means to determine their composition, including laser desorption/ionization mass spectrometry (LDI-MS), [43, 44] laser desorption/ionization time-of-flight (LDI-TOF), [191] matrix-assisted laser desorption ionization mass spectrometry (MALDI-MS), [45–47] fast atom bombardment ionization mass spectrometry (FAB-MS), [48] and electrospray ionization mass spectrometry (ESI–MS) in positive- or negative-ion modes.

3.2.7 *Nuclear Magnetic Resonance (NMR) Spectroscopy*

NMR gives the information on the structural and chemical environment of NCs, usually including proton (^{1}H) and carbon (^{13}C) NMR. and allows checking of the purity of the NCs and estimation of the binding interactions between the ligands and the metal core.

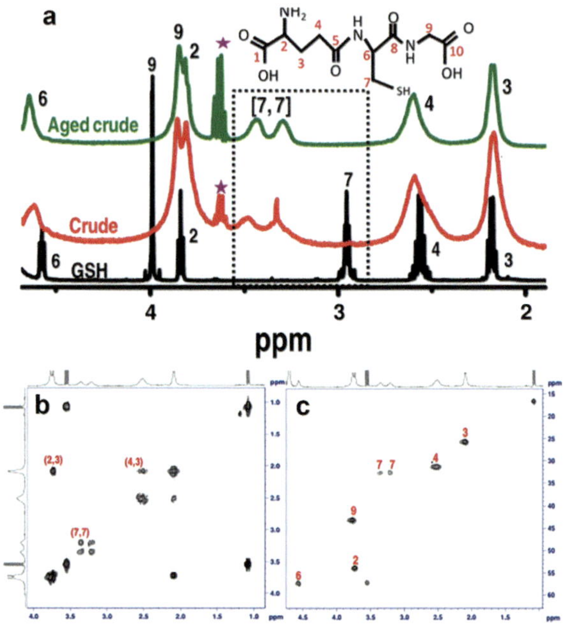

Scheme 3.21 **a** ^{1}HNMR spectra of GSH, crude cluster and aged crude. The peaks in dotted box are due to H-7. Peaks at ~3.6 ppm (marked with *) are due to residual EtOH. 2D NMR spectra of $Ag_{32}SG_{19}$ clusters. **b** COSY spectrum. **c** HSQC spectrum. Reprinted with permission from Ref. [49]. Copyright © 2009, Royal Society of Chemistry

Pradeep et al. proposed the Ag-thiolate structure of the cluster based on a detailed NMR study [49]. One-dimensional (1D) and two-dimensional (2D) NMR spectroscopic investigations of $Ag_{32}SG_{19}$ provided interesting spectral features which indicated the dominant –[SR-Ag-SR]– structural motif. This structural motif as the predominant entity is found for the first time in silver clusters. Protons H-7 (=CH–CH_2–S–) signals are significantly shifted downfield in the range of 3.2–3.5 ppm with broadening followed by splitting, noted as [7], because of its close proximity to the silver core. To know the cause of splitting and nature of the staple motif on the cluster surface in comparison to the structures presented above, homonuclear correlation spectroscopy (COSY) and heteronuclear single quantum correlation (HSQC) were performed (B and C, respectively). These results of the author's analysis suggest that the Ag-thiolate structure for $Ag_{32}SG_{19}$ is dominated by –[RS_c-Ag-S_cR]– (Scheme 3.21).

3.2.8 *Thermogravimetric Analysis (TGA)*

TGA is used to estimate the ratio of metal core and organic ligand by weight loss and to study the thermal stability of NCs.

Wu and co-workers reported the effect of thiolate ligands on the catalysis of CeO_2 rod supported $Au_{25}(SR)_{18}$ by using CO oxidation as a probe reaction [50]. The $Au_{25}(SR)_{18}/CeO_2$ rod sample was first calcined in a box oven in air at different temperatures (295, 423, 473, 498, 523, 573, and 673 K) for 1 h to make the temperature consistent with that used in the reactivity test in CO oxidation on the AMI system. These IR bands are most intense after 423 K treatment, then gradually decrease in intensity, and finally disappear at a treatment temperature of 523 K and above, implying complete removal of the thiolate ligands from the Au nanoclusters at this temperature. The temperature of 523 K is in good agreement with that for complete thiolate loss from the TGA result of bare and supported $Au_{25}(SR)_{18}$ nanoclusters [51].

3.2.9 X-Ray Absorption Spectroscopy (XAS)

This method gives the local structure, electronic characterization, and core-level energies of metal NCs and helps in understanding the NC structure. There are usually three steps, (1) identify the local structure. The wavelet transform of extended X-ray absorption fine structure (EXAFS) is used to roughly determine the types of coordination elements around the atom. (2) Fourier transform is applied to EXAFS to extract atomic spacing and coordination number information. (3) Identification of electronic and 3D structures. X-ray absorption near edge structure (XANES) analyzes the 3D configuration of atoms, builds a clear model based on existing conditions and experience, and performs theoretical calculations to iterate the atomic structure that best conforms to the actual and test data [52].

Previous studies on Cu-based NCs revealed remarkable XAS sensitivity to both ligand sets and Cu core [53, 54]. Therefore, Hayton and co-workers used Cu K-edge XAFS to compare structural features of complex 1 with its silica-supported analog, 2, as well as the recovered post-catalysis material, 2_{post} [52]. The absence of significant differences in the XANES profiles of 1 and 2 suggests that the nanocluster structure is preserved upon grafting onto SiO_2. In the EXAFS, two prominent features appear at 1.6 and 2.3 Å in the FT magnitude of the EXAFS of 1. The first peak represents light atoms directly coordinated to Cu, i.e., the donor atoms of the bidentate acetate and side-on bound acetylide ligands. The second peak represents the contributions of the Cu–Cu paths. EXAFS analysis suggests that immobilization of 1 results in little or no change to the cluster nuclearity, and only minor changes to the cluster structure.

3.3 Organic Reactions Catalyzed by MNCs

High catalytic efficiency has always been the target of industrial production, fortunately, MNCs satisfy this condition and have attracted much attention because of their high specific surface area and catalytic activity. Organic reaction as an ideal platform is provided to test the utility of MNCs with catalysis as their main application. So, a variety of reaction will be introduced from MNCs of monometallic, bimetallic alloy, and trimetallic alloy.

3.3.1 MNCs of Monometallic

3.3.1.1 Organic Reactions Catalyzed by Au NCs

Oxygen Reduction Reaction (ORR)

Chen et al. studied the electrocatalytic activity of $[Au_{25}(SC_{12}H_{25})_{18}]$ nanoclusters for the synthesis of H_2O_2 from the two-electron reduction of dioxygen and investigated the influence of the cluster charge states on the electrochemical production of H_2O_2 in alkaline media [55].

The electrochemical studies showed that maximum H_2O_2 production and the most efficient two-electron dioxygen reduction were obtained from the charge negative Au_{25}^- nanoclusters, followed by the neutral and cationic ones, indicating the significant effect of the charge state of a Au_{25} cluster on its catalytic properties. The present study has provided an insight into the highly selective two-electron reduction of dioxygen to hydrogen peroxide by controlling the charge states of gold nanoclusters.

Catalytic Hydrogenation Reaction

Zhu et al. studied the stereoselectivity of Au_{25} nanoparticle catalyst in the hydrogenation of cyclic ketones [56]. Firstly, Au_{25} nanoparticle catalysts with atomically precise structure were successfully prepared by the wet chemical method. Then, the crystal structure of Au_{25} nanoparticles was analyzed by X-ray crystallography, and their electronic properties were revealed. The results show that $Au_{25}(SR)_{18}$ catalysts offer a ~100% stereoselectivity toward the exo-alcohol, instead of the endo-alcohol, and the catalytic results are correlated with the atomic spatial construction and core–shell nature of the $Au_{25}(SR)_{18}$ nanoparticle (Scheme 3.22).

Li et al. reported on the Lewis acid-promoted hydrogenation of aldehyde to alcohol catalyzed by $Au_{25}(SR)_{18}/CeO_2$ in the presence of a base (pyridine or $NH_3{\cdot}H_2O$) under

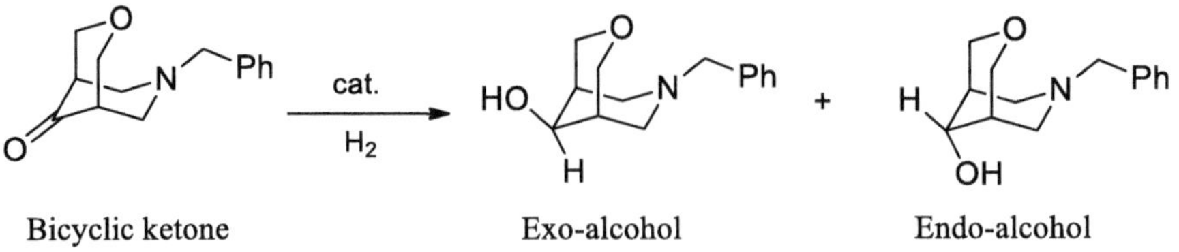

Scheme 3.22 Catalytic hydrogenation of bicyclic ketone to alcohol. Reprinted with permission from Ref. [56]. Copyright © 2010 Elsevier Inc. Published by Elsevier Inc. All rights reserved

Scheme 3.23 Catalytic activity of $Au_{25}(SR)_{18}$ nanoclusters ($R = C_2H_4Ph$) for aldehyde hydrogenation in the presence of base, ammonia or pyridine, and transition metal ion Mz^+. Reprinted with permission from Ref. [45]. Copyright © 2015, American Chemical Society

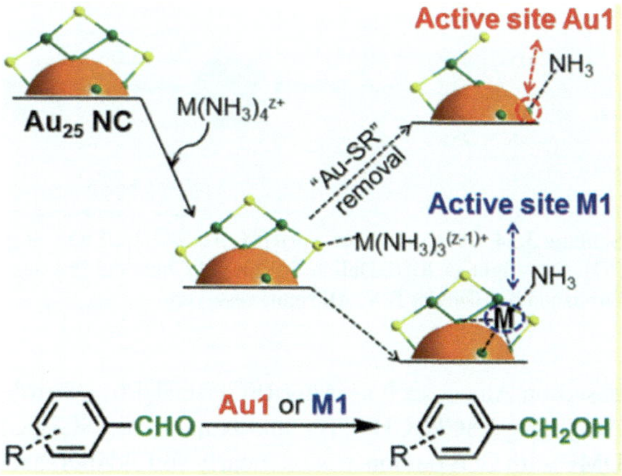

mild conditions (18 bar H_2 at 323 K) [45]. Since essentially all surface gold atoms of $Au_{25}(SC_2H_4Ph)18^-$ are protected by thiolate groups, the Au_{25} NC nanoclusters are inert toward adsorption of nonradical molecules on their surface, so, the author proposes bond activation of H_2 on these metal sites is the first step. Upon removal of the capping NH_3 from the metal site, the H–H bond of a H_2 molecule becomes activated with the aid of a base, and no hydrogenation reaction occurred in the absence of a base. PhCHO becomes physically adsorbed on the surface of the nanoclusters via interactions of its O and C (–CHO group) with NH_4^+ and H^-, respectively (Scheme 3.23).

The research idea is to prepare gold nanoclusters encapsulated in silica (Au_{25}/SiO_2) by a simple method and study its catalytic performance in the reduction of 4-nitrophenol [57]. Firstly, the gold nanocluster precursor $Au_{25}[SC_3H_6Si(OCH_3)_3]_{18}$ was prepared by 'size focusing' the polydisperse gold nanocluster in the presence of (3-mercaptopropyl) trimethoxysilane (MPTS).Then, the Au_{25}/SiO_2 catalyst was prepared by hydrolysis of MPTS surface ligands by tetraethyl orthosilicate (TEOS) to form a silica coating layer on the periphery of the silica core. Finally, it was found that Au_{25}/SiO_2 was stable in air at 400 °C and exhibited good catalytic activity in the reduction of 4-nitrophenol (TOF up to 343 h^{-1}). In this study, the difficulties of gold nanoclusters in catalytic reactions, such as aggregation and deactivation, were solved by preparing gold nanoclusters encapsulated in silica. This encapsulation approach can improve the stability and catalytic activity of gold nanoclusters, and also prevent the interaction between gold nanoclusters and other substances. Therefore, this study provides a new strategy for the application of gold nanoclusters in the field of catalysis (Scheme 3.24).

Electrocatalytic Reaction

There are few NCs applied to electrocatalytic CO_2 reduction, because the research is almost entirely focused on Au_{25} model catalysts, therefore, it is necessary to develop more NCs with high activity to reduce CO_2. Dass et al. synthesized a series of NCs

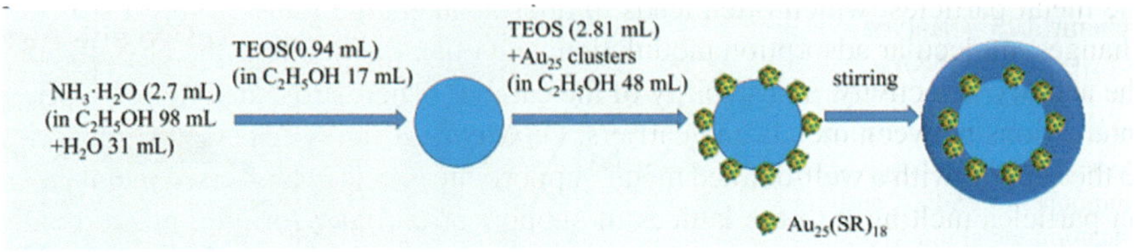

Scheme 3.24 Preparation of $Au_{25}[SC_3H_6SiO_3]_{18}/SiO_2$. Reprinted with permission from Ref. [57]. Copyright © 2016, Dalian institute of Chemical Physics, the Chinese Academy of Sciences. Published by Elsevier B.V. All rights reserved

based on Au_{137}, such as $Au_{137}(SCH_2CH_2Ph)_{56}$, $Au_{137}(SC_6H_{13})_{56}$, $Au_{137}(SC_4H_9)_{56}$, $Au_{137-x}Ag_x(SCH_2CH_2Ph)_{56}$ and $Au_{137-x}Pd_x(SCH_2CH_2Ph)_{56}$ [58]. $Au_{137}(PC_2)_{56}$ in DMF with CO_2 demonstrates a nearly sixfold increase in current during CV measurements when compared to $Au_{137}(PC_2)_{56}$ without CO_2. The observed asymmetric current increase is indicative of electrocatalytic reduction of CO_2. Importantly, a significant increase in current is observed at ~–0.55 V versus NHE (a normal hydrogen electrode, inset) followed by a dramatic current increase with an onset at −1.5 V. The initial increase in current is observed at an overpotential of <200 mV for 2 electron reduction of CO_2 to CO. Electrocatalytic CO_2 reduction in this work is at the beginning stage and needs to be further studied in detail (Scheme 3.25).

Abroshan et al. synthesized Au_{25} nanoclusters and characterized them by spectroscopic techniques [59]. Then, they studied the electrocatalytic activity of Au_{25} nanoclusters in CO_2 reduction reaction by electrochemical experiments and computational simulations. Researchers have found that there is a unique interaction between Au_{25} nanoclusters and CO_2, which can effectively catalyze the reduction of CO_2. They observed that the spectral characteristics of Au_{25} nanoclusters in CO_2 environment changed, indicating that there was an interaction between CO_2 and clusters.

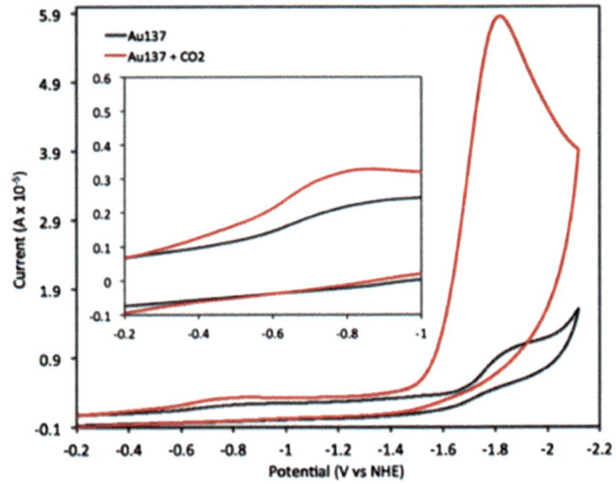

Scheme 3.25 CV of $Au_{137}(PC_2)_{56}$ under N_2 (black) and CO_2 (red) in DMF with 0.1 M Bu_4NPF_6 as the electrolyte and a scan rate of 100 mV s^{-1}. Reprinted with permission from Ref. [58]. Copyright © 2014, Royal Society of Chemistry

In addition, they also found that Au_{25} nanoclusters exhibited excellent electrocatalytic activity in CO_2 reduction reaction and produced highly selective CO products through electrochemical experiments.

Photocatalytic Reaction

Nano-carbon is a transition state between atoms and bodies, with unique discrete electronic structure and optical properties. They are considered as good models for the correlation between structure and high activity in many catalytic reactions. However, nanocarbons with precise atomic number and structure have few applications in nanomaterials. Kogo et al. prepared and deposited a GSH-protected Au_{25}/TiO_2 film on the electrode [60]. Photocatalytic performance of Au_{25} modified TiO_2 under visible light and near-infrared light. In this study, the electrons of Au_{25} clusters were excited, injected into the conduction band of TiO_2, and finally accepted by the oxidized electron donor on the counter electrode, realizing the oxidation of the donor and the reduction of the acceptor. It is found that even under the light of 860 nm, Au_{25} clusters can still produce photocatalytic reactions. The results of this study indicate that Au_{25} modified TiO_2 can be used as a visible and near-infrared light-driven photocatalyst, which can oxidize phenol derivatives and ferrocyanide, and reduce Ag^+, Cu^{2+} and dissolved oxygen. In addition, the study also found that Au_{25} cluster modified TiO_2 can achieve thermodynamic uphill reaction, such as the oxidation of phenol accompanied by the reduction of Cu^{2+}. In the presence of a donor (phenol derivative or ferrocyanide), the composite system exhibits a much higher photocurrent density (11.5 μAcm^{-2}) under visible light irradiation than that of bare TiO_2 (0.18 μAcm^{-2}) or the composite without the donor (0.16 μAcm^{-2}).

Hu et al. reported similar results in the photocatalytic degradation of rhodamine B (Rh B). These authors deposited Au_{25} into porous ZnO_{397} and optimized the loading of Au_{25} in the composite to balance the separation efficiency of e^-–H^+ and the active surface area [61]. When the loading was 1.16%, the catalytic activity of the composite was the highest, and the reduction rate was 95% within 15 min (Scheme 3.26).

Lee et al. first reported that the photocatalytic activity and electron transfer were controlled by the size of Au nanocarbons [62]. The main goal of this study is to explore how electron transfer and photocatalytic activity in metal nanoparticle-semiconductor composites are controlled by the size of metal nanoparticles. The researchers first synthesized different sizes of thiol-protected gold nanoparticles (AuNPs) and connected them to the surface of ZnO. They synthesized three different sizes of GSH-protected Au NCs (1.1 nm Au_{25}, 1.6 nm Au_{144} and 2.8 nm Au_{807}) and deposited them on the surface of ZnO by carboxyl binding to degrade thiophene under light irradiation. By adjusting the size of gold nanoparticles, researchers can control the charge transfer ability of gold nanoparticles. They found that smaller gold nanoparticles have higher charge transfer efficiency. In addition, they also found that the size of gold nanoparticles also has an effect on photocatalytic activity, and smaller size of gold nanoparticles can improve the efficiency of the photocatalytic reaction (Scheme 3.27).

Han et al. reported the same catalytic activity trend in Nile blue photoreduction [63]. However, in the photoreduction of azobenzene, this conclusion is no longer

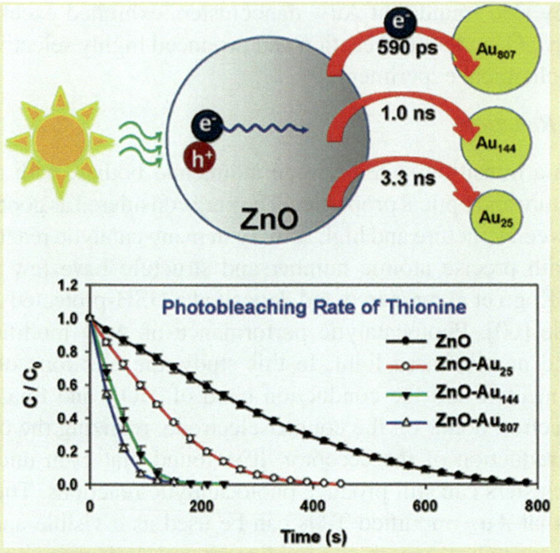

Scheme 3.26 Photobleaching rate of Thionine. Reprinted with permission from Ref. [62]. Copyright © 2011, American Chemical Society

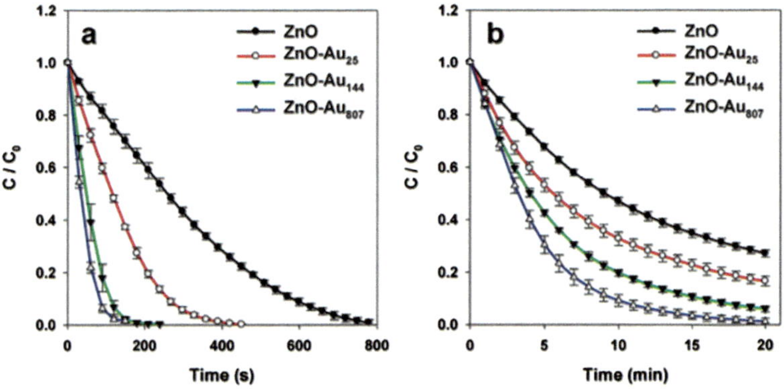

Scheme 3.27 Change in **a** TH and **b** R6G concentration (C) relative to the initial concentration (C_0) during photocatalysis with ZnO and ZnO–Au composites. Reprinted with permission from Ref. [62]. Copyright © 2011, American Chemical Society

applicable because the reaction is a multi-electron/multi-proton process involving proton-coupled electron transfer and is no longer directly related to charge transfer. In the photoreduction reaction of azobenzene, the ZnO–Au NC composite system has obvious size threshold and excitation intensity threshold, among which ZnO-Au_{807} exhibits the highest activity. The higher excitation intensity leads to the reduction of azobenzene aniline in the process of four electrons/holes, while the lower excitation intensity leads to the reduction of azobenzene hydrogenated azobenzene in the process of two electrons/holes. Liu et al. prepared 6-mercaptocaproic acid (MHA)-protected Au_{25} NCs as photosensitizers and loaded them on highly ordered TNA for the photocatalytic decomposition of tetracycline [64]. They changed the protective ligand from GSH to MHA based on two considerations: (1) The structure of MHA is simpler than that of GSH, which is more suitable for the synthesis of Au_{25}; (2) MHA with negative charge is more suitable to combine with $TiO_2$with positive charge. Color change, EDS and XPS tests confirmed that Au_{25} was successfully deposited on TNA. After Au_{25} deposition, the diffuse reflectance spectra (DRS) of TNA showed that Eg decreased significantly from 3.25–3.35 to 2.51–2.25 eV, indicating effective visible light capture. In the photodissociation of tetracycline, the Au_{25}-TNA complex showed 1.6-fold enhanced activity and 1.67~1.75-fold enhanced kinetic constants compared to bare TNA. A series of photoelectrochemical measurements (photocurrent test, EIS and linear sweep voltammograms) were performed, indicating that Au_{25} NC is an effective photosensitizer for promoting charge transfer and inhibiting charge recombination (Scheme 3.28).

Li et al. used $^1O_2$ produced by photoexcited Au NCs to selectively oxidize sulfides and amines [65]. They developed a new synthetic method to produce Au_{38} NC (1-adamantanethiolate (S-Adm) as ligand) in an improved yield (10%). Compared with Au_{38} (with thiol ligand), S-Adm-protected Au_{38} shows a larger HOMO–LUMO energy gap (about 1.57 eV), which is suitable for generating $^1O_2$. Au_{38} NCs exhibit higher activity than Au_{25} NCs in the selective photocatalytic oxidation of sulfides due to the higher efficiency of photoexcited Au_{38} to produce $^1O_2$. In addition, Au_{38} NCs have good reusability, and still maintain catalytic activity and selectivity after 3 cycles. The same Au_{38} core with different ligands shows different selective sulfide or

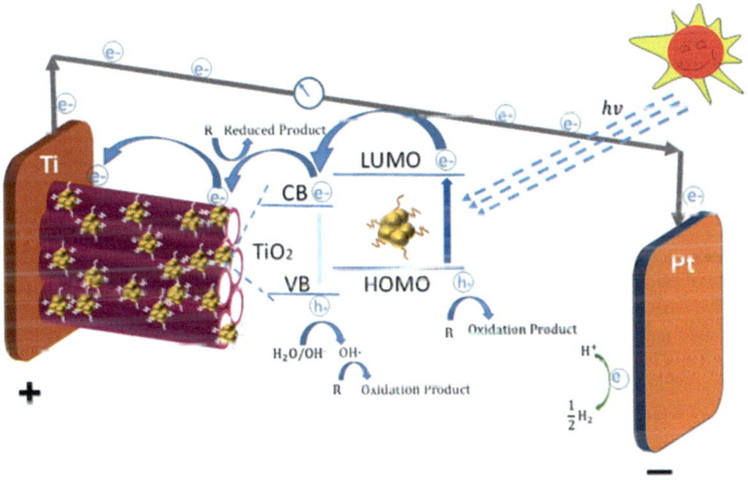

Scheme 3.28 The proposed mechanism for visible-light-driven photoelectrocatalytic oxidation of organic compounds. Reprinted with permission from Ref. [64]. Copyright © 2016, Royal Society of Chemistry

amine oxidation activity. $Au_{38}(PET)_{20}$($PET = SCH_2CH_2Ph$) is inactive under light irradiation. This indicates the effect of ligands on the photoreactivity of Au NC.

In addition, Zhu et al. achieved 100% conversion and selectivity of benzylamine to *N*-benzylidenebenzylamine under visible light irradiation using the alloy $Au_{25-x}Ag_x(PET)_{18}$ [66]. The advantages of alloy NC in this reaction (complete conversion and high selectivity) have been clearly demonstrated by a series of comparative experiments using Au_{25}^-, $Au_{25}0$ and Ag_{25}^- as catalysts. EPR and chemical probe tests confirmed the mechanism of the oxidation of benzylamine to *N*-benzylidenebenzylamine by $^1O_2$ generated by light irradiation of NCs (Schemes 3.29 and 3.30).

Solar photovoltaic conversion is the use of solar cells to directly convert light energy into electrical energy, which is a major way of solar energy utilization. Sakai and Tatsuma prepared metal cluster sensitized solar cells (MCSSC) by electrostatic attraction of a series of GSH-protected Au_n NC ($Au_{15}(SG)_{13}$, $Au_{18}(SG)_{14}$, $Au_{22}(SG)_{16}$, $Au_{22}(SG)_{17}$, $Au_{25}(SG)_{18}$, $Au_{29}(SG)_{20}$, $Au_{33}(SG)_{22}$ and $Au_{39}(SG)_{24}$) with precise structure to modify TiO_2 electrode [67]. In this system, hydroquinone was selected as the electron donor, because Au NCs are stable in this environment. All nanocrystals/TiO_2 MCSSC exhibited photoresponse, and the photocurrent behavior was consistent with the light absorption behavior of nanocrystals. This indicates that the photoexcited Au_n NC generates electrons into TiO_2 conduction bands through

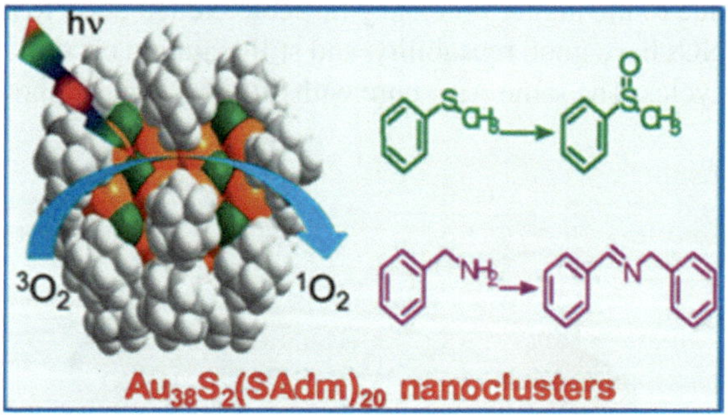

Scheme 3.29 Au_{38}-generated $^1O_2$ under photoexcitation and its catalyzed selective aerobic oxidation of sulfide to sulfoxide and benzylamine to imine. Reprinted with permission from Ref. [65]. Copyright © 2017, American Chemical Society

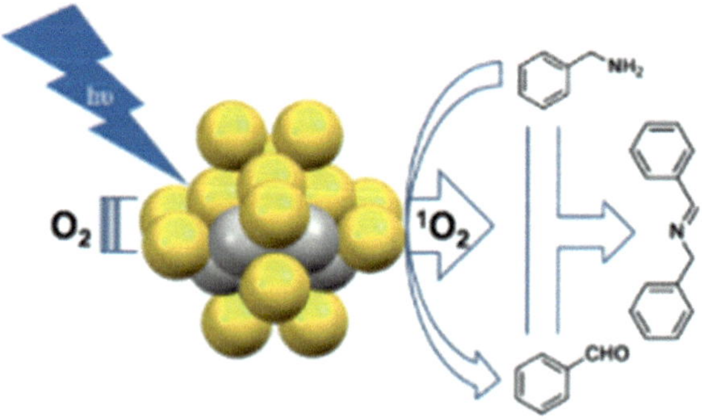

Scheme 3.30 The proposed mechanism of photocatalysis catalyzed by $Au_{25-x}Ag_x^-$. Reprinted with permission from Ref. [66]. Copyright © 2018 Wiley-VCH Verlag GmbH & Co. KGaA, Weinheim

the HOMO–LUMO transition. The Voc value (0.46 V), Jsc value (290 μA/cm^2) and IPCE value of Au_{25}/TiO_2 MCSC are the highest among all the tested Au NCs at an appropriate Au_{25} level. By comparing the electron donors with different redox potentials, the principle that the more positive the redox potential of the electron donor, the lower the IPCE of the MCSSC is proposed. These results indicate that Au NCs can be used as sensitizers for the dye-sensitized solar cell even if the performance is relatively poor. Therefore, the electronic energy level of the MCSSC system (e.g., Au NC, electron donor) is studied and the system is optimized to obtain a higher IPCE. They further explored a series of Au-GSH NCs (with 25, 30 or 90 Au atoms, corresponding to core diameters of 0.9, 1.0 or 1.4 nm) and various electron donors (e.g., thiocyanate ions, triethanolamine, hydroquinone, phenol, benzyl alcohol, and iodide ions) to gain insight into their energy level relationships [68].

In order to obtain high-performance MCSSC, they further combine Au NP and localized surface plasmon resonance (LSPR) as light-harvesting antennas, and use Au NC as a discrete energy sensitizer to fabricate Au NC-NP dual MCSSC systems [69]. In order to separate the influence of spacing (between NC and NP) and cluster size on electron transport, a two-dimensional (2D) ITO/Au NP/TiO_2/Au NC electrode is used instead of a 3D system. The ITO/Au NP/TiO_2/Au_{25} electrode shows a photocurrent of 2–9 times higher than that of the electrode without Au NP in the Au-NP-presented LSPR region. The enhanced photocurrent is attributed to the LSPR of Au NPs, because the number of Au NCs in the two systems (with and without Au NPs) is almost equal. The LSPR effect of Au NP is a near-field effect, so the photocurrent enhancement is related to the spacing between NP and NC (Scheme 3.31a, b). These authors found that the optimal spacing was 6 nm and the maximum enhancement was observed. They simulated the electric field distribution of a system with various spacings by finite-difference time-domain (FDTD) calculations (Scheme 3.31c)

Coupling Reaction

Li et al. used $Au_{25}(SR)_{18}$ nanoclusters to catalyze the homocoupling reaction of Ullmann-type aryl iodides [70]. The research question is how to use gold nanoclusters to catalyze the homocoupling reaction of aryl iodides and solve some limitations of traditional catalysts in this reaction. Firstly, $Au_{25}(SR)_{18}$ nanoclusters were synthesized and loaded on different oxide supports. Then by optimizing the selection of solvents and bases, the best reaction conditions were found. The experimental results show that under optimal conditions, the $Au_{25}(SR)_{18}$ nanoclusters catalyzed the homocoupling reaction of aryl iodides has high activity and selectivity, and the yield can reach 99.8%. In addition, the catalyst also showed good recyclability. The interaction between Au_{25} nanoclusters and CO_2 and its application in electrocatalytic activity (Scheme 3.32).

In the above examples, the main goal of the study is to regulate the electronic and catalytic properties of gold nanoclusters by changing the ligand structure. Recently, this group synthesized a new type of aromatic thiol-protected gold nanocluster $[Au_{25}(SNap)_{18}]^-[TOA]^+$, where SNap represents 1-naphthylthiolate ion and TOA represents tetraoctylammonium [71]. Compared with the previously reported $[Au_{25}(SC)_{18}]^-$, this aromatic thiol-protected nanoclusters show significant

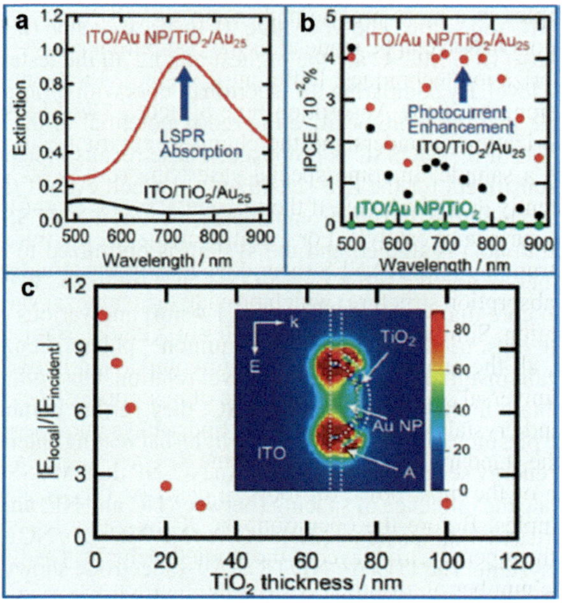

Scheme 3.31 **a** UV–vis absorption and **b** photocurrent spectra of ITO/Au NP/TiO_2/Au NC, ITO/TiO_2/Au NC, and ITO/Au NP/TiO_2. **c** Dependence of the electric field intensity on the TiO_2 thickness. Reprinted with permission from Ref. [69]. Copyright © 2013, Royal Society of Chemistry

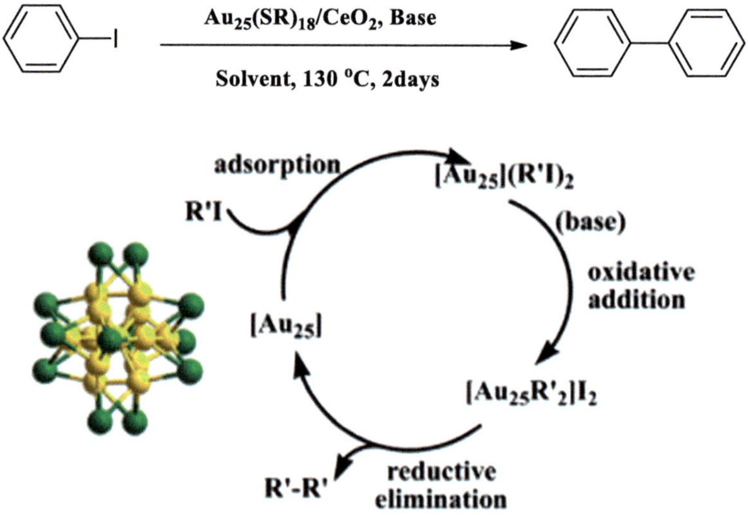

Scheme 3.32 Proposed mechanism for the Ullmann-type homocoupling reaction of iodobenzene catalyzed by $Au_{25}(SR)_{18}$ nanoclusters (green: surface Au atoms, yellow: core Au atoms, the thiol ligands are omitted for clarity). Reprinted with permission from Ref. [70]. Copyright © 2012, Royal Society of Chemistry

differences in electronic and catalytic properties, although their skeletons (i.e., the $Au_{25}S_{18}$ framework) are similar. It was found that the introduction of aromatic sulfate ions led to the expansion of the Au_{13} core of the nanoclusters, resulting in a red shift of the HOMO–LUMO electron absorption peak wavelength of the nanoclusters by about 10 nm. In addition, the unsupported $[Au_{25}(SNap)_{18}]^-$ nanoclusters exhibit high thermal stability and oxidation resistance, which is attributed to the stabilizing effect of aromatic ligands on the frontier orbitals (HOMO and LUMO) of nanoclusters. The catalytic activity of the $Au_{25}(SR)_{18}/CeO_2$ (R = Nap, Ph, CH_2CH_2Ph and n-C_6H_{13}) catalyst in the Ullmann cross-coupling reaction of 4-methyl-iodobenzene and 4-nitro-iodobenzene was also investigated. The results show that the activity and selectivity of the catalyst are greatly affected by the chemical properties of the protected thiol ligands (Scheme 3.33).

The complete structure of the $Au_{25}(SCH_2CH_2Ph)_{18}$ nanocluster catalyst, in-depth understanding of the mechanism of the Suzuki coupling reaction, and special attention to the promotion effect of imidazolium ionic liquids (ILs) on the coupling reaction between phenylboronic acid and p-iodotoluene [59]. Ultraviolet–visible spectroscopy, MALDI mass spectrometry, density functional theory (DFT) calculations and classical molecular dynamics (MD) simulations were used to study the catalytic mechanism. Through the interaction with imidazolium cations, the thiol ligand (–SR) and 'Au-SR' units are partially removed, which is considered to be the promotion

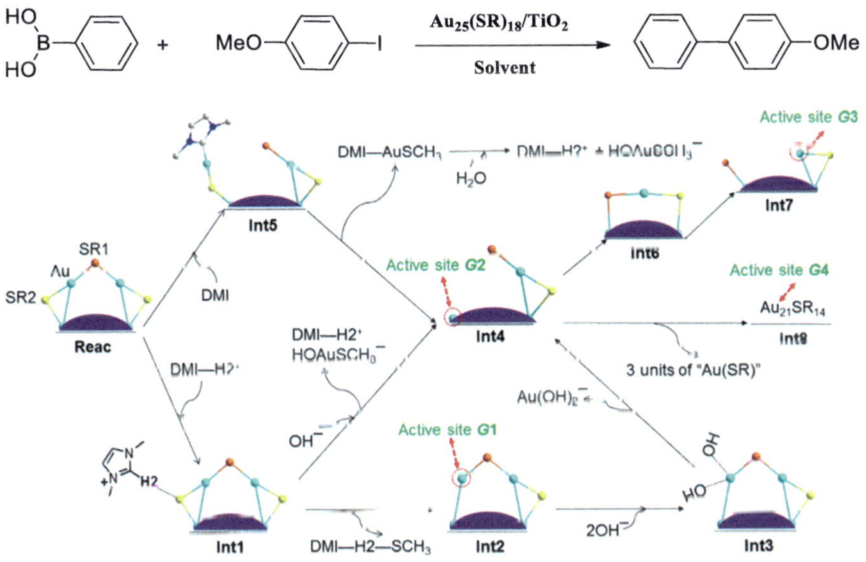

Scheme 3.33 Proposed reaction mechanism 1 for the formation of catalytically active sites on the Au_{25} cluster promoted by imidazolium-based ionic liquids. Placement of the intermediates in this schematic diagram is not based on their relative energies. Color code: Au, cyan; SR2, yellow; SR1, orange. The purple section represents the icosahedral Au_{13} core. The thiolate carbon tails are omitted for clarity. Reprinted with permission from Ref. [59].

mechanism of the cross-coupling reaction. Both experiments and DFT calculations show that the acidic protons at position 2 of the imidazolium ion play an important role in the detachment of ligands and Au atoms from the surface of the nanoclusters, so that the cross-coupling reaction occurs at a milder temperature than in the absence of ILs. The results show that the oxidation of active metal species can not only prevent further breakage of nanoparticles caused by ILs, but also increase the interaction between reactants and catalysts, thereby improving the conversion rate. Through DFT/PCM theoretical analysis, we predicted that weak acids (such as acetic acid) can better promote the cross-coupling reaction than conventional solvents. Experiments confirmed this prediction. Therefore, in this study, through the in-depth study of the structure and mechanism of the $Au_{25}(SCH_2CH_2Ph)_{18}$ nanocluster catalyst, the catalytic mechanism of the Suzuki coupling reaction was solved, and the promotion effect of imidazolium ionic liquid and the promotion effect of weak acid were proposed.

The Sonogashira cross-coupling reaction is an important organic synthesis reaction that can form new carbon–carbon bonds between terminal alkynes and aryl or alkenyl halides. However, the traditional metal complex catalyst has the problem of recovery and removal in this reaction. Therefore, Jin et al. used gold nanocluster catalysts to solve these problems. Specifically, this study first synthesized a gold nanocluster $Au_{25}(SR)_{18}$ with an atomic-level precise structure and loaded it on the surface of oxides (such as CeO_2, TiO_2, MgO and SiO_2) [72]. Then, the catalytic performance of gold nanocluster catalysts in the Sonogashira cross-coupling reaction was studied by experiments and density functional theory (DFT) simulations. The results show that the gold nanocluster catalyst supported on the oxide surface can efficiently catalyze the Sonogashira cross-coupling reaction, so that the conversion of aryl iodides is as high as 96.1% and the selectivity is as high as 88.1%. The precise structure of gold nanoclusters provides an important clue for the catalytic active sites: the unimpeded surface of the cluster surface facilitates the entry of reactants, and each surface consists of three independent 'U'-shaped surface protection groups. DFT simulation results show that the reactants are more inclined to adsorb on the open surface, where the benzene ring faces a surface gold atom. When the two reactants are co-adsorbed on the catalyst, the total adsorption energy reaches a relatively strong state, and the phenyl and iodine groups of the two reactants point to the third gold atom on the open surface, thereby achieving the coupling reaction (Scheme 3.34).

The catalytic potential of atomically precise quantum-sized gold nanoclusters $Au_{38}(SC_2H_4Ph)_{24}$ is explored by this group for the three-component coupling of an aldehyde, an alkyne and an amine to synthesize propargylamines [73]. A high catalytic efficiency with a very low loading (0.01 mol%) is achieved (Scheme 3.35).

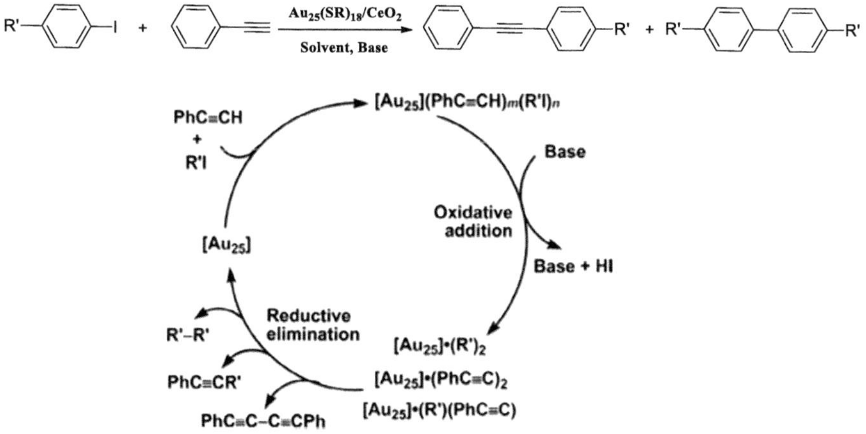

Scheme 3.34 Proposed mechanism for the Songashira-type cross-coupling reaction catalyzed by [Au_{25}] nanoclusters. For simplicity, [Au_{25}] represents $Au_{25}(SCH_2CH_2Ph)_{18}$. Reprinted with permission from Ref. [72] Copyright © 2013 Elsevier Inc. Published by Elsevier Inc. All rights reserved

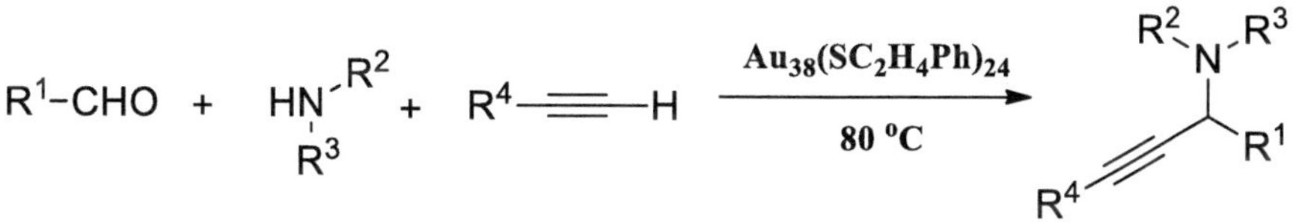

Scheme 3.35 Three-component coupling reactions of an aldehyde, alkyne, and amine catalyzed by $Au_{38}(SC_2H_4Ph)_{24}$. Reprinted with permission from Ref. [73]. Copyright © 2016, Royal Society of Chemistry

3.3.1.2 Organic Reactions Catalyzed by Pd NCs

CO Oxidation Reaction

Kunz et al. studied the oxidation mechanism of CO on selective Pd clusters under temperature-dependent conditions by using p-MBRS experiments and TPR and FTIR measurements [74] (Scheme 3.36).

The research questions mainly include two aspects: the activation of oxygen molecules and the adsorption strength of carbon monoxide. Through the experimental results, the activation temperature of oxygen molecules is determined. The experimental results show that the activation temperature of oxygen molecules on the surface of small Pd clusters and single crystal Pd is similar, which is about 360 K. Adsorption strength of carbon monoxide and CO poisoning effect: compared with single crystal Pd and nanoparticles with crystal faces on the surface, the adsorption strength of carbon monoxide on small Pd clusters is weaker, and no CO poisoning occurs in the temperature range above 300 K. This means that small Pd clusters still have catalytic activity for CO oxidation at higher temperatures, while single

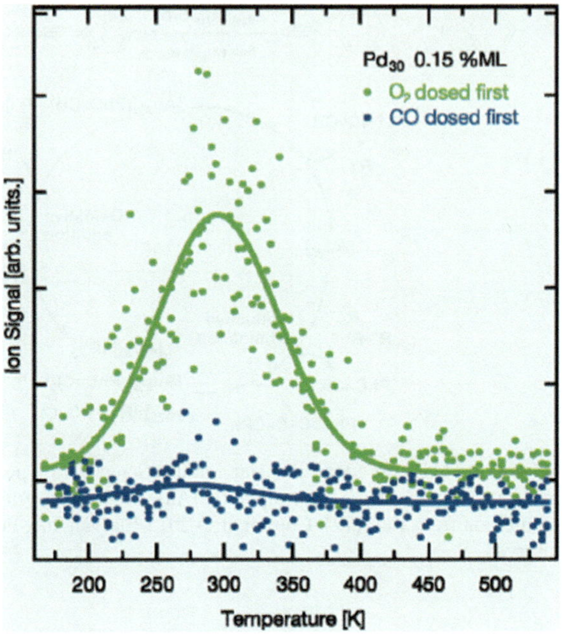

Scheme 3.36 TPR experiment of Pd_{30}. Reactants dosed in different order. Reprinted with permission from Ref. [74]. Copyright © 2010, American Chemical Society

crystal Pd and nanoparticles with crystal faces on the surface will be poisoned by CO adsorption in this temperature range. Through these experimental results, this study reveals the different behaviors of small Pd clusters and single crystal Pd and nanoparticles with crystal faces on the surface in the CO oxidation reaction.

Matthew et al. explored the effect of alumina thickness on the CO oxidation activity of Pd_{20}/alumina/Re (0001) catalyst, and studied the effects of alumina electronic properties and Pd_{20} geometry on the activity [75]. Firstly, the Pd_{20}/alumina/Re (0001) catalyst was characterized by surface science and technology, including the measurement of CO and O adsorbed on the surface and the characterization after the catalytic reaction. Through these characterization experiments, the researchers found that the adsorption state of CO on the surface of Pd_{20} is related to the thickness of alumina. The thinner alumina will lead to the change of the adsorption state of CO on the surface of Pd_{20}, thus affecting the activity of the catalyst. Next, the electronic structures of alumina films and Pd_{20} with different thicknesses were characterized by UPS and XPS (Scheme 3.37).

It is found that the thicker alumina film can keep the electronic structure of Pd_{20} stable, while the thinner alumina film will lead to the change of the electronic structure of Pd_{20}, thus affecting the activity of the catalyst. Finally, the researchers systematically studied the effect of alumina film with different thickness on the activity of Pd_{20}/alumina/Re (0001) catalyst through CO oxidation reaction experiments. The experimental results show that the thinner alumina film can improve the activity of the catalyst, while the thicker alumina film will reduce the activity of the catalyst. Through the analysis of the experimental results, in summary, this study solved the

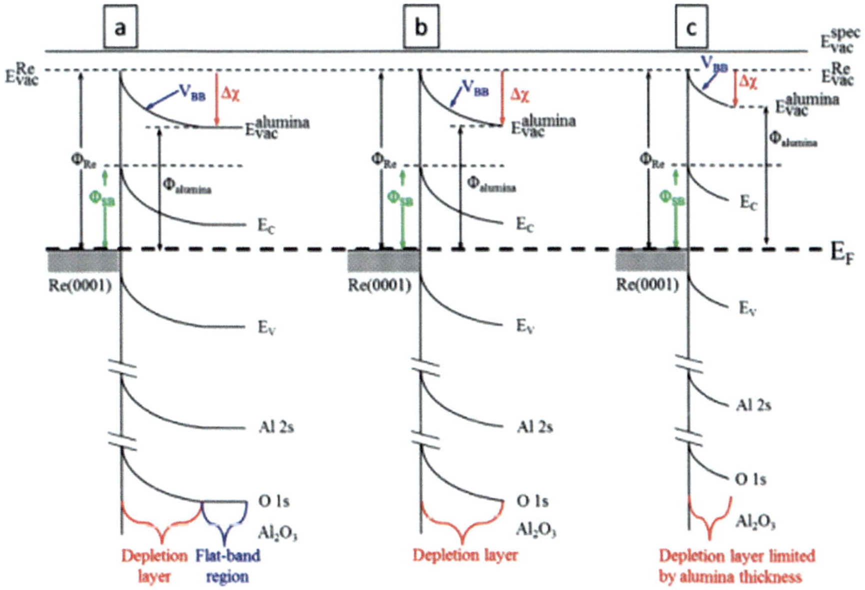

Scheme 3.37 Potential energy diagrams illustrating band bending expected for contact of Re with alumina films of thickness **a** much greater than the depletion layer thickness, **b** comparable to the depletion layer thickness, and **c** thinner than the depletion layer. Reprinted with permission from Ref. [75]. Copyright © 2015, American Chemical Society

influence mechanism of alumina thickness on CO oxidation activity by studying the characterization and activity of Pd_{20}/alumina/Re (0001) catalyst, and revealed the related effects of alumina electronic properties and Pd_{20} geometry on activity. This has important guiding significance for the design and optimization of CO oxidation catalysts.

In order to explore how the oxidation state and symmetry of the magnesium oxide-supported $Pd_{13}Ox$ nanocatalyst affect the activation energy of the CO oxidation reaction, Moseler et al. studied the combination of experiment and theoretical simulation to solve this problem [76]. First, Pd_{13} clusters were deposited on MgO films using a low kinetic energy (1 eV/atom) metal cluster beam. The catalytic properties of these clusters were then studied by temperature-programmed reaction (TPR) and isotope labeling experiments. The experimental results show that the $Pd_{13}O_4$ cluster forms a highly symmetrical structure under low temperature oxygen treatment, while the $Pd_{13}O_6$ cluster forms an asymmetric structure under high temperature oxygen treatment. The calculation results show that the highly symmetric cluster/reactant complex structure increases the activation energy of the catalytic reaction step, while the asymmetric cluster/reactant structure can accelerate the reaction process. This reduced activation energy is related to the destruction of local charge and orbital symmetry, resulting in a stronger bonding of the oxygen atom to one of the Pd atoms, which promotes the CO reaction with the shared Pd atom (Scheme 3.38).

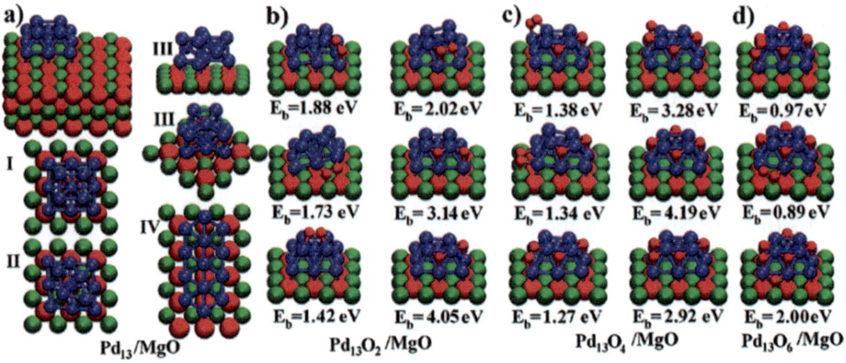

Scheme 3.38 Oxidation mechanisms of magnesia-supported Pd_{13}. **a** Ground-state structure and energetically higher lying isomers of the Pd_{13} clusters after soft-landing. The top left panel displays a side view of the full system for the ground state. The three lowermost MgO layers are not shown in the following panels. The middle left panel (I) shows a top view of the C_{4v} ground state, while a C_{2v} distorted energetically degenerate isomer is shown in the bottom panel (II). The top right and middle right panels show the first higher isomer (III) from two different perspectives, and the bottom left panel (IV) displays a 13 atom monolayer structure on the MgO substrate. **b** Left column: Adorption sites of a single oxygen molecule (shown from top to bottom) on a corner bridge (E_b = 1.88 eV), on top of Mg atoms (E_b = 1.73 eV), and on the Pd top square (E_b = 1.42 eV). Right column: The peroxo O_2 on a Pd side square (E_b = 2.02 eV), shown in the top row, dissociates with a barrier of 0.01 eV to form a nanooxide (middle row), with two atomic O on top of neighboring Pd atoms (E_b = 3.14 eV), which can evolve into the $Pd_{13}O_2$ ground state (E_b = 4.05 eV), shown at the bottom row, via a 0.1 eV barrier. **c** Left column: A second O_2 dimer can attach to corner bridge sites (E_b = 1.38 eV and E_b = 1.27 eV, shown at the top and bottom rows, respectively) or on top of Mg sites (E_b = 1.34 eV, shown at the middle row). Bottom row: The bridge-bonded O_2 can dissociate via barriers of the order 0.3 eV, crossing again peroxo states, into a metastable nanooxide (E_b = 3.28 eV and E_b = 2.92 eV in the sixth colunn). The E_b = 3.28 eV structure (top row) transforms easily (barrier 0.1 eV) into a $C_{4v}Pd_{13}O_4$ ground state with E_b = 4.19 eV (middle row), while the E_b = 2.92 eV structure (bottom row) has to cross barriers of the order 0.8 eV to isomerize into the ground state. **d** The third O_2 binds to a corner bridge (E_b = 0.97 eV, top row) or an Mg site (E_b = 0.89 eV, middle row) and dissociates (barrier 0.8 eV) in both cases to a $Pd_{13}O_6$ nanooxide (E_b = 2.00 eV, bottom row) with the two atomic O atoms on a corner bridge and on the Mg site. Color code: Mg atoms in green, oxygen in red (with adsorbed oxygen atoms depicted by the smaller spheres), and Pd atoms in blue. Reprinted with permission from Ref. [76]. Copyright © 2012, American Chemical Society

Methanol Steam Reforming Reaction

Iwasa et al. studied the effect of zinc addition on the performance of supported palladium catalyst in methanol steam reforming reaction [77]. In order to improve the efficiency of the methanol steam reforming reaction, the selectivity and activity of the supported palladium catalyst were improved by adding zinc. Firstly, the supported palladium catalyst without zinc addition was tested. It was found that methanol mainly underwent the decomposition reaction rather than the steam reforming reaction, resulting in poor selectivity. The addition of zinc to the supported palladium catalyst was studied and it was found that the selectivity was significantly improved.

Especially on the Pd/CeO_2 and Pd/AC catalysts with Zn addition, the selectivity reached 99%. Similar enhancement effects were also observed in Pd/CeO_2 catalysts with the addition of metals such as cadmium, indium and gallium. The activity of Zn-modified Pd/CeO_2 catalyst depends on the preparation method used. The Pd/Zn/CeO_2 catalyst prepared by the co-precipitation method showed the highest activity in hydrogen production.

Electrocatalytic Reaction

$Pd_6(SR)_{12}$ nanoclusters are used as electrocatalysts to solve the catalytic problem of the hydrogen oxidation reaction [78]. It has been found that Pd_6 nanoclusters have high catalytic performance, and the electronic state of the metal core shell plays a key role in mediating catalytic properties. Protective ligands play a dual role in catalytic reactions. In this study, $Pd_6(SR)_{12}$ nanoclusters were synthesized and loaded on activated carbon. Firstly, the synthesized products were characterized by UV–vis spectroscopy. The results showed that compared with PdC_{12} precursor, the prepared Pd_6 nanoclusters and Pd_6/AC composites exhibited three obvious absorption peaks near 275, 325 and 410 nm, indicating that Pd nanoclusters were successfully synthesized and loaded on activated carbon. The surface chemical properties of the product were further characterized by infrared spectroscopy. The results showed that Pd_6/AC had infrared absorption characteristics similar to $Pd_6(SR)_{12}$. Thermogravimetric analysis experiments show that the stability of $Pd_6(SR)_{12}$ nanoclusters is high. The morphology and size of the synthesized $Pd_6(SR)_{12}$ were studied by high-resolution transmission electron microscopy. The results show that $Pd_6(SR)_{12}$ nanoclusters have uniform spherical morphology (Scheme 3.39).

Catalytic Hydrogenation Reaction

The researchers conducted experimental studies on the effects of different support materials on the catalytic activity of palladium-based catalysts. Palladium/gallium oxide (Pd/Ga_2O_3) catalyst has higher activity than copper/zinc oxide (Cu/ZnO) catalyst in methanol synthesis [79]. By testing the activity and selectivity of the catalyst, they found that the Pd/Ga_2O_3 catalyst was superior to the Cu/ZnO catalyst in terms of methanol yield and methanol selectivity. In addition, the TOF (conversion frequency per unit time) of Pd/Ga_2O_3 catalyst is 20 times higher than that of Cu/ZnO catalyst. Therefore, this study solved the problem of catalytic activity in methanol synthesis reaction by developing Pd/Ga_2O_3 catalyst. Carbon nanotube (CNT) supported Pd-ZnO catalyst is used for the hydrogenation of CO_2 to methanol. A series of Pd-ZnO/CNTs catalysts with different compositions were prepared by the co-precipitation method, and their catalytic properties were evaluated and optimized [77]. By using CNTs as the support of the catalyst, the relative content of Pd^0 active species closely related to methanol formation on the catalyst surface can be increased compared with traditional activated carbon (AC) or γ-Al_2O_3 [80].

Coupling Reaction

In the past few decades, the active species of palladium catalysts have been an unsolved mystery (Scheme 3.40). In this study, the reaction was carried out by using

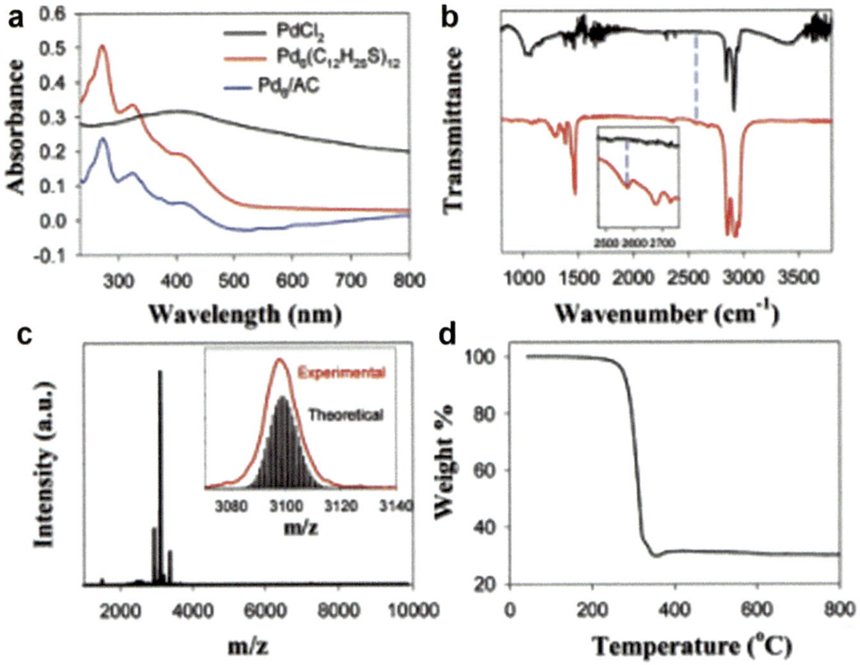

Scheme 3.39 **a** UV–vis absorption spectra of $PdCl_2$, $Pd_6(C_{12}H_{25}S)_{12}$ and Pd_6/AC. **b** Infrared spectra of 1-dodecanthiol (red line) and $Pd_6(C_{12}H_{25}S)_{12}$ (black line). The inset shows the zoomed-in spectra at 2470–2760 cm^{-1}. **c** MALDI-TOF MS of $Pd_6(C_{12}H_{25}S)_{12}$. The inset shows a comparison between the experimental and simulated patterns for the highest peak in the full spectrum. **d** Thermogravimetric curve of $Pd_6(C_{12}H_{25}S)_{12}$. Reprinted with permission from Ref. [78]. Copyright © 2017, Royal Society of Chemistry

different palladium sources (including salts, complexes and nanoparticles), and the reaction rate and product formation were observed [81]. It was found that the reaction could be carried out at the highest rate only when small clusters composed of three and four palladium atoms were formed. These small clusters are stable when heated in *N*-methylpyrrolidone (NMP) and can be stored for at least one month for subsequent use. In addition, other nucleophiles can also release palladium clusters from nanoparticles, and the properties of nucleophiles and the size of nanoparticles can affect the formation of clusters. Whether the starting palladium source is salt, complex or nanoparticle, the carbon–carbon coupling reaction can only proceed at a higher rate after the formation of small clusters composed of three and four palladium atoms. These palladium clusters can be stabilized and stored by water, and the catalytic activity can last for one month.

The Heck reaction is a commonly used carbon–carbon bond formation method in organic synthesis, which is achieved by palladium-catalyzed coupling reactions of olefins with aryl or vinyl halides. In recent years, the main reasons are the development of more active catalyst systems, the discovery of waste-free versions of reaction

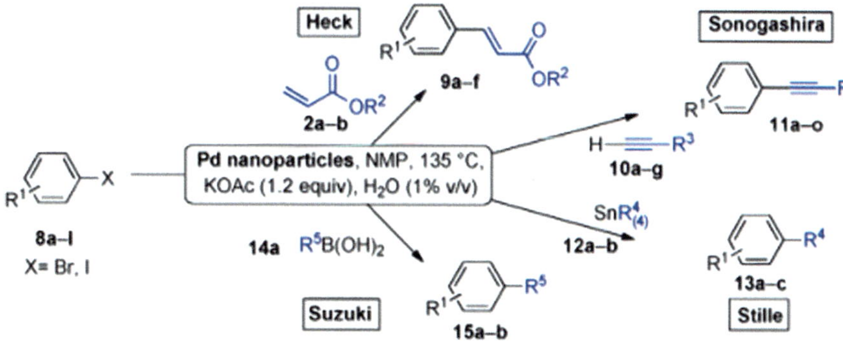

Scheme 3.40 Heck, Sonogashira, Stille, and Suzuki cross-coupling reactions of different iodo- and bromoarenes under palladium-cluster catalysis in the presence of water. Reprinted with permission from Ref. [81]. Copyright © 2013 WILEY–VCH Verlag GmbH & Co. KGaA, Weinheim

conditions, and the hope to build a large amount of empirical data on a solid mechanism. In the ligand-free Heck reaction of aryl bromides, the formation of palladium clusters exists, and the turnover frequency of the catalyst increases with the decrease of the catalyst concentration [82]. This indicates that at low concentrations, there is a balance between higher-order palladium species (such as palladium clusters) and lower-order species (such as monomer or dimer catalysts), and the balance will shift to lower-order species. Low concentrations of palladium can inhibit the formation of palladium black and make all metals available for catalytic reactions. The author also discussed the role of palladium clusters in the catalyst and proposed the possibility of the reaction mechanism. Finally, the researchers summarized the advantages and application prospects of ligand-free Heck reaction. Palladium clusters can be used as a repository of catalyst active species, thus achieving efficient reactions at low concentrations (Scheme 3.41).

The application of stable and efficient Pd_3 cluster catalyst in Suzuki reaction and its reaction mechanism. A Pd_3 cluster catalyst named Pd_3Cl was synthesized and characterized [83]. The experimental results show that Pd_3Cl is very stable at high temperature and in the presence of oxygen. Pd_3Cl can efficiently catalyze the Suzuki–Miyaura cross-coupling reaction of various aryl bromides and aryl boric acids at room temperature. Through experimental methods such as mass spectrometry and X-ray absorption fine structure, the researchers found that the structure of the Pd_3 cluster catalyst remained intact during the reaction and participated in the reaction through the Pd_3Ar intermediate. Under the reaction conditions, Pd_3OH species were also detected by mass spectrometry in the absence of substrates, which strongly supported the reaction mechanism of the 'oxidized Pd' pathway, in which the chlorine ligand was replaced by a hydroxide radical, and then the hydroxide radical was combined with the aryl boric acid (Scheme 3.42).

Polymer-stabilized and dendrimer-stabilized NPs have proved to be efficient catalysts for a variety of reactions. The advantage of dendrimers, besides their loose binding of the NP surface, is that a precise and predictable number of Pd atoms exist

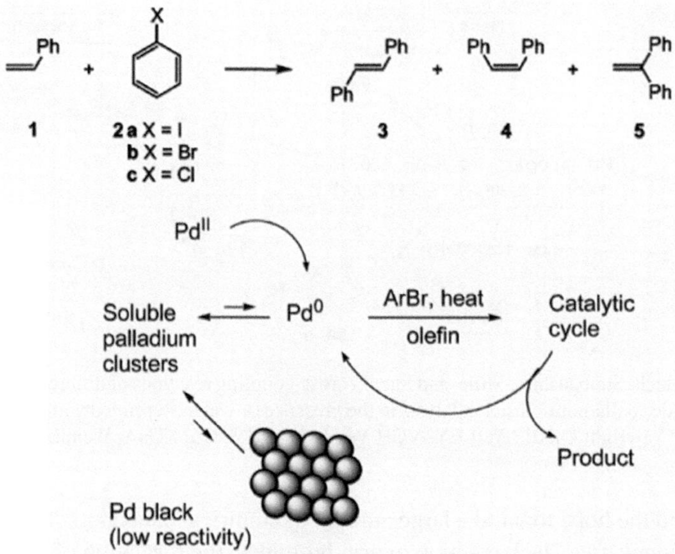

Scheme 3.41 General route for ligand-less Heck reactions at low Pd-loading. Reprinted with permission from Ref. [82]. Copyright © 2004, Royal Society of Chemistry

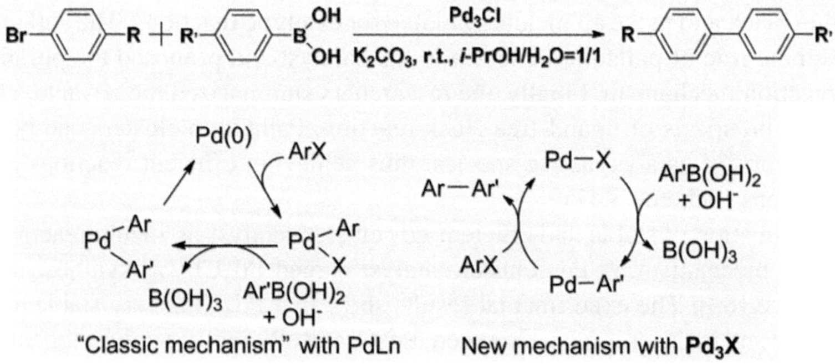

Scheme 3.42 Comparison between the "classic" mechanism of the Suzuki–Miyaura reaction and the mechanism shown with the Pd_3Cl cluster as catalyst. Reprinted with permission from Ref. [83]. Copyright © 2017, American Chemical Society

in the PdNP catalyst or precatalyst. Recently, Astruc et al. described dendrimers synthesized by click chemistry with 9, 27, and 81 ferrocenyl termini and 1,2,3-triazole ligands on the dendritic tethers: G0 with nine terminal ferrocenyltriazole units, G1 with 36 triazole units (9 interior and 27 terminal), and G2 with 117 triazole units (9 + 27 interior and 81 terminal). PdNPs that can be stabilized by these new dendrimers [84]. Firstly, the coordination ratio of triazole ligand to $Pd(OAc)_2$

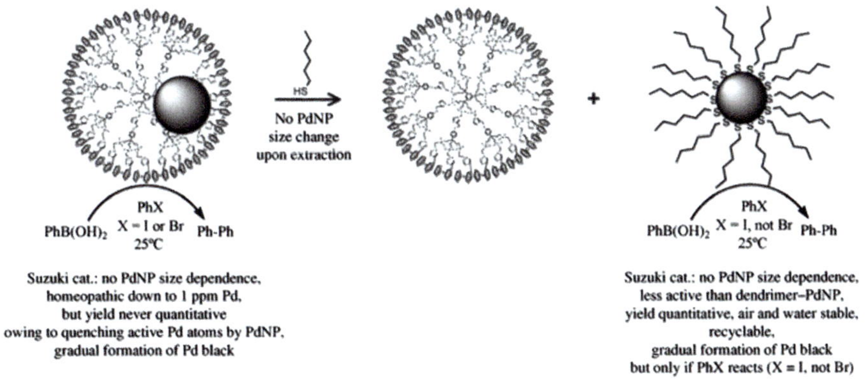

Scheme 3.43 Extraction of PdNPs from DSNs or DENs with hexanethiol without any change in the size of the PdNPs to produce air- and water-stable, catalytically active hexanethiolate PdNPs. Reprinted with permission from Ref. [84]. Copyright © 2007 WILEY–VCH Verlag GmbH&Co. KGaA, Weinheim

in the dendritic polymer was determined by electrochemical test, and the number of palladium atoms in the designed Pd nanoparticles was determined in advance. Then, the researchers successfully reduced Pd^{II} to Pd nanoparticles using methanol as a reducing agent. Through transmission electron microscopy (TEM) observation, it was found that G0 formed dendrimer-encapsulated NPs (DSN), while G1 and G2 formed dendrimer-encapsulated NPs (DEN). The researchers found that the size of Pd nanoparticles is closely related to the catalytic efficiency, and the DEN formed by G1 is smaller and more efficient than that formed by G2. Next, the researchers applied these Pd nanoparticles to the Suzuki reaction between phenyl halides and phenylboronic acid. The results showed that neither DSN nor DEN, nor two consecutive generations of DEN, could achieve a fully quantitative reaction. However, when the G0 catalyst was used, the conversion of the reaction increased as the concentration of Pd nanoparticles decreased to 1 ppm (Scheme 3.43).

3.3.1.3 Redox Reaction Catalyzed by Ag NCs

Zhang et al. prepared carbon nanofibers (CNFs) by electrospinning technology, and assembled uniformly dispersed silver nanoparticles (AgNPs) in situ on electrospun CNFs by hydrothermal treatment to catalyze the reduction reaction of 4-nitrophenol [85]. This study solves the following problems: 1. Preparation of efficient catalysts. By assembling AgNPs on electrospun CNFs, the high specific surface area of AgNPs and the synergistic effect between CNFs and AgNPs were utilized to achieve efficient catalytic reduction reaction. 2. The problem of controlling silver content. By properly changing the hydrothermal reaction conditions, the silver content on CNFs can be effectively controlled, thereby adjusting the performance of the catalyst. 3. The problem of separating and recovering catalysts. Due to the fiber morphology of

CNFs/AgNPs composite nanofibers, they can be separated from the liquid by precipitation to achieve the recycling of the catalyst through the above research ideas and methods. Specifically, $Au_{12}Ag_{32}(SR)_{30}$@ZIF-8 composite was successfully synthesized, and it was found that the catalyst had excellent performance for the conversion of phenylacetylene under mild conditions (50 °C and atmospheric CO_2), and its TON value was as high as 18,164, far exceeding most known catalysts. In addition, the catalyst has good stability and remains catalytic activity after 5 cycles. Through this synthesis method, researchers have successfully solved the synthesis problem of APNCs@ MOFs composites and obtained catalysts with precise structure, which provides new possibilities for further research in the field of heterogeneous catalysis [86].

3.3.1.4 Hydrogen Production Reaction Catalyzed by Co NCs

An electrocatalytic hydrogen production method at low overpotential was developed [87]. A series of cobalt-based complexes were synthesized and their catalytic properties in hydrogen production were studied by electrochemical methods. Numerical simulations were also carried out to explore the hydrogen production mechanism. The key problem in the research is to determine the optimal catalyst and reaction conditions to achieve efficient hydrogen production reaction at low overpotential. Through experiments and simulations, cobalt-based complexes can effectively catalyze the hydrogen production reaction in the presence of acidic conditions, and exhibit higher catalytic activity at low concentrations. In addition, the researchers also found that the hydrogen production reaction may be carried out through a bimetallic mechanism, rather than a simple single metal mechanism.

3.3.1.5 Organic Reactions Catalyzed by Ru NCs

Hydrogenation Reaction

Several methods for reducing aldehydes and ketones by using soluble catalysts have been reported. The use of mononuclear complexes or active complexes to participate in the reaction, and the introduction of suitable ligands into the molecule can greatly change the catalytic activity. $H_4Ru_4(CO)_{12}$ and its phosphine-substituted derivatives have been used as homogeneous hydrogenation catalysts [88]. Therefore, the author studied the catalytic hydrogenation of cyclohexanone. By using PPh_3, PBu_3 and (−)-DIOP to partially replace the carbonyl ligands in H and $H_4Ru_4(CO)_{12}$, the author found that the hydrogenation reaction rate of cyclohexanone is proportional to the substrate concentration and hydrogen partial pressure, and is also proportional to the catalyst concentration. At the same time, the solvent has a significant effect on the reaction rate. In addition, by gradually replacing the carbonyl ligand in $H_4Ru_4(CO)_{12}$ as the phosphine ligand, it was found that the activity of carbonyl ruthenium hydride increased with the increase of phosphine substitution. Finally, the ketone does not

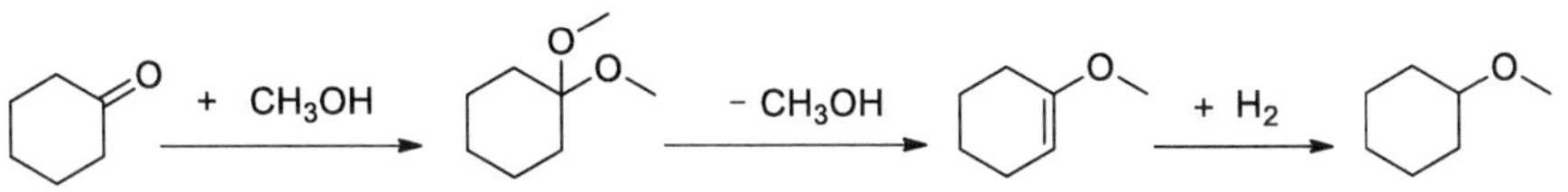

Scheme 3.44 Hydrogenation of cyclohexanone in the presence of $H_4Ru_4(CO)_{12}$. Reprinted with permission from Ref. [88]. Copyright © 1978 Published by Elsevier B.V

exist in the form of enol in the reduction step, but interacts with the catalyst through its carbonyl group (Scheme 3.44).

Cycloaddition Reaction

Transition-metal-catalyzed cycloaddition reactions using carbon monoxide as a one-carbon unit are very important methods to obtain cycloketones [89]. Pauson–Khand reaction is a typical example by a formal [2 + 2 + 1] cycloaddition to form cyclopentenones. But another kind of the construction of a five-membered ring system, the [4 + 1] mode, in which conjugated systems act as four-atom assembling units, is also attractive. The reaction of 1, 3-dienes with carbon monoxide would be expected to afford cyclopentenones, and the replacement of a terminal carbon by a heteroatom, such as oxygen and nitrogen, would give rise to unsaturated γ-lactones and γ-lactams, respectively. Especially, the latter catalytic transformation using α, β-unsaturated imines would be useful in synthetic organic chemistry, because γ-lactam skeleton is one of the most important nitrogen heterocycles for pharmaceutical agents. Murai et al. reported the first example of a [4 + 1] cycloaddition of structurally simple α, β-unsaturated imines with carbon monoxide. The standard reaction conditions established were 2 mol% of $Ru_3(CO)_{12}$, 10 atm of CO, in toluene, and at 180 °C. This reaction represents the first example of catalytic carbonylative [4 + 1] cycloaddition using a structurally simple 1, 3-diene system.

The reaction mechanism was proposed via a pathway shown in Scheme 3.45. The coordination of a nitrogen to ruthenium allows the complex to be easily converted to metallacycle **A** via an oxidative cyclization of α, β-unsaturated imine. The subsequent insertion of CO and reductive elimination of ruthenium initially produce the β, γ-unsaturated γ-lactam **B**. For the reaction of imines which contain a β-hydrogen, **B** is transferred to the thermally more stable α, β-unsaturated isomer **C**. The author speculated that the significant difference between the reactivity of different imines is due to the different direction of the resonance effect by the oxygen atom to the π-system. The nitrogen in imine partially loses, by the resonance effect, its sp^2-character which appears to be essential for the interaction between α, β-unsaturated imine and ruthenium.

Photocatalytic Reaction

In the presence of iodide ions, the dye molecules will form complexes with iodide ions, thereby promoting the charge regeneration process. The Clifford group studied the dye-dependent regeneration kinetics in dye-sensitized nanocrystalline solar cells

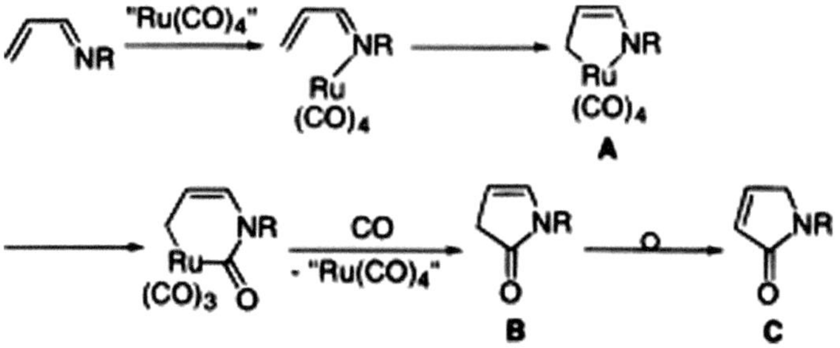

Scheme 3.45 The pathway of [4 + 1] cycloaddition reactions of α, β-unsaturated imines with CO catalyzed by $Ru_3(CO)_{12}$. Reprinted with permission from Ref. [89]. Copyright © 1999, American Chemical Society

[90]. The photogenerated charge recombination process of four different dye-sensitized TiO_2 films in the solution containing $LiClO_4$ and LiI electrolyte was observed by transient absorption spectroscopy. The experimental results show that the initial transient absorption signal of the dye-sensitized TiO_2 film increases in the presence of LiI electrolyte, and shows a long-life dynamic behavior on the microsecond time scale. Low concentration of iodide ions can promote the rate of charge regeneration, but high concentration of iodide ions will lead to competitive reaction of charge regeneration, thus reducing the efficiency of the battery. Further studies have shown that this enhanced signal is related to the formation of dye cations and iodide ion complexes.

3.3.1.6 Hydrogenation Reaction Catalyzed by Rh NCs

Supported metals are a widely investigated and industrially applied class of catalyst, and their properties are sensitive to subtle changes in structure. The synergistic effect between the support and metal catalyst is affected by the structural properties of the support, but when the supported species incorporate only very few metal atoms each, the supports take on the role of ligands, providing opportunities to fine-tune the catalytic properties that may rival those in organometallic solution catalysis. Furthermore, when the metals are highly dispersed on a support, the opportunities for involving the support in the catalysis are optimized.

Gates et al. reported how to tune the catalytic properties of small, nearly uniform clusters and site-isolated complexes of rhodium on a highly uniform (crystalline) support, zeolite HY, for control of the selectivity for alkene dimerization versus hydrogenation [91]. This is the first example of precise control of the selectivity of a solid catalyst by tuning of the structure of essentially molecular supported species. The complexes initially present as $Rh(C_2H_4)_2$ bonded to the zeolite catalyze both the dimerization and hydrogenation of ethene, with a high selectivity for the former, and that treatment of the catalyst to make rhodium clusters of only a few atoms each,

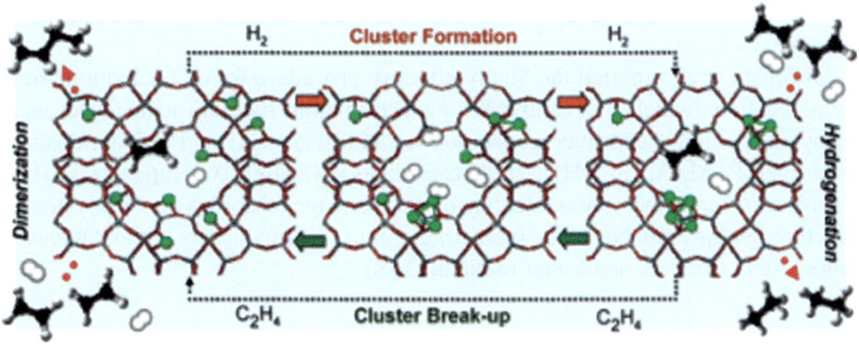

Scheme 3.46 Simplified schematic representation of the state of the rhodium (green) in the catalyst attained at each step in the cycle, as the catalytic performance of the solid is tuned. Reprinted with permission from Ref. [91]. Copyright © 2011, American Chemical Society

on average, boosts the activity for hydrogenation and essentially eliminates that for dimerization. Moreover, these changes are reversible (Scheme 3.46).

The data shown that the clusters were highly active catalysts for C–H bond formation, much more active than the complexes. They believe that a quantification of the changes in the catalyst behavior at the very early stages of cluster formation, when only few Rh–Rh bonds had been formed and only apportion of the rhodium had been converted into clusters. When hydrogenation was the predominant reaction, it was so fast in comparison with dimerization that the latter was swamped.

Selective hydrogenation reactions of 1, 3-butadiene were also studied by this team [92, 93]. They investigated the influence of the following catalyst design variables on the activity and selectivity of supported rhodium for 1, 3-butadiene hydrogenation: (a) the metal nuclearity, ranging from one to several; (b) the electron-donor properties of the support and (c) other ligands on the rhodium, including reactive hydrocarbons and CO. The selectivity for *n*-butene formation increases as the nuclearity of the metal species decreases from several Rh atoms to one, but these catalysts form the undesired product *n*-butane even at low diene conversions. The *n*-butene selectivity increases when the rhodium is selectively poisoned with CO ligands, and it is highest when the support is the electron-donor MgO and the rhodium is in the form of clusters that are well approximated as dimers. The electron-donor support is crucial for stabilization of the rhodium carbonyl dimer sites, as it limits the oxidative fragmentation of the clusters—which is facilitated when the support is HY zeolite (a poor electron donor)—that leads to decreased catalytic activity and selectivity. The selective MgO-supported rhodium carbonyl dimers suppress the catalytic routes that yield butane, limiting the activity for H_2 dissociation to avoid butane formation via primary reactions and also favoring the bonding of 1, 3-butadiene over butenes to limit secondary reactions giving butane. With this catalyst, selectivities to *n*-butene > 99% were achieved at 1, 3-butadiene conversions as high as 97%. This selectivity matches that of any reported for this reaction, and the catalyst works under milder conditions (313 K and 1 bar) than others that are selective for this reaction (Scheme 3.47).

Nishihara et al. reported the highly effective procedure for olefin and nitroarene hydrogenations which was catalyzed by encapsulating rhodium nanocluster using phenylazomethine dendrimer generation 4 (TPP-DPA G4) and polyamidoamine dendrimer (PAMAM G4-OH) as the encapsulants [94]. High TOF (up to 17,520 h^{-1}) was observed and the important feature of the nanocage catalyst is that substrates can pass through the branches of the protecting groups of nanoclusters without releasing nanoclusters from the dendrimer (Scheme 3.48).

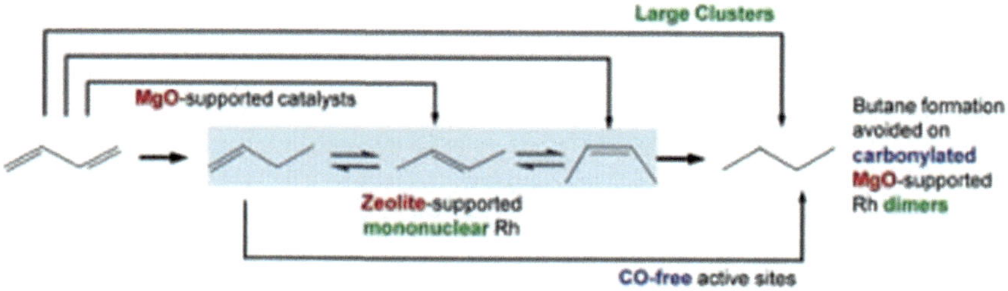

Scheme 3.47 Reaction pathways for the hydrogenation of 1, 3-butadiene. Reprinted with permission from Ref. [93]. Copyright © 1998, American Chemical Society

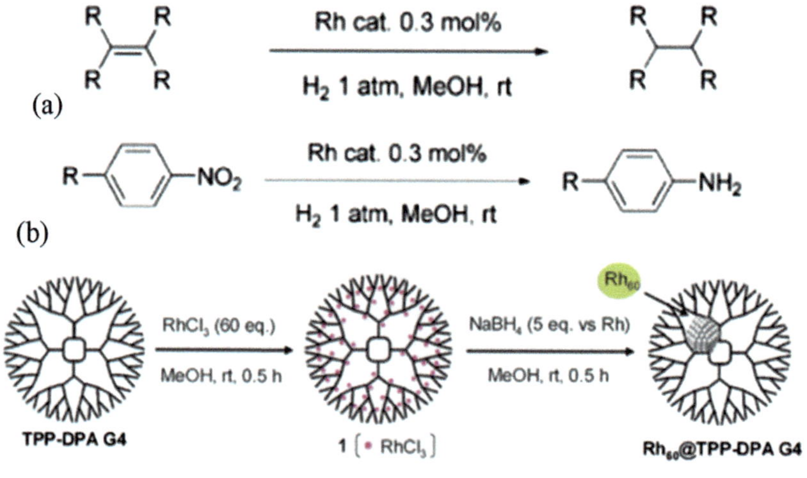

Scheme 3.48 The reactions of olefin hydrogenation (**a**), nitroarene hydrogenation (**b**), and synthetic procedure of Rh_{60}@TPP-DPA G4. Reprinted with permission from Ref. [94]. Copyright © 2008, Royal Society of Chemistry

3.3.1.7 Hydrogenation Reaction Catalyzed by Ni NCs

p-Nitrophenol is one of the most recalcitrant pollutants that often exists in industrial effluents, while *p*-aminophenol is of great commercial importance as an intermediate for the manufacture of pharmaceuticals (paracetamol, acetanilide, phenacetin and antipyretic drugs, polymers, dyes, and urethanes). Although noble metal catalysts have high catalytic activity, low selectivity and difficult separation are still prominent problems. In addition, high reaction temperature and high pressure of H_2 limited the application of the non-metallic catalyst.

Wu et al. prepared two-dimensional (2D) heterostructured Ni/graphene nanocomposites, which were constructed via electrostatic-induced spread by following in situ-reduction growth process for magnetically recyclable catalysis of *p*-nitrophenol to *p*-aminophenol [95].

A proposal mechanism was illustrated in Scheme 3.49 because each Ni atom includes a 0.6d band hole (unpaired electrons), these unpaired electrons could work with the adsorbed *p*-nitrophenol to form an adsorption bond in the chemical adsorption. During the hydrogenation reduction of *p*-nitrophenol, the generated *p*-nitrophenolate ions and BH_4^- were adsorbed by the electronic hole in the surface of the Ni nanocomposites via chemical adsorption and immediately followed the formation of *p*-aminophenol by hydrogenation reduction of *p*-nitrophenol on the surface of Ni nanocomposites. Besides enhanced electron transfer, graphene can also adsorb hydrogen and supply the favorable environment for the hydrogenation reduction reaction. In order to understand the hydrogen adsorption behavior of graphene, the catalytic dehydrogenation reaction of the $NaBH_4$ aqueous solution was evaluated by using Ni nanocomposites and Ni/graphene nanocomposites as catalysts. Ni/graphene generates a relative lower amount of H_2 than Ni nanocomposites, while graphene cannot produce H_2, indicating the hydrogen adsorption of graphene. Thus the synergistic effect of Ni/graphene nanocatalysts finally exhibits highly catalytic performance for the hydrogenation reduction of *p*-nitrophenol to *p*-aminophenol with the rate constant of $11.7 \times 10^{-3}\ s^{-1}$, which is about twofold more when compared to that of Ni nanocomposites ($5.45 \times 10^{-3}\ s^{-1}$). The heterostructured Ni/graphene nanocomposites with super catalytic activity and the long lifetime develop a new green chemistry method for the non-noble metal/graphene nanocatalysts applied in the nano-catalytic synthesis of *p*-aminophenol.

3.3.2 *MNCs of Bimetallic Alloy*

Hydrogen Evolution Reaction (HER) catalyzed by Pt-Au Alloy MNCs

Platinum-based catalysts are widely recognized as an alternative energy carrier, facilitating the development of artificial catalytic systems. These systems often involve the synthesis of precisely composed molecular-like clusters through size-controlled methods. Examples of such clusters include $Au_{25}(SR)_{18}$, $Au_{102}(SR)_{44}$, and

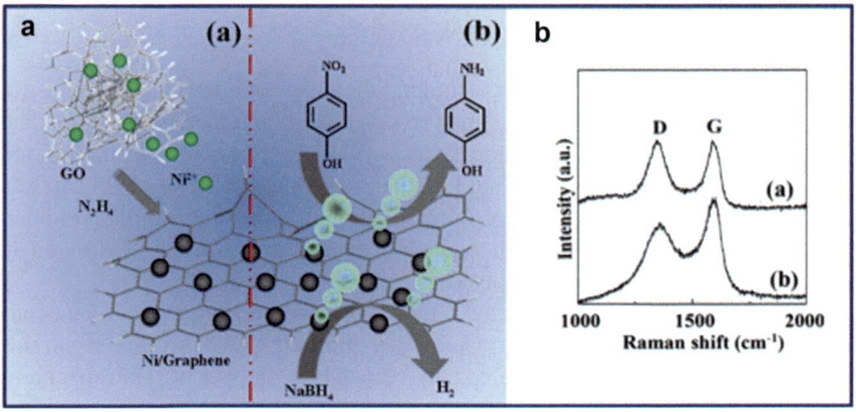

Scheme 3.49 **A** Schematic representation of the fabrication of Ni/graphene nanocomposites (a) and its catalytic hydrogenation mechanism of *p*-nitrophenol to *p*-aminophenol (b). **B** Raman spectra of Ni/graphene nanocomposites (a) and graphene oxides (b). Reprinted with permission from Ref. [95]. Copyright © 2014, American Chemical Society

$Ag_{44}(SR)_{30}$, where SR represents the thiolate ligand. The stability of these clusters is believed to arise from the closure of superatomic electron shells. This unique electrocatalyst exhibits efficient performance in both homogeneous and heterogeneous conditions.

Hydrogen-metal interactions are used in many reactions, especially in hydrogenation catalysis and water splitting. But the role of hydrogen in gold nanoclusters remains unknown. Kwak's team doped platinum atoms into gold nanoclusters of similar molecules to produce a thermally neutral catalyst for electrocatalytic reactions for preparing hydrogen [96].

First, NCs were applied in tetrahydrofuran (THF) electrolyte containing 1.0 M trifluoroacetic acid (TFA) to study the HER performance. Pt_1Au_{24} showed better catalytic activity than Au_{25} (−1.10 V) with a higher onset potential (−0.89 V) (Scheme 3.50a). The thermodynamic reduction potential of protons in this electrolyte is approximately −0.82 V; therefore, Pt_1Au_{24} greatly reduces the difference between the actual and theoretical potentials. Furthermore, the quasi-first-order rate constant (kobs) of Pt_1Au_{24} for HER (121,000 s^{-1} at 1.5 V; Scheme 3.50b) is higher than that of previously reported molecular electrocatalysts. As the TFA concentration increased, 362–364 LSV was measured (Scheme 3.50c).

The previous examples demonstrate the significant impact of changing or doping one atom in ultra-small metal nanoclusters on the hydrogen evolution reaction. In their study, Hu et al. utilized first principles to investigate the interactions between hydrogen and $[Au_{25}(SR)_{18}]^q$ clusters, as well as single-atom doped Au $[M_1Au_{24}(SR)_{18}]^q$ clusters (M = Pt, Pd, Ag, Cu, Hg, Cd) [97]. The findings revealed that hydrogen behaves as a metal in these clusters and contributes its 1 s electrons to the superatomic free electron count. The analysis of doping energetics indicated that when H is doped to form an eight-electron superatom, the binding energy of

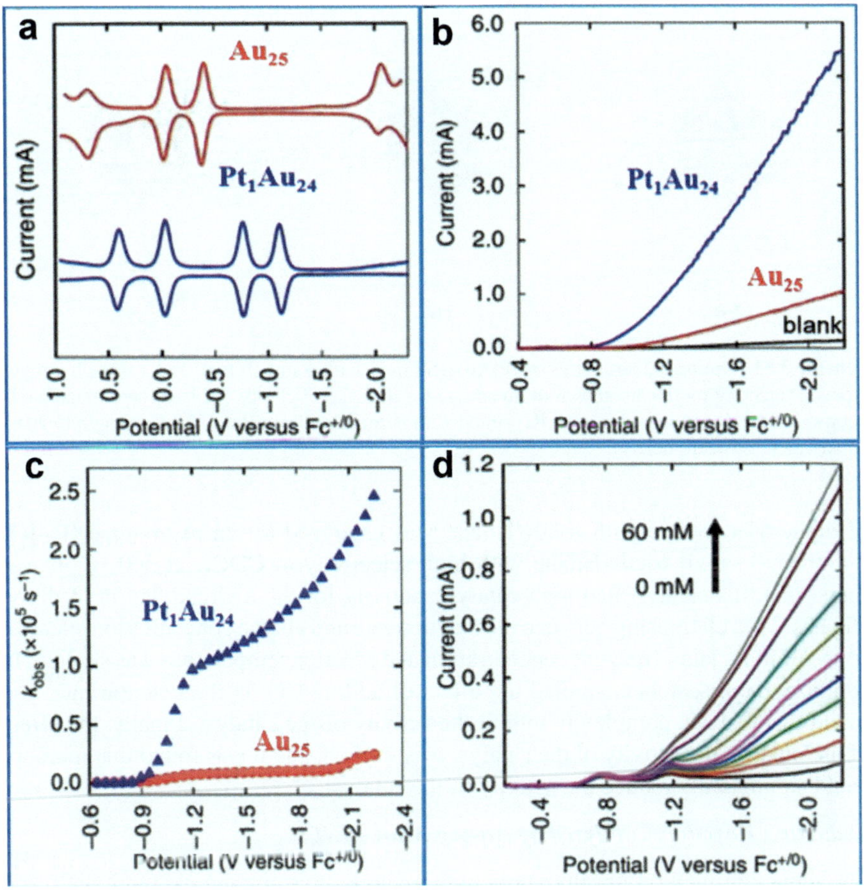

Scheme 3.50 **a** SWVs ofAu_{25} and Pt_1Au_{24} NCs in CH_2Cl_2, **b** LSVs of Au_{25},Pt_1Au_{24} NCs, and without NCs in TFA. **c** Plots of Kobs versus potential for the Au_{25} and Pt_1Au_{24} NCs. **d** LSVs of the Pt1Au_{24} NCs in solutions containing various concentrations of TFA. Reprinted with permission from Ref. [96]. Copyright © 2017, The Author(s)

H becomes significantly stronger, whereas the binding energy of H to a superatom that already possesses eight electrons weakens. Consequently, Pd_1Au_{24} and center-doped Cu_1Au_{24} should also be considered as promising catalysts for the hydrogen evolution reaction. Furthermore, the authors calculated the Gibbs free energy of hydrogen adsorption and concluded that $PtAu_{24}(SR)_{18}$, $PdAu_{24}(SR)_{18}$, and center-doped $CuAu_{24}(SR)_{18}$ exhibit potential as effective electrocatalysts for the hydrogen evolution reaction, as depicted in Scheme 3.51.

Hydrosilylation Reaction of Diarylacetylenes Catalyzed by Pt-Ru Alloy MNCs

Adams et al. studied the application of a platinum-ruthenium mixed metal cluster catalyst in the catalytic hydrosilylation of diaryl alkynes [98]. The hydrosilylation

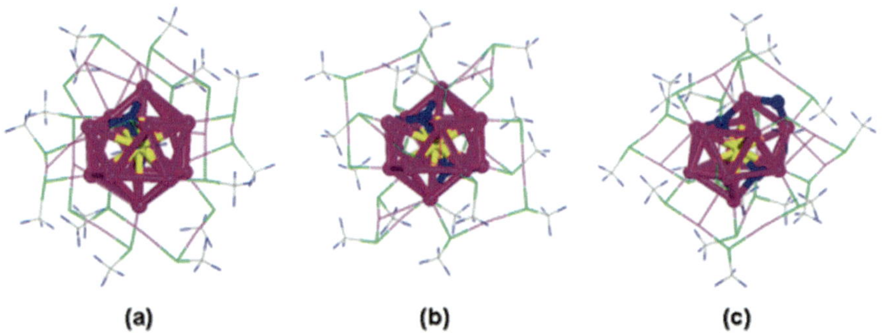

Scheme 3.51 Optimized structures of $[PtAu_{24}(SR)_{18}]^0$ with **a** one, **b** two, and **c** three hydrogen atoms. Protecting motifs are shown in line mode for the sake of clarity. Au is colored magenta, Pt yellow, H blue, S green, and C gray. Reprinted with permission from Ref. [97]. Copyright © 2017, American Chemical Society

of diphenylacetylene with triethylsilane was catalyzed by 2a to produce (E)-[(1, 2-diphenyl) vinyl] triethylsilane with high selectivity in $CDCl_3$ at 300 °C. It was found that this catalyst had high catalytic activity for the hydrosilylation of diaryl alkynes. The relationship between the reaction rate and alkyne concentration, catalyst concentration, silane reagent concentration and reaction temperature was studied. In addition, the researchers studied the effect of adding CO on the reaction rate, and found that CO can completely inhibit the activity of the catalyst. Finally, the effect of mercury on the activity of the catalyst was studied, and it was found that mercury could significantly reduce the activity of the catalyst.

Reduction Reaction Catalyzed by Au–Ag Alloy MNCs

The reduction method usually starts with a single crystal nano as a template, and when adding foreign metal complexes or metal salt precursors, foreign metal atoms replace the original metal atoms in the template nano to form a new alloy NC. Because considering the redox potential, reverse current reduction is harmful in principle.

In 2010, Choi et al. observed this phenomenon. $[Au_{25}(SC_2H_4Ph)_{18}]^-$ can react with Ag^+ to form $[(Au_{24}Ag)\ (SC_2H_4Ph)_{18}]^{2+}$, $[(Au_{23}Ag_2)(SC_2H_4Ph)_{18}]^{2+}$ and $[(Au_{22}Ag_3)(SC_2H_4Ph)_{18}]^{2+}$ alloy NC, and control the ratio of original NC to Ag ions [99].

However, on the basis of Choi, the reaction mechanism was not clear until nearly two years later, when Wu et al. proposed that gold and silver nanoparticles could be converted to each other by anti-cell reduction (AGR). Gold nanoparticles can reduce silver ions to silver metal and replace gold atoms in gold nanoparticles. This finding is contrary to the traditional battery theory, but the existence of AGR is confirmed by experiments [100]. In addition, the study also found that thiol-protected gold and silver nanoparticles showed high activity in the AGR reaction. At the same time, they also proposed the basic principle of the reverse current reduction method. They believe that metal ions with higher redox potential will be reduced by metal atoms with lower redox potential. However, when the size of the nanoparticles is less than 3 nm, the ordered redox potential of the metal is no longer the only decisive thermodynamic parameter. By this method, $Au_{23}(SC_6H_{11})_{16}$ NC is transformed into $Au_{25-n}Ag_n(SC_6H_{11})_{16}$ NC ($n \leq 19$). They also successfully doped Cd and Hg atoms into $Au_{25}(SC_2H_4Ph)_{18}$ to prepare $Au_{24}Cd(SC_2H_4Ph)_{18}$ with a Cd inner shell and $Au_{24}Hg(SC_2H_4Ph)_{18}$ with a Hg outer shell [101, 102].

Oxidation Reaction Catalyzed by Au–Ag Alloy MNCs

The surface ligands can be removed and the metal nanoclusters with intact metal skeleton structure can study the catalytic effect of surface ligands on metal nanoparticles in the hydrolysis and oxidation of organosilanes of bimetallic gold and silver into silanol by Wang's team [103]. The surface ligands were studied by synthesizing a gold-silver nanocluster $Au_{34}Ag_{28}(PhCCH)_{34}$ containing 62 metal atoms and using a removablephenylalkynyl (PA) group as the surface ligand, which was found to be important in metal nanoparticle catalysis. It was found that nanoclusters were completely wrapped by PA ligands showed higher catalytic activity than nanoclusters with partial or complete removal of surface ligands. This indicates that surface ligands have an important influence on the catalytic performance of metal nanoparticles.

While studying the styrene oxidation process, Wang's team obtained an achiral bimetallic $Ag_2@Au_{18}@Ag_{20}$ core and a chiral $Ag_{24}Au_6(SR)_{32}$ shell in 2015 [104]. X-ray crystal structure of magic gold (Au–Ag) nanoclusters $[Ag_{46}Au_{24}(SR)_{32}](BPh_4)_2$, which fills the vacancies between Ag_{44} and Au_{102} nanoclusters. Surface-doped nanoclusters (NCs) exhibit high selectivity toward benzaldehyde, while core–shell structured nanoclusters exhibit high selectivity toward epoxides (Scheme 3.52).

Jin et al. also used bimetallic nanoclusters to catalyze the oxidation of styrene, and found that the dopants of Ag and Cu have higher selectivity and catalytic activity than

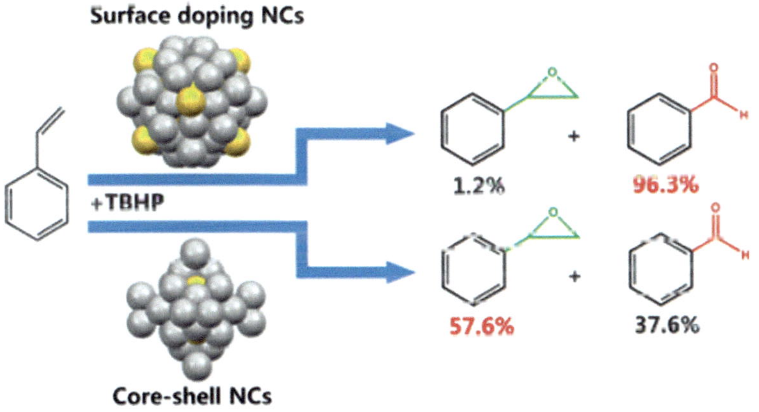

Scheme 3.52 Two different kinds of nanocluster catalysts are present in the styrene oxidization. The surface-doped nanocluster (NC) shows high selectivity for benzaldehyde, whereas the core–shell structured nanocluster shows high selectivity for epoxide. Reprinted with permission from Ref. [104]. Copyright © 2015, The Authors

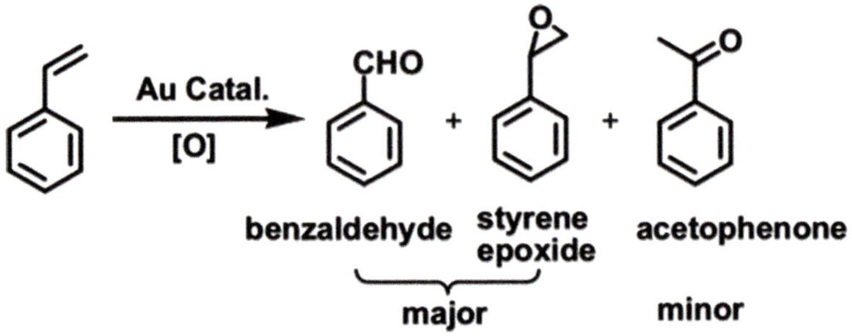

Scheme 3.53 Oxidation of styrene to yield styrene epoxide, benzaldehyde, and acetophenone. Reprinted with permission from Ref. [36]. Copyright © 2015 Elsevier B.V. All rights reserved

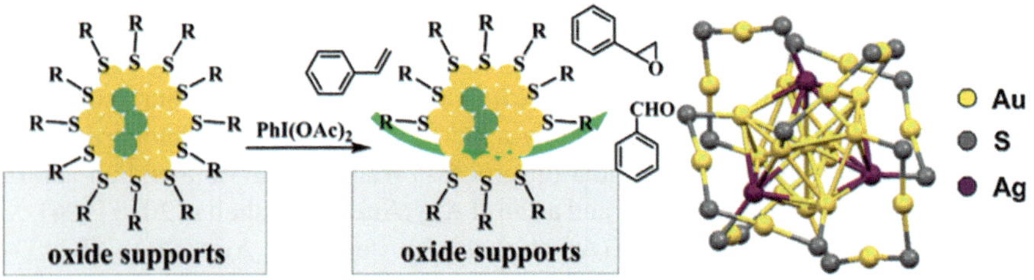

Scheme 3.54 A possible catalytic mechanism of oxidation of styrene catalyzed by gold nanoclusters with $PhI(OAc)_2$ as the oxidant. The right panel shows the low-energy configuration of $Ag_3Au_{22}(SR)_{18}$ (the R groups are omitted in the drawing). Reprinted with permission from Ref. [36]. Copyright © 2015 Elsevier B.V. All rights reserved

the homometallic clusters for styrene, and briefly described the catalytic mechanism of styrene oxidation reaction, as shown in the following figure (Schemes 3.53 and 3.54) [36].

3.3.3 *MNCs of Trimetallic Alloy*

Aerobic Oxidation of Hydrocarbons Reaction Catalyzed by Cu-Pt-Au Alloy MNCs

Yamamoto and co-workers reported the synthesis of finely controlled Cu-Pt-Au alloy NCs (around 1 nm) using a macromolecular template with coordination sites arranged in a gradient of basicity which supported on graphitized mesoporous carbon [105]. Excitingly, aerobic oxidation of hydrocarbons to ketones at room temperature was achieved as a heterogeneous catalytic reaction by this alloy NCs catalyst for the first time, using small amounts of a radical initiator and under solvent-free condition.

The realization of the reaction is related to the characteristics of the alloy NCs catalyst. Fourth-generation dendritic polyphenylazomethine with a tet-raphenylmethane core (TPM-DPA G4) has a defined number of coordination sites for metal ions, and the dendrimer prevents aggregation of the alloy NCs because of its dense tree-like structure, which encapsulates the NCs, and its coordination sites, which stabilize them but with enough surface accessibility for reactants in the catalytic reaction process.

Direct conversion of peroxides to ketones has been demonstrated by studying the mechanism of the reaction. However, the origin of the remarkable catalytic activity and the acceleration of catalytic activity under O_2 after the addition of a third metal are still unclear, and additional mechanistic studies are ongoing (Scheme 3.55).

Oxygen Reduction Reaction (ORR) Catalyzed by Pd-Au-Pt Alloy MNCs

Pt-shelled catalysts have significantly less Pt loading than commercial alternatives, and can show higher activities with stable state in the ORR.

Crooks et al. predicted optimal compositions for different alloy-core@Pt nanoparticles. Subsequent electrochemical measurements of ORR activities of AuPd@Pt dendrimer-encapsulated nanoparticles (DENs) are in a good agreement with the theoretical prediction that the peak of activity is achieved for a 28% Au/72% Pd alloy core supporting a Pt shell [106].

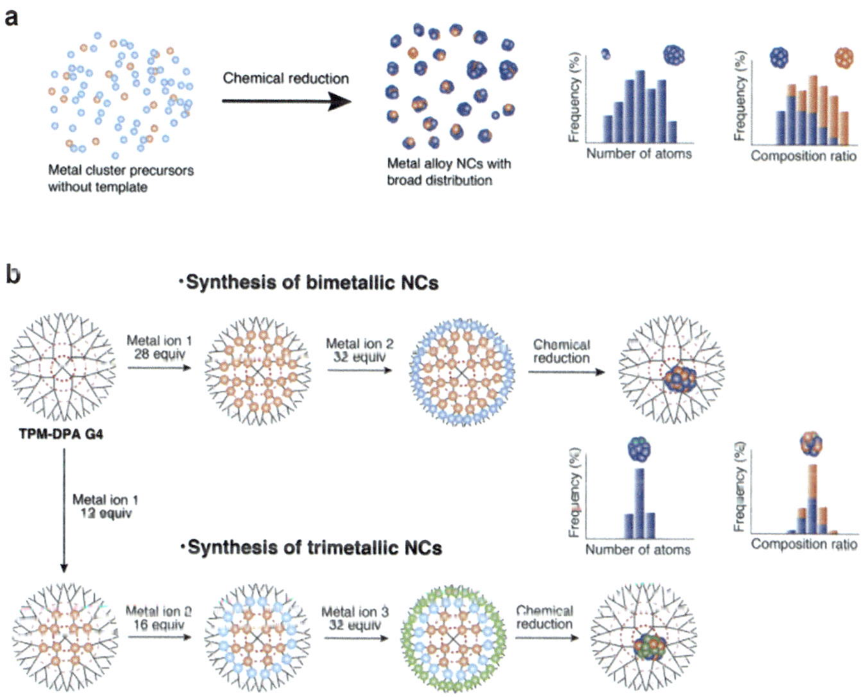

Scheme 3.55 Synthetic methods for MNCs in solution. **a** Typical synthetic method for MNCs resulting in a statistical distribution in size and metal composition. **b** New synthetic approach using a dendrimer template whose coordination sites have a basicity gradient. Reprinted with permission from Ref. [105]. Copyright © 2017 The Authors

This research was carried out by DFT calculation. A target O binding energy was chosen, calculated to be -1.51 eV on a five-layer 3×3 Pt(111) slab model, using a gas-phase molecular O_2 reference. It's worth noting that, this is an unusual case of first-principles theory leading to nearly perfect agreement with experimental results.

3.4 Conclusion and Outlook

NCs have become one of the most active research frontiers in many fields, such as catalysis, chemical sensors, electronics, biological labeling and biomedicine. In particular, in catalysis by NCs, surprisingly good performances are often displayed compared with conventional catalysts. In this chapter, some recent works on nanoclusters have been summarized, including synthesis methods, characterization techniques, and excellent performance in important catalytic reactions.

The NCs tend to agglomerate, because they have high surface free energy, it is particularly important to obtain stable metal nanoclusters. So, some methods have been carried out, such as direct synthesis, ligand-exchange, chemical etching, metal-exchange, separation, electrochemical synthesis, support, atomic layer deposition, photo-induction reduction, heat-induction synthesis, etc. But each approach has its own advantages, researchers can choose synthesis methods according to different using environments.

With the deepening of research and advance in science and technology, the characterization techniques of nanoclusters are also developing. To confirm the properties of NCs, especially for the size, composition, morphology and surface properties, which affect optical, electronic and catalytic activities, is very necessary. Understanding the microstructure of nanoclusters is an effective method to improve their properties, and it can make the research content more targeted.

As an important application of nanoclusters, catalytic activity has attracted much attention. In addition to monometallic, bimetallic and trimetallic nanoclusters are also used in organic reactions. However, most of the current research is theoretical and far from industrialization, such as carbon dioxide hydrogenation to methanol, methanol to olefin (MTO), etc., further exploration is still needed.

In future, simple and effective synthesis methods, fast and accurate characterizations, these factors will make the study of nanoclusters more in-depth. With the emergence of more methods to stabilize nanoclusters, the influence of the electronic environment around metal atoms on their catalytic activity will also be diversified, and more organic reactions will use nanoclusters as catalysts. The exploration of the reaction mechanism will increase the possibility of its industrial application.

References

1. Negishi Y, Nobusada K, Tsukuda T (2005) Glutathione-protected gold clusters revisited: bridging the gap between gold (I)—thiolate complexes and thiolate-protected gold nanocrystals. J Am Chem Soc 127(14):5261–5270. https://doi.org/10.1021/ja042218h
2. Chen Y, Zeng C, Kauffman DR, Jin R (2015) Tuning the magic size of atomically precise gold nanoclusters via isomeric methylbenzenethiols. Nano Lett 15(5):3603–3609. https://doi.org/10.1021/acs.nanolett.5b01122
3. Zeng C, Chen Y, Iida K, Nobusada K, Kirschbaum K, Lambright KJ, Jin R (2016) Gold quantum boxes: on the periodicities and the quantum confinement in the Au(2)(8), Au(3)(6), Au(4)(4), and Au(5)(2) magic series. J Am Chem Soc 138(12):3950–3953. https://doi.org/10.1021/jacs.5b12747
4. Woehrle GH, Warner MG, Hutchison JE (2002) Ligand exchange reactions yield subnanometer, thiol-stabilized gold particles with defined optical transitions. J Phys Chem B 106(39):9979–9981. https://doi.org/10.1021/jp025943s
5. Shichibu Y, Negishi Y, Tsukuda T, Teranishi T (2005) Large-scale synthesis of thiolated Au_{25} clusters via ligand exchange reactions of phosphine-stabilized Au_{11} clusters. J Am Chem Soc 127(39):13464–13465. https://doi.org/10.1021/ja053915s
6. Jia X, Li J, Wang E (2013) Cu nanoclusters with aggregation induced emission enhancement. Small 9(22):3873–3879. https://doi.org/10.1002/smll.201300896
7. Han Y, Wang Y-G, Chen W, Xu R, Zheng L, Zhang J, Luo J, Shen R-A, Zhu Y, Cheong W-C (2017) Hollow N-doped carbon spheres with isolated cobalt single atomic sites: superior electrocatalysts for oxygen reduction. J Am Chem Soc 139(48):17269–17272. https://doi.org/10.1021/jacs.7b10194
8. Qiu HJ, Ito Y, Cong W, Tan Y, Liu P, Hirata A, Fujita T, Tang Z, Chen M (2015) Nanoporous graphene with single-atom Nickel dopants: an efficient and stable catalyst for electrochemical hydrogen production. Angew Chem Int Ed Engl 54(47):14031–14035. https://doi.org/10.1002/anie.201507381
9. Li Q, Luo T-Y, Taylor MG, Wang S, Zhu X, Song Y, Mpourmpakis G, Rosi NL, Jin R (2017) Molecular "surgery" on a 23-gold-atom nanoparticle. Sci Adv 3(5):e1603193. https://doi.org/10.1126/sciadv.1603193
10. Negishi Y, Kurashige W, Niihori Y, Iwasa T, Nobusada K (2010) Isolation, structure, and stability of a dodecanethiolate-protected Pd(1)Au(24) cluster. Phys Chem Chem Phys 12(23):6219–6225. https://doi.org/10.1039/b927175a
11. Niihori Y, Matsuzaki M, Pradeep T, Negishi Y (2013) Separation of precise compositions of noble metal clusters protected with mixed ligands. J Am Chem Soc 135(13):4946–4949. https://doi.org/10.1021/ja4009369
12. Kumar S, Bolan MD, Bigioni TP (2010) Glutathione-stabilized magic-number silver cluster compounds. J Am Chem Soc 132(38):13141–13143. https://doi.org/10.1021/ja105836b
13. Ghosh A, Hassinen J, Pulkkinen P, Tenhu H, Ras RH, Pradeep T (2014) Simple and efficient separation of atomically precise noble metal clusters. Anal Chem 86(24):12185–12190. https://doi.org/10.1021/ac503165t
14. Reetz MT, Helbig W (1994) Size-selective synthesis of nanostructured transition metal clusters. J Am Chem Soc 116(16):7401–7402. https://doi.org/10.1021/ja00095a051
15. Rodriguez-Sanchez L, Blanco MC, Lopez-Quintela MA (2000) Electrochemical synthesis of silver nanoparticles. J Phys Chem B 104(41):9683–9688. https://doi.org/10.1021/jp001761r
16. Blanco MC, Lourido R, Vázquez-Vázquez C, Pastor E, Planes GA, Rivas J, López-Quintela MA (2008) Synthesis of atomic gold clusters with strong electrocatalytic activities. Langmuir 24(21):12690–12694. https://doi.org/10.1021/la8018474
17. Vilar-Vidal N, Blanco MC, Lopez-Quintela MA, Rivas J, Serra C (2010) Electrochemical synthesis of very stable photoluminescent copper clusters. J Phys Chem C 114(38):15924–15930. https://doi.org/10.1021/jp911380s

18. Wang C, Lv P, Xue D, Cai Y, Yan X, Xu L, Fang J, Yang Y (2018) Zero-dimensional/two-dimensional $Au_{25}(Cys)18$ nanoclusters/g-C_3N_4 nanosheets composites for enhanced photocatalytic hydrogen production under visible light. ACS Sustain Chem Eng 6(7):8447–8457. https://doi.org/10.1021/acssuschemeng.8b00643
19. Yang S, Ding S, Zhao C, Huo S, Yu F, Fang J, Yang Y (2021) Ligand-free Au nanoclusters/g-C_3N_4 ultra-thin nanosheets composite photocatalysts for efficient visible-light-driven photocatalytic H_2 generation. J Mater Sci 56(24):13736–13751. https://doi.org/10.1007/s10853-021-06184-7
20. Du XL, Wang XL, Li YH, Wang YL, Zhao JJ, Fang LJ, Zheng LR, Tong H, Yang HG (2017) Isolation of single Pt atoms in a silver cluster: forming highly efficient silver-based cocatalysts for photocatalytic hydrogen evolution. Chem Commun (Camb) 53(68):9402–9405. https://doi.org/10.1039/c7cc04061j
21. Ji S, Chen Y, Wang X, Zhang Z, Wang D, Li Y (2020) Chemical synthesis of single atomic site catalysts. Chem Rev 120(21):11900–11955. https://doi.org/10.1021/acs.chemrev.9b00818
22. Sun S, Zhang G, Gauquelin N, Chen N, Zhou J, Yang S, Chen W, Meng X, Geng D, Banis MN, Li R, Ye S, Knights S, Botton GA, Sham T-K, Sun X (2013) Single-atom catalysis using Pt/graphene achieved through atomic layer deposition. Sci Rep 3(1). https://doi.org/10.1038/srep01775
23. Cheng N, Stambula S, Wang D, Banis MN, Liu J, Riese A, Xiao B, Li R, Sham TK, Liu LM, Botton GA, Sun X (2016) Platinum single-atom and cluster catalysis of the hydrogen evolution reaction. Nat Commun 7:13638. https://doi.org/10.1038/ncomms13638
24. Yan H, Lin Y, Wu H, Zhang W, Sun Z, Cheng H, Liu W, Wang C, Li J, Huang X, Yao T, Yang J, Wei S, Lu J (2017) Bottom-up precise synthesis of stable platinum dimers on graphene. Nat Commun 8(1):1070. https://doi.org/10.1038/s41467-017-01259-z
25. Liu P, Zhao Y, Qin R, Mo S, Chen G, Gu L, Chevrier DM, Zhang P, Guo Q, Zang D (2016) Photochemical route for synthesizing atomically dispersed palladium catalysts. Science 352(6287):797–800. https://doi.org/10.1126/science.aaf5251
26. Wang G, He CT, Huang R, Mao J, Wang D, Li Y (2020) Photoinduction of Cu single atoms decorated on UiO-66-NH_2 for enhanced photocatalytic reduction of CO_2 to liquid fuels. J Am Chem Soc 142(45):19339–19345. https://doi.org/10.1021/jacs.0c09599
27. Lang R, Xi W, Liu JC, Cui YT, Li T, Lee AF, Chen F, Chen Y, Li L, Li L, Lin J, Miao S, Liu X, Wang AQ, Wang X, Luo J, Qiao B, Li J, Zhang T (2019) Non defect-stabilized thermally stable single-atom catalyst. Nat Commun 10(1):234. https://doi.org/10.1038/s41467-018-08136-3
28. Zhang E, Wang T, Yu K, Liu J, Chen W, Li A, Rong H, Lin R, Ji S, Zheng X, Wang Y, Zheng L, Chen C, Wang D, Zhang J, Li Y (2019) Bismuth single atoms resulting from transformation of metal-organic frameworks and their use as electrocatalysts for CO_2 reduction. J Am Chem Soc 141(42):16569–16573. https://doi.org/10.1021/jacs.9b08259
29. Lisiecki I, Pileni M (1995) Copper metallic particles synthesized "in situ" in reverse micelles: influence of various parameters on the size of the particles. J Phys Chem 99(14):5077–5082. https://doi.org/10.1021/j100014a030
30. Wei W, Lu Y, Chen W, Chen S (2011) One-pot synthesis, photoluminescence, and electrocatalytic properties of subnanometer-sized copper clusters. J Am Chem Soc 133(7):2060–2063. https://doi.org/10.1021/ja109303z
31. Mott D, Galkowski J, Wang L, Luo J, Zhong C-J (2007) Synthesis of size-controlled and shaped copper nanoparticles. Langmuir 23(10):5740–5745. https://doi.org/10.1021/la0635092
32. Zhu M, Aikens CM, Hollander FJ, Schatz GC, Jin R (2008) Correlating the crystal structure of a thiol-protected Au_{25} cluster and optical properties. J Am Chem Soc 130(18):5883–5885. https://doi.org/10.1021/ja801173r
33. Rao TUB, Nataraju B, Pradeep T (2010) Ag9 quantum cluster through a solid-state route. J Am Chem Soc 132(46):16304–16307. https://doi.org/10.1021/ja105495n
34. Zaluzhna O, Brightful L, Allison TC, Tong YJ (2011) Spectroscopic evidence of a bidentate-binding of meso-2,3-dimercaptosuccinic acid on silver nanoclusters. Chem Phys Lett 509(4–6):148–151. https://doi.org/10.1016/j.cplett.2011.04.087

35. Adhikari B, Banerjee A (2010) Facile synthesis of water-soluble fluorescent silver nanoclusters and HgII sensing. Chem Mater 22(15):4364–4371. https://doi.org/10.1021/cm1001253
36. Li G, Jin R (2016) Atomic level tuning of the catalytic properties: doping effects of 25-atom bimetallic nanoclusters on styrene oxidation. Catal Today 278:187–191. https://doi.org/10.1016/j.cattod.2015.11.019
37. Wu Z, Chen J, Jin R (2010) One-pot synthesis of $Au_{25}(SG)_{18}$ 2- and 4-nm gold nanoparticles and comparison of their size-dependent properties. Adv Funct Mater 21(1):177–183. https://doi.org/10.1002/adfm.201001120
38. Yang S, Chai J, Song Y, Fan J, Chen T, Wang S, Yu H, Li X, Zhu M (2017) In situ two-phase ligand exchange: a new method for the synthesis of alloy nanoclusters with precise atomic structures. J Am Chem Soc 139(16):5668–5671. https://doi.org/10.1021/jacs.7b00668
39. Yun Y, Sheng H, Yu J, Bao L, Du Y, Xu F, Yu H, Li P, Zhu M (2018) Boosting the activity of ligand-on atomically precise Pd3Cl cluster catalyst by metal-support interaction from kinetic and thermodynamic aspects. Adv Synth Catal 360(24):4731–4743. https://doi.org/10.1002/adsc.201800603
40. Du Y, Sheng H, Astruc D, Zhu M (2020) Atomically precise noble metal nanoclusters as efficient catalysts: a bridge between structure and properties. Chem Rev 120(2):526–622. https://doi.org/10.1021/acs.chemrev.8b00726
41. Ge X, Zhou P, Zhang Q, Xia Z, Chen S, Gao P, Zhang Z, Gu L, Guo S (2020) Palladium single atoms on TiO_2 as a photocatalytic sensing platform for analyzing the organophosphorus pesticide chlorpyrifos. Angew Chem Int Ed Engl 59(1):232–236. https://doi.org/10.1002/anie.201911516
42. Whetten RL, Khoury JT, Alvarez MM, Murthy S, Vezmar I, Wang ZL, Stephens PW, Cleveland CL, Luedtke WD, Landman U (2004) Nanocrystal gold molecules. Adv Mater 8(5):428–433. https://doi.org/10.1002/adma.19960080513
43. Schaaff T, Shafigullin M, Khoury J, Vezmar I, Whetten R, Cullen W, First P, Gutierrez-Wing C, Ascensio J, Jose-Yacaman M (1997) Isolation of smaller nanocrystal Au molecules: robust quantum effects in optical spectra. J Phys Chem B 101(40):7885–7891. https://doi.org/10.1021/jp971438x
44. Alvarez MM, Khoury JT, Schaaff TG, Shafigullin MN, Vezmar I, Whetten RL (1997) Optical absorption spectra of nanocrystal gold molecules. J Phys Chem B 101(19):3706–3712. https://doi.org/10.1021/jp962922n
45. Li G, Abroshan H, Chen Y, Jin R, Kim HJ (2015) Experimental and mechanistic understanding of aldehyde hydrogenation using Au_{25} nanoclusters with Lewis acids: unique sites for catalytic reactions. J Am Chem Soc 137(45):14295–14304. https://doi.org/10.1021/jacs.5b07716
46. Niihori Y, Matsuzaki M, Uchida C, Negishi Y (2014) Advanced use of high-performance liquid chromatography for synthesis of controlled metal clusters. Nanoscale 6(14):7889–7896. https://doi.org/10.1039/c4nr01144a
47. Dass A, Stevenson A, Dubay GR, Tracy JB, Murray RW (2008) Nanoparticle MALDI-TOF mass spectrometry without fragmentation: $Au_{25}(SCH_2CH_2Ph)$ 18 and mixed monolayer Au_{25} $(SCH_2CH_2Ph)_{18-x}(L)_x$. J Am Chem Soc 130(18):5940–5946. https://doi.org/10.1021/ja710323t
48. Dass A, Dubay GR, Fields-Zinna CA, Murray RW (2008) FAB mass spectrometry of Au_{25} (SR) 18 nanoparticles. Anal Chem 80(18):6845–6849. https://doi.org/10.1021/ac801259j
49. Udayabhaskararao T, Bootharaju MS, Pradeep T (2013) Thiolate-protected Ag_{32} clusters: mass spectral studies of composition and insights into the Ag-thiolate structure from NMR. Nanoscale 5(19):9404–9411. https://doi.org/10.1039/c3nr03463a
50. Wu Z, Jiang DE, Mann AK, Mullins DR, Qiao ZA, Allard LF, Zeng C, Jin R, Overbury SH (2014) Thiolate ligands as a double-edged sword for CO oxidation on CeO_2 supported $Au_{25}(SCH_2CH_2Ph)_{18}$ nanoclusters. J Am Chem Soc 136(16):6111–6122. https://doi.org/10.1021/ja5018706
51. Nie X, Qian H, Ge Q, Xu H, Jin R (2012) CO oxidation catalyzed by oxide-supported $Au_{25}(SR)_{18}$ nanoclusters and identification of perimeter sites as active centers. ACS Nano 6(7):6014–6022. https://doi.org/10.1021/nn301019f

52. Cook AW, Jones ZR, Wu G, Scott SL, Hayton TW (2018) An organometallic Cu_{20} nanocluster: synthesis, characterization, immobilization on silica, and "Click" chemistry. J Am Chem Soc 140(1):394–400. https://doi.org/10.1021/jacs.7b10960
53. Nguyen T-AD, Jones ZR, Goldsmith BR, Buratto WR, Wu G, Scott SL, Hayton TW (2015) A Cu_{25} nanocluster with partial Cu_0 character. J Am Chem Soc 137(41):13319–13324. https://doi.org/10.1021/jacs.5b07574
54. Nguyen T-AD, Jones ZR, Leto DF, Wu G, Scott SL, Hayton TW (2016) Ligand-exchange-induced growth of an atomically precise Cu_{29} nanocluster from a smaller cluster. Chem Mater 28(22):8385–8390. https://doi.org/10.1021/acs.chemmater.6b03879
55. Lu Y, Jiang Y, Gao X, Chen W (2014) Charge state-dependent catalytic activity of $[Au_{25}(SC_{12}H_{25})_{18}]$ nanoclusters for the two-electron reduction of dioxygen to hydrogen peroxide. Chem Commun (Camb) 50(62):8464–8467. https://doi.org/10.1039/c4cc01841a
56. Zhu Y, Wu Z, Gayathri C, Qian H, Gil RR, Jin R (2010) Exploring stereoselectivity of Au_{25} nanoparticle catalyst for hydrogenation of cyclic ketone. J Catal 271(2):155–160. https://doi.org/10.1016/j.jcat.2010.02.027
57. Chen H, Liu C, Wang M, Zhang C, Li G, Wang F (2016) Thermally robust silica-enclosed Au_{25} nanocluster and its catalysis. Chin J Catal 37(10):1787–1793. https://doi.org/10.1016/s1872-2067(16)62478-6
58. Jupally VR, Dharmaratne AC, Crasto D, Huckaba AJ, Kumara C, Nimmala PR, Kothalawala N, Delcamp JH, Dass A (2014) $Au_{137}(SR)_{56}$ nanomolecules: composition, optical spectroscopy, electrochemistry and electrocatalytic reduction of CO_2. Chem Commun (Camb) 50(69):9895–9898. https://doi.org/10.1039/c4cc03424d
59. Abroshan H, Li G, Lin J, Kim HJ, Jin R (2016) Molecular mechanism for the activation of $Au_{25}(SCH_2CH_2Ph)_{18}$ nanoclusters by imidazolium-based ionic liquids for catalysis. J Catal 337:72–79. https://doi.org/10.1016/j.jcat.2016.01.011
60. Kogo A, Sakai N, Tatsuma T (2010) Photocatalysis of Au_{25}-modified TiO_2 under visible and near infrared light. Electrochem Commun 12(7):996–999. https://doi.org/10.1016/j.elecom.2010.05.021
61. Hu J, Yuan H, Li P, Wang J, Liu Q, Wang H, Wang Q, Yuan X (2015) Synthesis and photocatalytic activity of ZnO-Au_{25} nanocomposites. Sci China Chem 59(3):277–281. https://doi.org/10.1007/s11426-015-5487-6
62. Lee J, Shim HS, Lee M, Song JK, Lee D (2011) Size-controlled electron transfer and photocatalytic activity of ZnO–Au nanoparticle composites. J Phys Chem Lett 2(22):2840–2845. https://doi.org/10.1021/jz2013352
63. Han NS, Kim D, Lee JW, Kim J, Shim HS, Lee Y, Lee D, Song JK (2016) Unexpected size effect observed in ZnO-Au composite photocatalysts. ACS Appl Mater Int 8(2):1067–1072. https://doi.org/10.1021/acsami.5b09987
64. Liu Y, Yao Q, Wu X, Chen T, Ma Y, Ong CN, Xie J (2016) Gold nanocluster sensitized TiO_2 nanotube arrays for visible-light driven photoelectrocatalytic removal of antibiotic tetracycline. Nanoscale 8(19):10145–10151. https://doi.org/10.1039/c6nr01702a
65. Li Z, Liu C, Abroshan H, Kauffman DR, Li G (2017) $Au_{38}S_2(SAdm)_{20}$ photocatalyst for one-step selective aerobic oxidations. ACS Catal 7(5):3368–3374. https://doi.org/10.1021/acscatal.7b00239
66. Hanbao C, Guiqi G, Jinsong C, Sha Y, Bo R, Guang L, Manzhou Z (2018) Photoinduced oxidation catalysis by $Au_{25-x}Ag_x(SR)_{18}$ nanoclusters. Chem Nano Mat 4(5):482–486. https://doi.org/10.1002/cnma.201700336
67. Sakai N, Tatsuma T (2010) Photovoltaic properties of glutathione-protected gold clusters adsorbed on TiO_2 electrodes. Adv Mater 22(29):3185–3188. https://doi.org/10.1002/adma.200904317
68. Kogo A, Sakai N, Tatsuma T (2012) Photoelectrochemical analysis of size-dependent electronic structures of gold clusters supported on TiO_2. Nanoscale 4(14):4217–4221. https://doi.org/10.1039/c2nr30480e
69. Kogo A, Takahashi Y, Sakai N, Tatsuma T (2013) Gold cluster-nanoparticle diad systems for plasmonic enhancement of photosensitization. Nanoscale 5(17):7855–7860. https://doi.org/10.1039/c3nr02420b

70. Li G, Liu C, Lei Y, Jin R (2012) Au25 nanocluster-catalyzed Ullmann-type homocoupling reaction of aryl iodides. Chem Commun (Camb) 48(98):12005–12007. https://doi.org/10.1039/c2cc34765b
71. Li G, Abroshan H, Liu C, Zhuo S, Li Z, Xie Y, Kim HJ, Rosi NL, Jin R (2016) Tailoring the electronic and catalytic properties of Au_{25} nanoclusters via ligand engineering. ACS Nano 10(8):7998–8005. https://doi.org/10.1021/acsnano.6b03964
72. Li G, Jiang D-e, Liu C, Yu C, Jin R (2013) Oxide-supported atomically precise gold nanocluster for catalyzing Sonogashira cross-coupling. J Catal 306:177–183. https://doi.org/10.1016/j.jcat.2013.06.017
73. Li Q, Das A, Wang S, Chen Y, Jin R (2016) Highly efficient three-component coupling reaction catalysed by atomically precise ligand-protected $Au_{38}(SC_2H_4Ph)_{24}$ nanoclusters. Chem Commun (Camb) 52(99):14298–14301. https://doi.org/10.1039/c6cc07825g
74. Kunz S, Schweinberger FF, Habibpour V, Röttgen M, Harding C, Arenz M, Heiz U (2010) Temperature dependent CO oxidation mechanisms on size-selected clusters. J Phys Chem C 114(3):1651–1654. https://doi.org/10.1021/jp911269z
75. Kane MD, Roberts FS, Anderson SL (2015) Effects of alumina thickness on CO oxidation activity over Pd_{20}/Alumina/Re(0001): correlated effects of alumina electronic properties and Pd20 geometry on activity. J Phys Chem C 119(3):1359–1375. https://doi.org/10.1021/jp5093543
76. Moseler M, Walter M, Yoon B, Landman U, Habibpour V, Harding C, Kunz S, Heiz U (2012) Oxidation state and symmetry of magnesia-supported $Pd_{13}O_x$ nanocatalysts influence activation barriers of CO oxidation. J Am Chem Soc 134(18):7690–7699. https://doi.org/10.1021/ja211121m
77. Iwasa N, Mayanagi T, Nomura W, Arai M, Takezawa N (2003) Effect of Zn addition to supported Pd catalysts in the steam reforming of methanol. Appl Catal A 248(1–2):153–160. https://doi.org/10.1016/s0926-860x(03)00184-4
78. Gao X, Chen W (2017) Highly stable and efficient $Pd_6(SR)_{12}$ cluster catalysts for the hydrogen and oxygen evolution reactions. Chem Commun (Camb) 53(70):9733–9736. https://doi.org/10.1039/c7cc04787h
79. Fujitani T, Nakamura I (2002) Methanol synthesis from CO and CO_2 hydrogenations over supported palladium catalysts. Bull Chem Soc Jpn 75(6):1393–1398. https://doi.org/10.1246/bcsj.75.1393
80. Liang X-L, Dong X, Lin G-D, Zhang H-B (2009) Carbon nanotube-supported Pd–ZnO catalyst for hydrogenation of CO_2 to methanol. Appl Catal B 88(3–4):315–322. https://doi.org/10.1016/j.apcatb.2008.11.018
81. Leyva Perez A, Oliver-Meseguer J, Rubio-Marques P, Corma A (2013) Water-stabilized three- and four-atom palladium clusters as highly active catalytic species in ligand-free C–C cross-coupling reactions. Angew Chem Int Ed Engl 52(44):11554–11559. https://doi.org/10.1002/anie.201303188
82. Reetz MT, de Vries JG (2004) Ligand-free Heck reactions using low Pd-loading. Chem Commun (Camb) 14:1559–1563. https://doi.org/10.1039/b406719n
83. Fu F, Xiang J, Cheng H, Cheng L, Chong H, Wang S, Li P, Wei S, Zhu M, Li Y (2017) A robust and efficient Pd_3 cluster catalyst for the Suzuki reaction and its odd mechanism. ACS Catal 7(3):1860–1867. https://doi.org/10.1021/acscatal.6b02527
84. Diallo AK, Ornelas C, Salmon L, Ruiz Aranzaes J, Astruc D (2007) Homeopathic" catalytic activity and atom leaching mechanism in Miyaura-Suzuki reactions under ambient conditions with precise dendrimer-stabilized Pd nanoparticles. Angew Chem Int Ed Engl 46(45):8644–8648. https://doi.org/10.1002/anie.200703067
85. Zhang P, Shao C, Zhang Z, Zhang M, Mu J, Guo Z, Liu Y (2011) In situ assembly of well-dispersed Ag nanoparticles (AgNPs) on electrospun carbon nanofibers (CNFs) for catalytic reduction of 4-nitrophenol. Nanoscale 3(8):3357–3363. https://doi.org/10.1039/c1nr10405e
86. Sun L, Yun Y, Sheng H, Du Y, Ding Y, Wu P, Li P, Zhu M (2018) Rational encapsulation of atomically precise nanoclusters into metal–organic frameworks by electrostatic attraction for CO_2 conversion. J Mater Chem A 6(31):15371–15376. https://doi.org/10.1039/c8ta04667k

87. Hu X, Brunschwig BS, Peters JC (2007) Electrocatalytic hydrogen evolution at low overpotentials by cobalt macrocyclic glyoxime and tetraimine complexes. J Am Chem Soc 129(29):8988–8998. https://doi.org/10.1021/ja067876b
88. Frediani P, Matteoli U, Bianchi M, Piacenti F, Menchi G (1978) Cluster ruthenium hydrogenation catalysts. J Organomet Chem 150(2):273–278. https://doi.org/10.1016/S0022-328X(00)84730-0
89. Morimoto T, Chatani N, Murai S (1999) The first catalytic carbonylative [4+ 1] cycloaddition using a 1, 3-conjugated system. A new transformation of α, β-unsaturated imines to unsaturated γ-lactams catalyzed by $Ru_3(CO)_{12}$. J Am Chem Soc 121(8):1758–1759. https://doi.org/10.1021/ja983546i
90. Clifford JN, Palomares E, Nazeeruddin MK, Grätzel M, Durrant JR (2007) Dye dependent regeneration dynamics in dye sensitized nanocrystalline solar cells: evidence for the formation of a ruthenium bipyridyl cation/iodide intermediate. J Phys Chem C 111(17):6561–6567. https://doi.org/10.1021/jp067458t
91. Serna P, Gates BC (2011) Zeolite-supported rhodium complexes and clusters: switching catalytic selectivity by controlling structures of essentially molecular species. J Am Chem Soc 133(13):4714–4717. https://doi.org/10.1021/ja111749s
92. Yardimci D, Serna P, Gates BC (2012) A highly selective catalyst for partial hydrogenation of 1,3-butadiene: MgO-supported rhodium clusters selectively poisoned with CO. Chem Cat Chem 4(10):1547–1550. https://doi.org/10.1002/cctc.201200033
93. Yardimci D, Serna P, Gates BC (2012) Tuning catalytic selectivity: zeolite- and magnesium oxide-supported molecular rhodium catalysts for hydrogenation of 1,3-butadiene. ACS Catal 2(10):2100–2113. https://doi.org/10.1021/cs300475c
94. Nakamula I, Yamanoi Y, Yonezawa T, Imaoka T, Yamamoto K, Nishihara H (2008) Nanocage catalysts-rhodium nanoclusters encapsulated with dendrimers as accessible and stable catalysts for olefin and nitroarene hydrogenations. Chem Commun (Camb) 44:5716–5718. https://doi.org/10.1039/b813649a
95. Wu Y-g, Wen M, Wu Q-s, Fang H (2014) Ni/graphene nanostructure and its electron-enhanced catalytic action for hydrogenation reaction of nitrophenol. J Phys Chem C 118(12):6307–6313. https://doi.org/10.1021/jp412711b
96. Kwak K, Choi W, Tang Q, Kim M, Lee Y, Jiang DE, Lee D (2017) A molecule-like $PtAu_{24}(SC_6H_{13})_{18}$ nanocluster as an electrocatalyst for hydrogen production. Nat Commun 8:14723. https://doi.org/10.1038/ncomms14723
97. Hu G, Tang Q, Lee D, Wu Z, Jiang D-e (2017) Metallic hydrogen in atomically precise gold nanoclusters. Chem Mater 29(11):4840–4847. https://doi.org/10.1021/acs.chemmater.7b00776
98. Adams RD, Barnard TS (1998) Catalytic hydrosilylation of diarylalkynes by layer-segregated platinum—ruthenium cluster complexes $Pt_3Ru_6(CO)_{20}(\mu_3\text{-}RC_2R)(\mu_3\text{-}H)(\mu\text{-}H)$. Organometallics 17(12):2567–2573. https://doi.org/10.1021/om980117k
99. Choi J-P, Fields-Zinna CA, Stiles RL, Balasubramanian R, Douglas AD, Crowe MC, Murray RW (2010) Reactivity of $[Au_{25}(SCH_2CH_2Ph)_{18}]1$—nanoparticles with metal ions. J Phys Chem C 114(38):15890–15896. https://doi.org/10.1021/jp9101114
100. Wu Z (2012) Anti-galvanic reduction of thiolate-protected gold and silver nanoparticles. Angew Chem Int Ed Engl 51(12):2934–2938. https://doi.org/10.1002/anie.201107822
101. Tian S, Yao C, Liao L, Xia N, Wu Z (2015) Ion-precursor and ion-dose dependent anti-galvanic reduction. Chem Commun (Camb) 51(59):11773–11776. https://doi.org/10.1039/c5cc03267a
102. Li Q, Wang S, Kirschbaum K, Lambright KJ, Das A, Jin R (2016) Heavily doped $Au_{25-x}Ag_x(SC_6H_{11})_{18}^-$ nanoclusters: silver goes from the core to the surface. Chem Commun (Camb) 52(29):5194–5197. https://doi.org/10.1039/c6cc01243d
103. Wang Y, Wan XK, Ren L, Su H, Li G, Malola S, Lin S, Tang Z, Hakkinen H, Teo BK, Wang QM, Zheng N (2016) Atomically precise alkynyl-protected metal nanoclusters as a model catalyst: observation of promoting effect of surface ligands on catalysis by metal nanoparticles. J Am Chem Soc 138(10):3278–3281. https://doi.org/10.1021/jacs.5b12730

104. Wang S, Jin S, Yang S, Chen S, Song Y, Zhang J, Zhu M (2015) Total structure determination of surface doping $[Ag_{46}Au_{24}(SR)_{32}](BPh_4)_2$ nanocluster and its structure-related catalytic property. Sci Adv 1(7):e1500441. https://doi.org/10.1126/sciadv.1500441
105. Takahashi M, Koizumi H, Chun W-J, Kori M, Imaoka T, Yamamoto K (2017) Finely controlled multimetallic nanocluster catalysts for solvent-free aerobic oxidation of hydrocarbons. Sci Adv 3(7):e1700101. https://doi.org/10.1126/sciadv.1700101
106. Zhang L, Iyyamperumal R, Yancey DF, Crooks RM, Henkelman G (2013) Design of Pt-shell nanoparticles with alloy cores for the oxygen reduction reaction. ACS Nano 7(10):9168–9172. https://doi.org/10.1021/nn403788a

Chapter 4
Organic Synthesis Catalyzed by Supported Metal Single-Atom Catalysts

Abstract Single-atom catalysts (SACs) provide the ultimate value in metal usage efficiency: each isolated atom is an active site. Single-atom catalysts also possess unique catalytic activities that differ from those of their particle and cluster counterparts. This chapter mainly describes supported metal SACs and discusses several relevant topics ranging from their preparation, characterization to their catalytic applications for organic synthesis. Organic synthesis reactions involving oxidation, hydrogenation, addition, coupling, and hydroformylation reactions catalyzed by supported metal SACs are then summarized.

Keywords Single-atom catalyst · Support · Preparation and characterization · Application · Organic synthesis

4.1 Introduction

Single-atom catalysts (SACs) present the ultimate advantage in metal usage efficiency: each isolated atom is an active site [1]. Single-atom catalysts also possess unique catalytic activities that differ from those of their cluster and particle counterparts. The main advantages of SAC are the following aspects: (1) SACs realize the utmost usage of metal atoms. (2) SACs have the same isolated active centers, and thus producing a similar coordination environment which can afford desired product in high selectivity. (3) New reaction pathway and different reaction mechanism might be built compared with the cluster or particle catalysts.

The history of the idea that single atoms have unique catalytic properties can be dated back to nearly a century ago. In 1925, Taylor pointed out that the uppermost Ni atoms (Fig. 4.1) may behave like gaseous metal atoms due to their extremely unsaturated coordination, providing a theoretical prediction for single atom catalysis [2]. Until 1964, Rooney and coworkers verified that the surface atoms in metals possess chemical properties which are identical with those of the free atoms [3, 4].

In 1979 Yates et al. reported that single Rh atoms can catalyze the decomposition of formaldehyde to produce carbonyl hydride species, but lack of valid evidence for

M. Bao et al., *Nanometal Catalysis in Organic Synthesis*, Molecular Catalysis 3,
https://doi.org/10.1007/978-981-97-4573-9_4

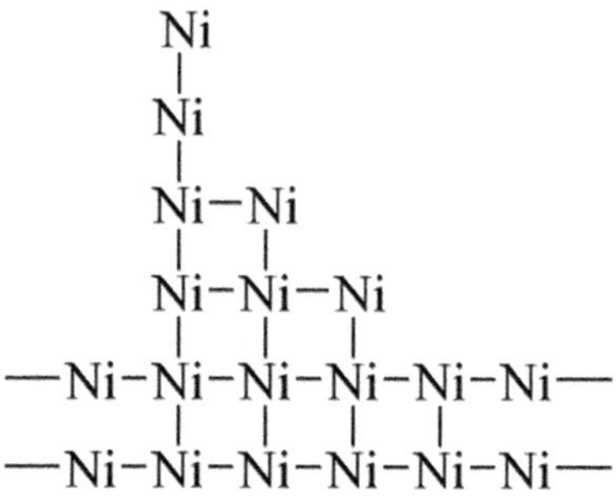

Fig. 4.1 Diagrammatic representation of a portion of a granule surface in cross-section [2]. Copyright 1925, Royal Society

the single atom state [5]. Then in 1995 Maschmeyer et al. produced a catalyst for the epoxidation of cyclohexene by attachment of a titanocene-derived catalyst precursor to the pore walls of MCM-41 [6]. According to the EXAFS and FTIR analysis, the Ti^{4+} ion is solely anchored on the support.

In 1999 Asakura et al. prepared atomically dispersed Pt species at MgO surface which were found to be as active as metallic Pt particles supported on MgO for propane combustion [7]. At the turn of the century Abbet et al. studied the cyclotrimerization of acetylene on size-selected Pd_n clusters ($1 \leq n \leq 30$) supported on thin MgO(100) [8]. Density functional theory (DFT) calculations showed that single Pd atoms can be catalytically activated when adsorbed on basic defect sites of MgO and when charge is transferred to the metal atoms. So they proposed that "one atom is enough". In 2003 Fu et al. found that when the metal nanoparticles of Au(Pt) were etched from Au(Pt)/CeO_2, the catalysts still exhibited excellent catalytic activity toward water–gas shift reaction [9]. Therefore it is not the metal nanoparticles but the metal atoms associated with surface cerium-oxygen groups who are responsible for the activity. In 2005 Zhang et al. prepared isolated Au^{3+} ions supported on a zirconia support [10]. These heterogenized Au^{3+} ions are highly active for the selective hydrogenation of 1,3-butadiene to butene. But still the direct observation of single atoms has not been realized.

Until 2007, Hackett et al. successfully observed a lot of Pd single atoms distributed across the Al_2O_3 framework [11]. This is a milestone since the single atom state is confirmed convincingly for the first time (Fig. 4.2). Two years later Tierney et al. prepared Pd single atoms embedded in Cu(111), and observed them by STM (Fig. 4.3) [13, 14]. Then in 2011 the concept "single-atom catalyst (SAC)" was proposed by Qiao and coworkers (Fig. 4.4) [12]. After that a new "SAC" era has dawned. They also proposed to use "M_1/support" (such as Pt_1/FeO_x) to represent M single atom dispersing on a support.

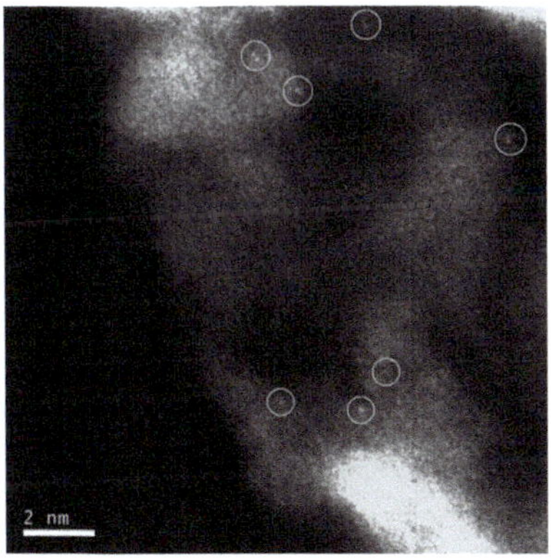

Fig. 4.2 HAADF-STEM image of a 0.03 wt% Pd/meso-Al_2O_3 sample. Reproduced with permission from Ref. [11], Copyright 2007, Wiley–VCH Verlag GmbH & Co. KGaA, Weinheim

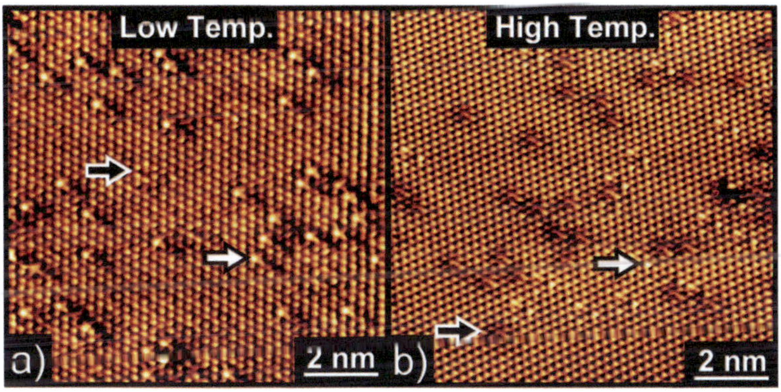

Fig. 4.3 STM images show that both surface and subsurface Pd atoms are observed in the Cu(111) surface at a wide range of deposition temperatures. Surface Pd appears as single atom protrusions (white arrows). **a** Depositions at 350 K; **b** deposition at 500 K. Reproduced with permission from Ref. [13]. Copyright 2009, American Chemical Society

4.2 Preparation of Supported Metal SACs

To prepare SACs, special strategies need to be developed to keep the formed single atoms stable. Thus support is indispensable for the single metal atom catalyst. Moreover, the support must be elaborately chosen so that they can tightly bond the single atom by strong electronic or covalent interactions.

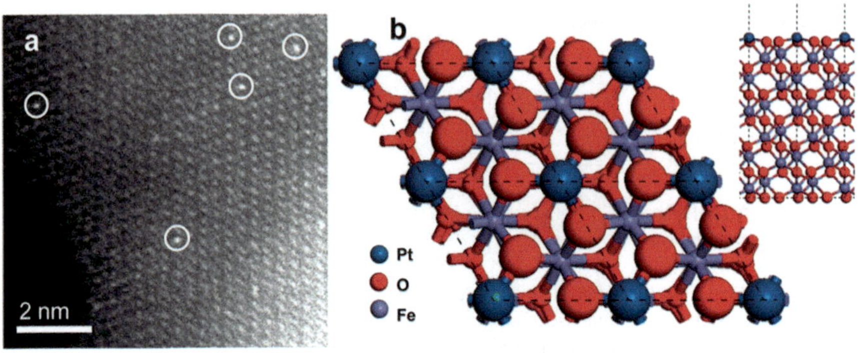

Fig. 4.4 **a** HAADF-STEM image and **b** density functional modeling of Pt_1/FeO_x. Reprinted with permission from Ref. [12]. Copyright 2011, Springer Nature Limited

4.2.1 Mass-Selected Soft-Landing Techniques

The mass-selected soft-landing technique is powerful in preparing supported metal SACs (Fig. 4.5), which can exactly control the size of the metal species by using mass-selected molecular or atom beams and precisely regulate the surface structure of the support by combining these with ultrahigh vacuum surface science procedures [15]. However, such an expensive and low yield fabrication method limits its wide application and is clearly not suitable for practical industrial applications of heterogeneous catalysis.

4.2.2 Atomic Layer Deposition (ALD)

The ALD method was previously developed to fabricate metal oxide thin films with atomically precise control. ALD allows one to control the size, morphology, density, and the loading of the deposited metal. The deposition is split into a sequence of two self-limiting surface reactions between a gaseous precursor and a substrate. The self-limiting nature of this method allows the uniform deposition of catalytic materials on a high surface area catalyst support at the atomic level. It shows great potential in the design and fabrication of SACs through delicate tuning of the support. For example, Li et al. prepared Pt single atoms on CeO_2, Co_3O_4, ZrO_2 and graphene by the ALD method [16]. Nonetheless, the ALD method still has some shortcomings, such as high cost, demanding operating conditions, and limited scope of application.

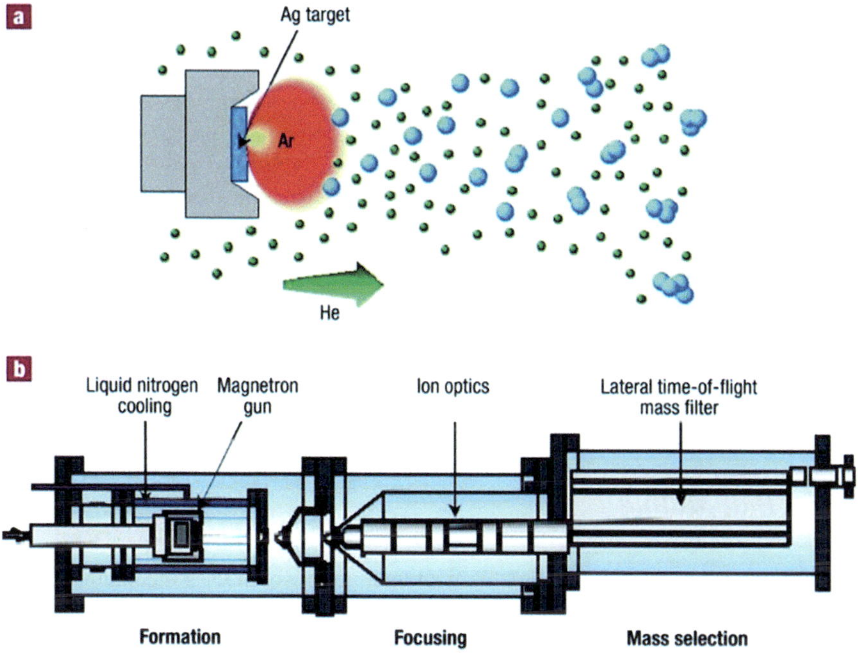

Fig. 4.5 Production of size-selected cluster beams. **a** Cluster formation by plasma sputtering and gas aggregation. **b** Overview of the complete cluster beam source [15]. Copyright 2003, Springer Nature Limited

4.2.3 Wet-Chemistry Approaches

Wet chemistry, is a common way to prepare SACs due to its simple operation and the ease of access to the appropriate raw materials. This method includes co-precipitation, impregnation which is carried out with mild heat, or even at room temperature. In the process of co-precipitation, the oxide support is simultaneously formed with the metal single atoms, while the as obtained support is mixed with metal precursors by the impregnation method. The precursors employed in the synthesis of SACs are organometallic complexes or inorganic salt. For example, Ir_1/FeO_x catalyst can be synthesized by co-precipitation of an aqueous solution of H_2IrCl_6 and $Fe(NO_3)_3$ with NaOH at 80 °C and pH 8 [17]. Qiao et al. successfully prepared CeO_2-supported Au single atoms (Au_1/CeO_2) by the impregnation method [18]. It is noteworthy that to obtain single atoms, the metal loading is usually limited. Thus how to increase the metal loading through wet-chemistry approaches is still a big challenge.

4.2.4 High-Temperature Pyrolysis

The pyrolysis method is a recent technology that has been widely used to synthesize high-loading SACs in situ. This method involves decomposing precursors at elevated temperatures under a gas atmosphere, such as N_2, NH_3, and Ar. The precursors include metal salts, metal complexes, metal oxides, and various carbon substrates such as metal organic frameworks (MOFs), polymers, and so on. Over the past decade, a series of metal-nitrogen-carbon SACs have been successfully synthesized by this method. For example, Yin et al. synthesized Co–N-C SACs using bimetallic Zn/Co MOF as precursor. Zn can be selectively evaporated away at high temperatures above 800 °C during the pyrolysis process, while Co can be reduced by the organic linker carbonization [19].

4.2.5 Ball Milling

Ball milling method is a mechanochemical technique which uses mechanical interactions to convert reactants into products. For example, Gan et al. used a scalable ball milling method to synthesize a Co supported Pt SACs (Pt_1/Co) in kg scale [20]. Results show that the Pt atoms were finely dispersed on Co, indicating that the scale-up production of SACs can be successfully realized by the ball milling method. Moreover, they also extend the synthesis method to fabricate Ru-, Pd-, Ir-, and Rh-based SACs (Fig. 4.6).

4.2.6 Metal Leaching Method

The metal leaching method is generally used to immerse the traditional supported metal catalyst in a dilute NaCN aqueous solution, and then rinse it in a high pH NaOH solution at room temperature in the presence of O_2. In the procedure, the metal nanoparticles are selectively eluted and the metal cations remain intact. The method was first applied to extract gold from gold mines, and was further developed to prepare metal oxide-supported metal cation catalysts. By this method, Yang et al. prepared atomically dispersed Au–$(OH)_x$ species on TiO_2 [21]. However, the metal leaching method is only suitable for preparing few metals and support, and is not suitable for industrial production due to the involvement of toxic cyanide solution.

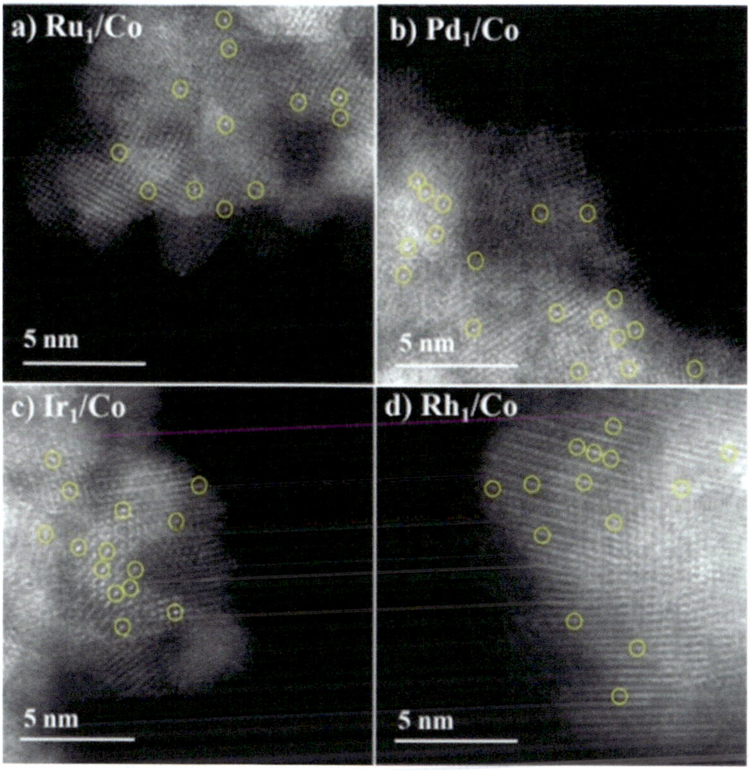

Fig. 4.6 HAADF-STEM images of **a** Ru_1/Co, **b** Pd_1/Co, **c** Ir_1/Co, and **d** Rh_1/Co [20]. Copyright 2020, American Chemical Society

4.2.7 Photoreduction Method

Photoreduction is a method of preparing SACs by reducing metal salts to metal state under the drive of light. For example, Zheng et al. prepared Pd_1/TiO_2 by the photochemical reduction assisted wet chemical method. The surface of TiO_2 is positively charged under suitable pH and attracts large number of $[PdCl_4]^{2-}$ anions which can be reduced to Pd with light [22]. It is worth noting that SACs with metal loading up to 1.5 wt% can be obtained with this UV irradiation method.

4.3 Characterization

The discovery and characterization of SACs are entirely based on the use of different characterization tools. A large number of complex analytical techniques have been used to unambiguously describe and identify the states (geometry, electron configuration, and local environment) of single atom locations. On the other hand, in order to understand the catalytic mechanism of SACs, it is necessary to characterize the reaction behavior of substrates or intermediates at the active sites. Here we will briefly introduce some widely used and representative characterization techniques.

4.3.1 Aberration-Corrected High-Resolution Transmission Electron Microscopy

Though the concept that single atoms have unique catalytic properties has been proposed nearly a century ago, the direct evidence, i.e., the direct observation of single atoms has not been realized until the development of high angle annular dark field scanning transmission electron microscopy technique.

Aberration-corrected high-resolution transmission electron microscopy has the advantages of high resolution and high sensitivity to chemical composition. Using high angle annual dark field (HAADF) detector to detect the distribution of single atoms has become the most direct characterization of metal single atoms loaded on the support surface. The intensity of the high-angle annual dark filed scanning transmission electron microscopy (HAADF-STEM) image is proportional to the square of the atomic number (Z^2), i.e., the brighter part of the image indicates the greater number of atoms present. As shown in Figs. 4.2, 4.3a, and 4.6, the bright spots in the white circles are the metal SACs. In the characterization process, it is necessary to ensure a low acceleration voltage, avoid the occurrence of single atom agglomeration caused by the electron beam, and ensure that the single atom catalyst will not be damaged in the characterization [23].

In addition to HAADF-STEM, the three-dimensional morphology resolution map of the atomic distribution on the catalyst surface can also be obtained by atomic force microscopy (AFM), and the distribution of single atoms can be directly observed, and the interaction force between single atoms and the support can be measured by the micro-cantilever probe [24].

4.3.2 X-Ray Absorption Fine Structure Spectroscopy (XAFS)

In order to understand the excellent catalytic performance of SACs, we need to understand the interaction between single atom and support at the atomic scale. We not only need to intuitively characterize the dispersion degree of the single atom on the support surface, but also need to understand the microstructure of the single atom, including atomic structure and electronic structure information. Because XAFS is sensitive to the local structure and chemical environment of the central atom, it becomes the most powerful tool for describing the local structure [25]. Information such as the distance between the central atom and neighboring atoms, the number and type of coordination atoms, the degree of disorder and the oxidation state of atoms can be obtained, which are all parameters for determining the local geometry and electronic structure.

XAFS can be divided into two regions, the near-edge region and the extended region. The near-edge XAFS, or X-ray absorption near edge structure (XANES) spectroscopy, provides information on the oxidation state, coordination symmetry, and can be used for identification of different species by fingerprinting. In contrast, extended XAFS, or EXAFS, measures the coordination structure including bond distance and coordination number of neighboring atoms.

In the Pt_1/FeO_x catalyst sample prepared by Zhang et al. [12], due to the extremely low Pt loading capacity of single atom, no Pt crystal phase could be detected by XRD. In order to verify the existence of the dispersed single atom in the catalyst, EXAFS characterization technology was used. The EXAFS characterization was performed on catalysts with different loading loads (0.17 wt% and 2.5 wt% Pt/FeO_x), and the EXAFS data fitting was compared with standard Pt foils and PtO_2. Different white line strengths reflected different oxidation states of metals. The white line strength of 0.17 wt% of the sample was between those of Pt foil and PtO_2, indicating that the single atom Pt in the sample was positively charged, and the comparison before and after the reaction of the catalyst indicated that Pt was further oxidized during the catalytic reaction. The same group analyzed the EXAFS spectra of the L_{III}-edge of Au and the K-edge of Pd in the Ullman-type C–C coupling reaction of the AuPd alloy catalyst, and obtained information such as the coordination number (N), bond length (R), Deby-Walker factor (σ^2) of the SACs [26].

4.3.3 Fourier Transform Infrared (FTIR) Spectroscopy

FTIR is one of the most powerful techniques for characterizing catalytic systems. It can be used to distinguish and quantify the concentration of metal species, local geometry, stability, reactivity, and uniformity of the supporting metal species. The infrared absorption peak can reflect the molecular structure, identify the structural composition, and determine the presence of chemical groups. Moreover, the content of chemical groups can also be tested by the absorption intensity. With CO as the probe molecule, the catalytic reaction mechanism, oxidation state, and geometry structure of SACs can be confirmed according to the change of vibration frequency and intensity.

4.3.4 X-Ray Photoelectron Spectroscopy (XPS)

XPS is widely used to study the composition of surface elements and the electronic states of specific elements. It has been widely used to analyze the valence states of metal centers in SACs. In recent years, a series of in-situ XPS techniques have been developed to characterize the valence evolution of catalysts and the electron transfer behavior of the active site during the reaction. However, due to limited detection depth, this technique can only detect elements near the surface.

By using the advanced characterization technology, we can combine the visually observed images with the electronic and structural information of the microenvironment where the single atom is located, more comprehensively reflecting the real microscopic details of the single atom catalyst, obtain reliable structural information, and accurately get the characteristics of the single atom catalyst. It provides sufficient theoretical reference for understanding the reaction process of the single atom catalyst and further designing the single atom catalyst.

XRD is a powerful tool for characterizing the crystal phase of nanoparticles. However, judging from the Debye–Scherrer formula, the full width at half maximum (FWHM) β is inversely proportional to the particle size D. Therefore, with the decrease of the particle size, the peaks will become wider and wider, and finally disappear completely. Actually when the particles enter the size of the cluster regime, no peaks will show any more. Thus XRD is seldom used for SACs.

4.4 Catalysis

4.4.1 Oxidation

Oxidation reaction is very important in organic synthesis. SACs have been used to catalyze oxidation reactions and have shown excellent catalytic performance. One typical example is the selective oxidation of methane to C1 oxygenates (Fig. 4.7).

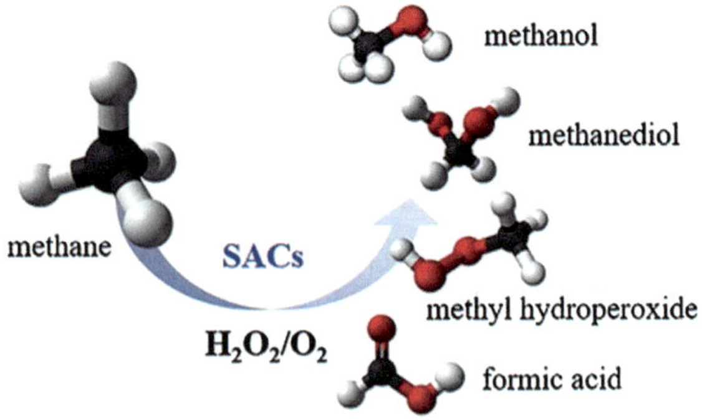

Fig. 4.7 Schematic illustration of selective oxidation of methane to C1 oxygenates. Reproduced with permission from Ref. [27]. Copyright 2022, Wiley–VCH GmbH

It is reported that the Rh SACs dispersed on ZrO_2 can oxidize methane directly to methanol with hydrogen peroxide under mild conditions [28]. According to EXAFS data, Rh_1/ZrO_2 SACs have a dominant Rh-O peak at 1.5 Å and no Rh-Rh peak at 2.4 Å, while Rh NPs/ZrO_2 has a strong Rh-Rh peak at 2.4 Å. The XANES data show that a single Rh atom has a higher oxidation state. The methyl group ($CH_3\cdot$) at the free Rh site has the lowest energy during methane oxidation, which is conducive to the adsorption and dissociation of methane. Rh SACs dispersed on CeO_2 nanowires have also been reported for direct methane conversion to oxygenates [29]. Compared with Rh clusters, the C1 oxidation yield of Rh SACs (1231.7 vs 189.4 $mmol \cdot g_{Rh}^{-1} \cdot h^{-1}$) is increased by a factor of 6.5, and the selectivity of CH_3OH and CH_3OOH is maintained at 94%. Characterization and calculations suggest that CeO_2 promotes the formation of ·OOH and ·OH radicals which interact with $\cdot CH_3$ to produce CH_3OH and CH_3OOH.

Pd SACs also can promote the direct oxidation of methane to methanol. Huang et al. anchored Pd SACs to the inner surface of the ZSM-5 micropore to form a highly uniform Pd_1O_4 active site with high activity for methane activation [30]. A selectivity of about 86% for methanol can be obtained with the help of CuO, which is significantly higher than the reaction with PdO nanoparticles.

Au SACs are also active in the selective oxidation of methane to methanol. Under light irradiation, the selectivity of methane oxidation to methanol on Au_1/BP (black phosphorus) reached >99% [31]. According to DRIFTs data, one can only observe the linear adsorption peak of CO on M_1/BP nanosheets (M = Pt, Pd and Rh), while on MNPS/BP nanosheets one can observe the bridge adsorption peak and linear adsorption peak. The mechanism study shows that water helps O_2 to activate to form active hydroxyl and ·OH free radicals under light irradiation. The hydroxyl group reacts with methane at the Au SACs to form water and CH_3*, which is then oxidized to methanol by the ·OH radical.

Due to the high price of the noble metals, non-noble metal SACs are receiving more and more attention. For example, Fe SACs embedded in graphene exhibit some activity in converting methane directly to C1 oxygenates at room temperature [32]. The C–H bond of methane is activated at the O–FeN_4–O site, and the resulting $\cdot CH_3$ is first converted to CH_3OH or CH_3OOH, and then CH_3OH is transformed to $HOCH_2OOH$ and HCOOH. Zhu et al. found that Fe_1/ZSM-5 has excellent efficiency and selectivity in the direct conversion of methane to formic acid [33]. The formic acid yield reached 383.2 $\mu mol\ g_{cat}^{-1}$ at 80 °C, with a selectivity of 91%. EPR data and DFT calculations demonstrate that C–H bonds can easily split on reactive oxygen species produced at Fe_1 sites. With the dissociation of C–H bonds, $\cdot CH_3$, CH_3OH, $\cdot CH_2OH$ and HCHO are continuously formed, and finally HCOOH is formed.

Cu_1/ZSM-5 containing highly uniform Cu_1-O_4 sites is reported to be able to preferentially activate CH_4 and prevent CH_3OH peroxidation [34]. The yield of C1 oxygenate can reach 4800 $\mu mol\ g_{cat}^{-1}$ in 30 min at 50 °C, and the selectivity is up to 99%. However, Cu NPs showed low activity and did not promote the selectivity of methanol. According to the DFT calculation, ·OH can extract the hydrogen in the adsorbed OH at the Cu_1O_4 site and form desorption H_2O with CH_3O, thus blocking the oxidative dehydrogenation of ·OH to CH_3OH. In addition, in the Cu_1-O_4 entity,

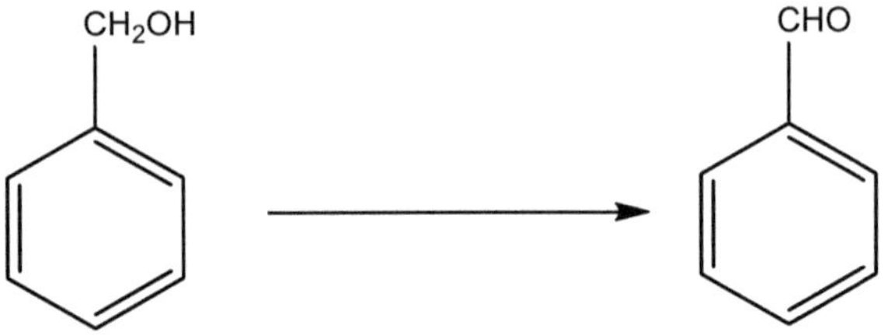

Fig. 4.8 Selective oxidation of benzyl alcohol to benzaldehyde

the C–H bond of CH_4 is more easily activated than CH_3OH in aqueous solution, which inhibits further oxidation of CH_3OH.

Cr_1/TiO_2 SACs were also used to catalyze the direct oxidation of methane to C1 oxygenate by H_2O_2 under mild conditions [35]. The selectivity is about 93%, and the yield of C1 oxygenate reaches 57.9 mol·mol_{Cr}^{-1}. EPR experiments show that a single Cr atom promotes the formation of $\cdot CH_3$ and ·OH, which can react to form CH_3OH. In addition, CH_3 can interact with –OOH on the TiO_2 surface to generate CH_3OOH, and the generated CH_3OH can be further oxidized to $HOCH_2OOH$ and HCOOH.

The oxidation of alcohol to aldehyde is an important organic reaction. However, it is very challenging to avoid the total oxidation of alcohol to carbon oxides in the gas phase oxidation process. In view of the low efficiency of traditional supported metal catalysts, the selective oxidation method of alcohol catalyzed by SACs was studied.

The M-N-C site is considered to be an important active site in alcohol oxidation. The Co_1/P-NC catalyst showed excellent activity and selectivity for the selective oxidation of benzyl alcohol due to the doping of heteratomic N and P at the atomic level (Fig. 4.8) [36]. Li et al. used Co_1@NG (nitrogen-doped graphene) with 4.1 wt% Co content to oxidize benzyl alcohol by activating peroxonosulfate under mild conditions [37]. This work sheds new light on the potential applications of transition metal-based SACs in selective oxidation of alcohols.

Co_1/NG was prepared by heat treatment of cobalt salt and graphene oxide in ammonia atmosphere for selective alcohol oxidation [38]. After the reaction for 5 h, the Co_1/NG catalyst achieved 94.8% benzyl alcohol conversion, 97.5% selectivity for benzaldehyde, and an average TOF over 500 h^{-1}. The DFT calculation shows that the oxygen atom is weakly adsorbed by the Co center, the electrons are transferred from the 3d orbital of Co to the 2p antibonding orbital of O_2, the superoxide is formed, and the hydrogen is extracted from benzyl alcohol to produce benzaldehyde. Huang et al. reported another kind of highly active Co-N-C SACs synthesized by one-pot method for selective aerobic oxidation of benzyl alcohol to benzaldehyde [39]. The conversion of benzyl alcohol and the selectivity of benzaldehyde over Co-N-C SAC were 93.8% and 97% respectively.

Wei et al. used a Mott-Schottky type nitrogen-doped carbon supported single-atom iron (Fe_1/N–C) catalyst for the selective oxidation of benzyl alcohol to aldehyde [40]. Fe and N atom coordinate to form FeN_4 active sites, and electrons transfer from Fe to N at the Mott-Schottky interface, resulting in a decrease in the electron density of Fe, which is conducive to the selective oxidation of benzyl alcohol to benzaldehyde.

Xin and coworkers compared the catalytic activity of CeO_2-supported Pd_1 single atoms and Pd_6 clusters for the oxidation of benzyl alcohol, and found that Pd_1/CeO_2 has higher catalytic activity and better selectivity, while Pd_6/CeO_2 has no activity [41]. The simulation results show that Pd_6 cluster not only blocks the interaction between OH and CeO_2 surface, but also has a high barrier of H transfer from cluster to CeO_2, resulting in its lower activity.

The active sites in three different Au species (SACs, nanoclusters, and nanoparticles) for alcohol oxidation were also studied by Lei and colleagues [42]. It was found that the size of Au species and the concentration of oxygen vacancy (O_v) significantly affect the activity of Au/CeO_2 catalysts. The activity of Au SACs is much higher compared to nanoclusters and nanoparticles due to the abundance of [O-O_v-Ce-O-Au] active sites, where oxygen vacancies can promote the adsorption of benzyl alcohol and the splitting of O–H bonds.

4.4.2 Hydrogenation

Hydrogenation of unsaturated organic compounds (such as nitroaromatics, alkenes, alkynes, aldehydes, ketones, imines, etc.) catalyzed by transition metals is one of the most important organic reactions, which has the advantages of high atomic economy, simple operation, clean green, etc., and has been widely used in organic synthesis.

Aromatic amines are the key intermediates for fine chemicals such as agrochemicals, pharmaceuticals and dyes, which can be synthesized by reducing the corresponding nitroaromatics. Many nitroaromatics have other reducible groups, such as –C≡C, –C=C, –C–X, –C≡N, –C=O, etc. Therefore, selective hydrogenation of -NO_2 groups without reducing other reducible groups (Fig. 4.9) is challenging. Many attempts have been made to develop efficient and selective catalysts for selective hydrogenation of nitroaromatics.

It is noteworthy that the Pt SACs tends to adsorb the –NO_2 group, thus decreasing the selectivity for the C=C double bond [43]. Wei et al. report that FeO_x-supported Pt SACs are very active, chemically selective, recyclable for hydrogenation of nitroaromatics [44]. For the hydrogenation of 3-nitrostyrene, Pt SACs show a selectivity of approximately 99% for 3-amino-styrene at mild conditions. The strong interaction between Pt SACs and FeO_x support is the reason for the excellent performance. The presence of separated positively charged Pt centers, as well as a suitably reduced

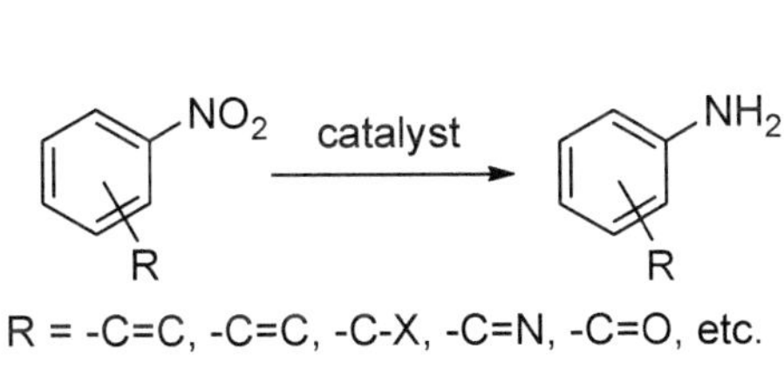

Fig. 4.9 Selective hydrogenation of –NO_2 groups without reducing other reducible groups

metal oxide surface, favors preferential adsorption of -NO_2 groups, thus significantly enhancing catalytic performance. It is worth noting that Pt nanoparticles would accelerate the C=C hydrogenation side reaction and decrease the selectivity.

Pt_1/Ni with Pt single atoms supported on Ni nanocrystals shows high activity and selectivity for the hydrogenation of 3-nitrostyrene [45]. The TOF of Pt_1/Ni nanocrystals is much higher than that of Pt clusters/Ni nanocrystals. It is found that only -NO_2 groups are hydrogenated on Pt SACs, with selectivity higher than 99% for 3-vinylaniline. However, both -NO_2 groups and C=C bonds are hydrogenated over Pt clusters/Ni nanocrystals.

Recently non-noble metal SACs loaded on nitrogen-doped carbon support are widely studied for selective hydrogenation of aromatic compounds. Wang et al. prepared Co SACs supported on NC ($Zn_{24}Co_1$-BMOF) which exhibit excellent performance in the selective hydrogenation of nitrobenzene, with >99% aniline selectivity at 100% conversion after 4 h [46], while Co NPs supported on NC show lower activity (97.4% selectivity and 84.8% conversion at 8 h). They found that atomically dispersed Co species are the key to obtain the high activity and selectivity. The doped N atoms in carbon not only decrease the dissociation energy of H_2, but also preferentially adsorb -NO_2 groups through adjustment of surface polarity and basicity of the catalyst [47]. He and his coworkers found that Co SAs/NC-800 (pyrolysis at 800 °C) shows excellent hydrogenation activity due to the Co-N_4 coordination. The higher electron density of Co SACs in the Co SAs/NC-800 promotes the donation of electrons from Co to H_2 and thus facilitates the dissociation of H_2 molecules [48].

Fe SACs (Fe_1/N–C) also shows higher activity and selectivity than that of Fe NP/N–C for transfer hydrogenation of nitroarenes with hydrazine hydrate [49]. The energy barrier in the rate-determining step is much lower on Fe SACs. DFT calculations show that the positively charged Fe SACs could effectively lowered energy barrier, leading to the superior performances for transfer hydrogenation of nitroarenes.

The SACs could effectively prevent the reduction of C=C bonds. The H^+/H^- produced from H_2 heterolytic cleavage tend to hydrogenate the polar NO_2 groups rather than the nonpolar C=C bond [50]. Since the heterolysis of H_2 can be realized on SACs, thus SACs show higher selectivity than their NP counterparts.

Selective hydrogenation of unsaturated aldehydes to unsaturated alcohols is another challenging reaction (Fig. 4.10) because the C=C bond is thermodynamically more prone to be hydrogenated than the C=O bond [51]. The homogeneous catalyst is effective for selective hydrogenation of C=O bond through the formation of ionic hydrogen. However, the homogeneous catalyst cannot be separated and recycled. Therefore, the heterogeneous catalysts for selective hydrogenation of unsaturated aldehydes are highly desirable.

Fortunately, SACs are reported to show remarkable selectivity in the hydrogenation of α, β-unsaturated aldehydes. For example, Lou et al. use Rh_1/MoS_2 SACs to convert crotonaldehyde to crotyl alcohol [52]. The Rh_1/MoS_2 SAC exhibits 100% selectivity and TOF of 26.6 h^{-1} at 80 °C and 5 bar H_2, while the nano Rh/MoS_2 catalyst only shows 47% selectivity with a lower TOF. It is revealed that the special

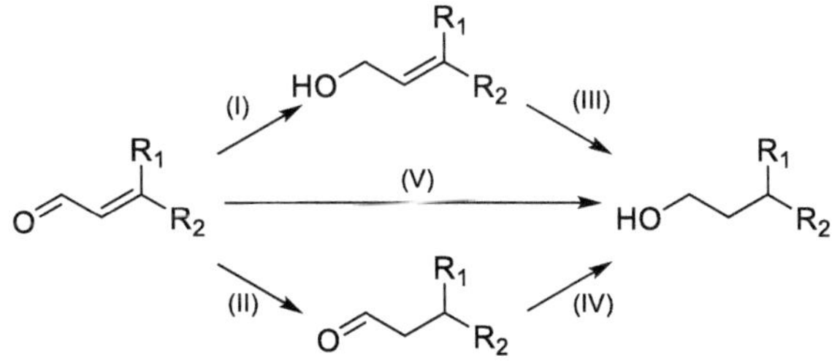

Fig. 4.10 Hydrogenation of α, β-unsaturated aldehydes

geometry and electron configuration of Rh_1/MoS_2 SACs limit the adsorption mode of crotonaldehyde molecules through spatial effects and lead to an excellent selectivity.

In 2021, Lou and coworkers reported that Pt_1/MoS_2 SACs also have remarkable selectivity in crotonaldehyde hydrogenation (Fig. 4.11) [53]. 100% selectivity to crotyl alcohol can be obtained at a 75 °C and 6 bar H_2. However, the activity of the nano-Pt/MoS_2 catalyst is only 41%, and the main product is 1-butanol. The unique bag-like active site of Pt_1/MoS_2 SACs limits the adsorption configuration of the reactant molecules and promotes H_2 dissociation, thereby improving catalytic selectivity and activity.

The hydrogenation of furfural to furfuryl alcohol is also very important (Fig. 4.12). Hu and colleagues synthesized Pd_1/C_3N_4 SACs for selective hydrogenation of furfural by confining a single Pd atom to a six fold N-coordination cavity of g-C_3N_4 [54]. The selectivity of Pd_1/C_3N_4 to furfuryl alcohol is 99%, the conversion rate is 64% when the reaction time is 4 h, and the activity is significantly high, and the TOF is 146 h^{-1}. In contrast, the Pd NPs/C_3N_4 catalyst has a lower activity with a TOF of 67 h^{-1}, a conversion rate of 39% to furfuryl alcohol and a selectivity of 99% under the same conditions. In addition, Pd_1/C_3N_4 SACs exhibit excellent durability and thermal stability. In 2021, Fan et al. used Ni_1/NC (Ni SACs supported on nitrogen doped carbon) in the catalytic transfer hydrogenation of furfural to furfuryl alcohol [55]. The atomically dispersed Ni–N_4 sites change the electron density of the Ni center and produce an excellent activity with TOF of 832 h^{-1} and 97.1% selectivity at 130 °C, which is nine times higher than that of Ni nanoparticles.

Ru SACs loaded onto mesoporous carbon nitride (Ru_1/C_3N_4) was used for the hydrogenation of vanillin (Fig. 4.13) [56]. It was found that the selectivity of vanillin hydrogenation to vanillin alcohol can reach ≈100% at 60 °C and the conversion rate can reach 95%. The conversion and selectivity of Ru NPs, which are 29% and 95% respectively, are much lower than that of the Ru_1/C_3N_4 catalyst. The reaction

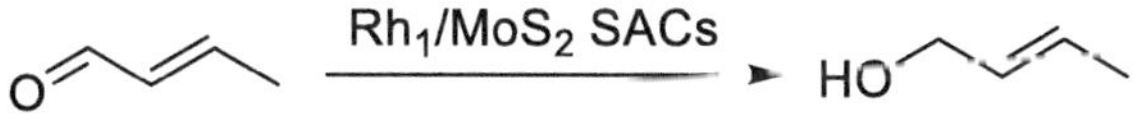

Fig. 4.11 Crotonaldehyde hydrogenation

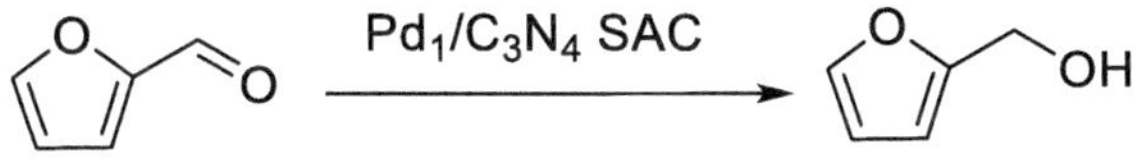

Fig. 4.12 Hydrogenation of furfural to furfuryl alcohol

mechanism of Ru SACs in vanillin hydrogenation was systematically studied by first principles calculation. They found that Ru SACs are mobilized and the active sites are induced by reactant adsorption. Interestingly, 2-methoxy-p-cresol as the deoxygenation product can also be obtained provided the reaction is carried out at 140 °C.

Alkyne semi-hydrogenation is an important and challenging step. Pd-based catalysts are considered to be one of the most effective semi-hydrogenation catalysts for acetylene due to their excellent hydrogenation activity. Ethylene has three adsorption modes on Pd catalysts, namely, ethylidyne adsorption mode on Pd trimers, di-σ mode on bridged Pd dimers and π-bonded adsorption mode on isolated Pd SACs (Fig. 4.14) [27]. When ethylene is adsorbed by acetylene or σ-mode, the selectivity of semi-hydrogenation to ethylene is low. At this mode, deep hydrogenation is preferred, because the energy barrier of desorption is higher than that of hydrogenation. On the contrary, when ethylene is adsorbed by the π-bond mode, the selectivity for ethylene is higher than other modes. In this mode, the energy required for desorption is lower than that required for further hydrogenation, so ethylene is easily desorbed from the surface of the supported metal catalyst rather than further hydrogenation. This means that Pd SACs with isolated active sites can improve ethylene selectivity through π-bond mode adsorption due to its characteristics of single site.

Huang et al. fabricated Pd_1/ND@G SACs by anchoring the isolated Pd atoms onto the defective nanodiamond-graphene which exhibited excellent performance in the selective hydrogenation of acetylene, with >90% ethylene selectivity at 100% conversion at 180 °C [57]. DFT calculations show that the energy barrier of adsorbed C_2H_4 intermediate on further hydrogenation to C_2H_6 is much higher than the desorption energy of surface C_2H_4 species to the gas phase. The competitive desorption of ethylene at the active site of Pd_1/ND@G is the reason for the high selectivity of semi-hydrogenated acetylene. Feng et al. also found that the ethylene selectivity of Pd_1/MPNC (mesoporous N-doped carbon) catalysts (82%) is much higher than that of Pd NP/MPNC catalysts (17%) [58]. DFT calculations also show that it is easier

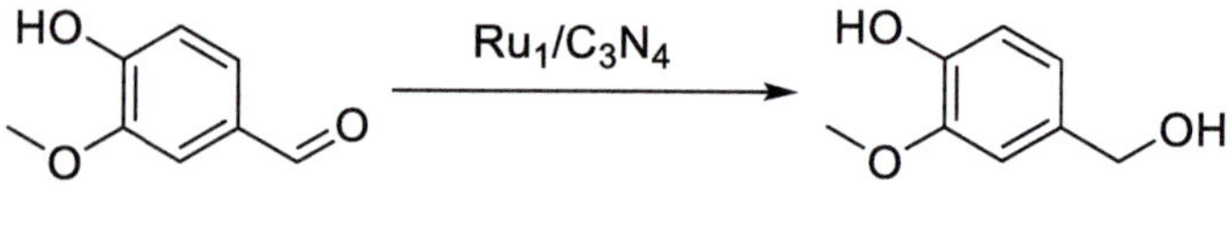

Fig. 4.13 Hydrogenation of vanillin

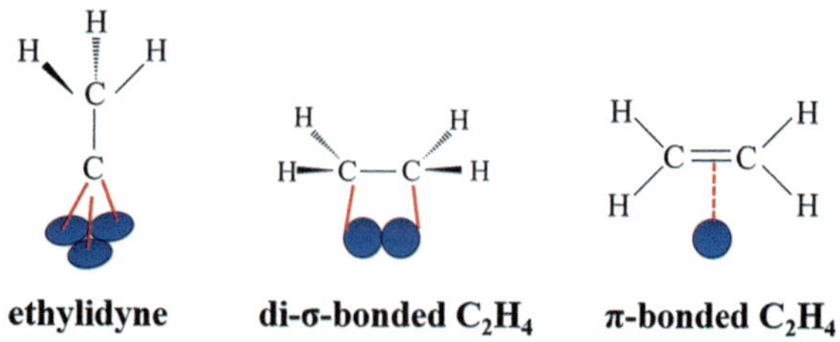

Fig. 4.14 Different adsorption patterns of ethylene on Pd catalysts with different structures. Reproduced with permission from Ref. [27]. Copyright 2022, Wiley–VCH GmbH

for ethylene to desorb rather than to be further hydrogenated to ethane over Pd_1/MPNC.

Wei et al. also found that Pd_1/CN showed remarkable catalytic performance in the semi-hydrogenation of acetylene [59]. At 120 °C, the acetylene conversion rate was as high as 96.0% and the ethylene selectivity was as high as 93.4%. In contrast, the acetylene conversion of the Pd-NPs/CN catalyst is only 70.1%, and the selectivity is 71.8%. Similar to other catalysts, Pd SACs need to overcome two obstacles in the hydrogenation of ethylene to ethane, with one transition state (0.89 eV) located above the ethylene. As a result, further hydrogenation over the Pd-SACs site is prevented.

Au-based catalysts are another kind of promising catalysts for selective hydrogenation of acetylene to ethylene. DFT simulations show that only the C≡C bond is adsorbed and activated on the Au catalyst, while the adsorption strength of the C=C bond is too weak to be activated [60]. As a result, molecules containing C=C bond leave easily, and only the C≡C bond can be hydrogenated on the Au catalyst, leading to selective hydrogenation of alkanes even in the presence of large quantities of alkenes.

Besides C=C bonds, the hydrogenation of nitriles can also be realized over SACs (Fig. 4.15). Liu et al. used Pd_1/ND@G (defect-rich nanodiamond-graphene) in the catalytic transfer hydrogenation of nitriles [61]. The Pd_1/ND@G catalyst preferentially generates secondary amines while the Pd_n/ND@G catalyst exhibits high primary amines selectivity. Characterizations and DFT calculations (Fig. 4.16) show that the structure of atomically dispersed Pd catalysts governs the dissociative adsorption pattern of H_2 and the hydrogenation pathway of the benzylideneimine intermediate, resulting in a different product selectivity over Pd_1/ND@G and Pd_n/ND@G, respectively.

In summary, besides the improvement of atomic utilization efficiency, the SAC active sites can also alter the reaction path by influencing the adsorption pattern of reactants or intermediates on the catalyst, resulting in high selectivity.

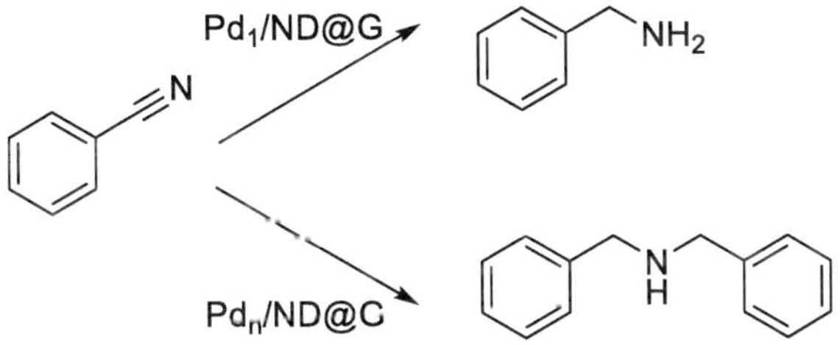

Fig. 4.15 The hydrogenation of nitriles

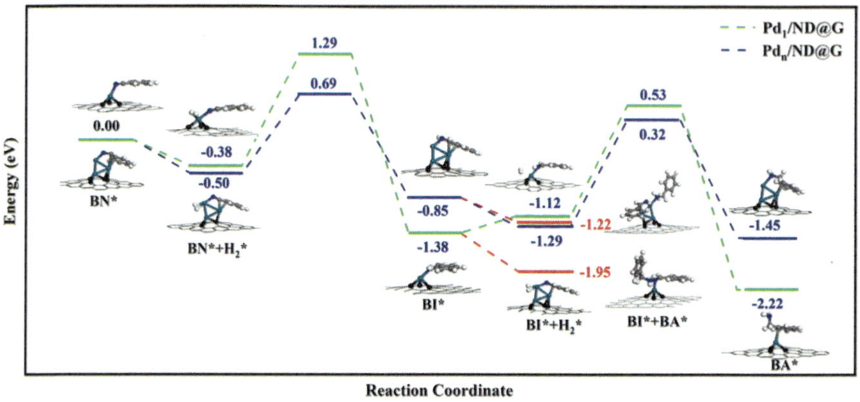

Fig. 4.16 Hydrogenation mechanism of benzonitrile to benzylamine on Pd_1/ND@G and Pd_n/ND@G. Reproduced with permission from Ref. [61]

4.4.3 Addition

Addition reaction of unsaturated substrates is one of the most widely used organic synthesis reactions in fine chemical production. Recently, SACs has been widely used in addition reactions, including hydrogenation, hydrosilylation, hydrochlorination, distannylation, hydroboration, and so on. Since hydrogenation reactions have been discussed in detail in Sect. 4.4.2, we will mainly focus on other addition reactions in this part.

Pt, Ru, and Au SACs have been proved that they can provide efficient catalytic performance for the production of vinyl chloride through acetylene hydrochlorination [62–64]. Chen et al. found that the N coordinated and positively charged Au_1 site has remarkable catalytic performance for acetylene hydrochlorination [65].

Kaiser and coworkers show that Pt SACs on carbon supports are much more stable than Au SACs, and their coordination environment can be precisely adjusted by simple temperature control [66]. By combining kinetic analysis and DFT calculation, Pt(II)−Cl was identified as the active site which is three times more active than Pt nanoparticles.

Li et al. designed and synthesized Cu_1/TiC SACs with strong electronic metal–support interaction for selective hydroboration of terminal alkynes (Fig. 4.17) [67]. This work vividly proves that the SACs can not only improve the catalytic efficiency, but also break through the limitation of regional selectivity, which provides inspiration for the development of organic synthesis methods.

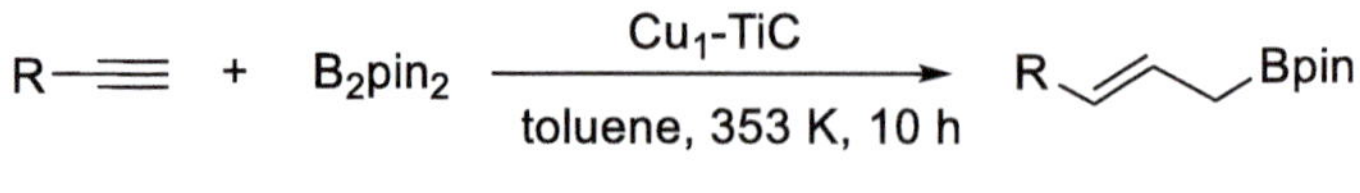

Fig. 4.17 Selective hydroboration of terminal alkynes

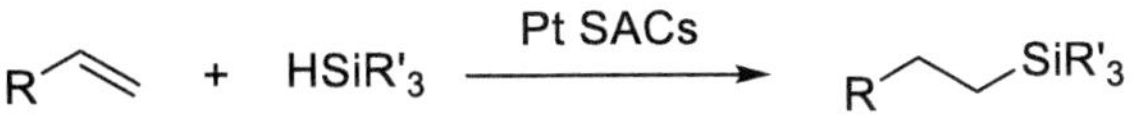

Fig. 4.18 Alkene hydrosilylation

A single-atom Pt catalyst ($Pt_1/Ni(OH)_x$) containing up to 2.3 wt% Pt was obtained by a simple wet impregnation method using nickel hydroxide nanoplates with abundant Ni^{2+} vacancy defects as support [68]. It is found that Ni^{2+} vacancy has a strong stabilizing effect on Pt SACs. The $Pt_1/Ni(OH)_x$ catalyst has high catalytic efficiency for diboration of alkynes and alkenes, and the total turnover rate of phenylacetylene after the reaction is ~3000 h^{-1}, which is much higher than other reported heterogeneous catalysts.

According to the reaction mechanism of distannylation of terminal alkynes, a porous organic polymer catalyst with triarylphosphonate as the backbone was designed and prepared [69], which has the advantages of pore enrichment, ligand regulation and single atom selective regulation. The catalyst solves the problems of narrow substrate range, poor selectivity and low catalytic efficiency in this reaction.

Alkene hydrosilylation, the addition reaction of a silicon hydride (Si–H) to the C=C bond (Fig. 4.18), can also be catalyzed by SACs. Anti-Markovnikov hydrosilylation of terminal alkenes with silanes can be catalyzed by Pt SACs. Pt SACs supported on anatase TiO_2 ($Pt_1^{\delta+}/TiO_2$) can catalyze almost full conversion of hydrosilylation with high selectivity [70]. DFT calculations show that the unexpected properties of $Pt_1^{\delta+}/TiO_2$ are due to the atomic dispersion of the active substance and the characteristic partial positive electronic structure of $Pt^{\delta+}$, which is not available in a traditional nanocatalyst. Zhu et al. used a Pt_1/NG (N-doped graphene) with Pt loading up to 5.3 wt% for anti-Markovnikov hydrosilylation of terminal alkenes to tertiary silanes under mild conditions [71]. Characterization showed that single Pt species on graphene support were stabilized by N in the form of $Pt–N_4$. An overall turnover frequency of 180 h^{-1} was obtained in hydrosilylation of 1-octene, fourfold enhancement than that over Pt/C. As shown in Table 4.1, the Pt_1/NG catalyst exhibits good universality for anti-Markovnikov hydrosilylation of diverse alkenes with triethoxysilane.

In 2021, Liu et al. used electron-deficient Pt_1@AHA_U_400 (humic matter) in hydrosilylation reactions and obtained superior activity (TOF 3.0×10^7 h^{-1}) and selectivity (>99%) [72]. DFT calculations reveal that the high performance of the catalyst results from the atomic dispersion and the electron deficiency of the Pt_1 atoms.

4.4.4 Coupling

The Suzuki–Miyaura coupling reaction is a reaction in which organoborate reagents are cross-coupled with organohalides or sulfonates (Fig. 4.19), and is one of the most effective ways to form C–C bonds.

Table 4.1 Hydrosilylation of Various Alkenes with Triethoxysilane by Pt_1/NG^a. Reproduced with permission from Ref. [71]. Copyright © 2018, American Chemical Society

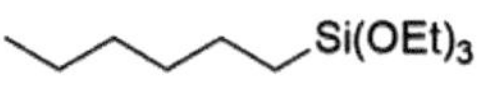 **3b**, 50 °C, 2 h 99%[b] (99%[c])	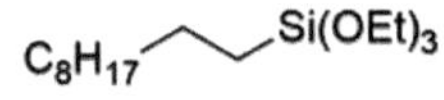 **3c**, 60 °C, 3 h 95% (99%)	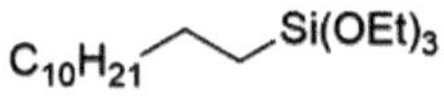 **3d**, 60 °C, 6 h 97% (99%)
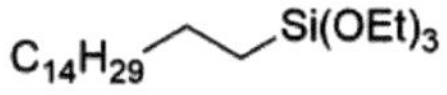 **3e**, 60 °C, 6 h 94% (99%)	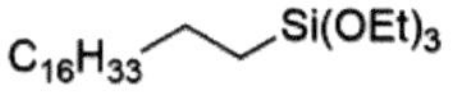 **3f**, 60 °C, 6 h 92% (99%)	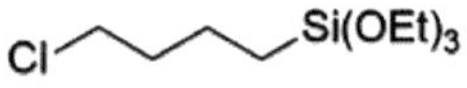 **3g**, 60 °C, 6 h 93% (98%)
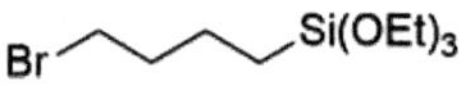 **3h**, 60 °C, 6 h 97% (98%)	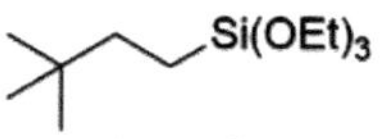 **3i**, 40 °C, 2 h 97% (99%)	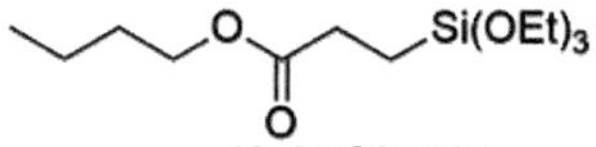 **3j**, 60 °C, 16 h 94% (99%)
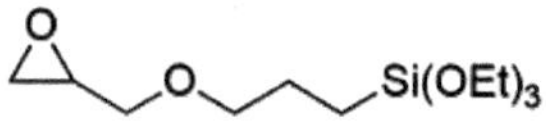 **3k**, 60 °C, 18 h 94% (98%)	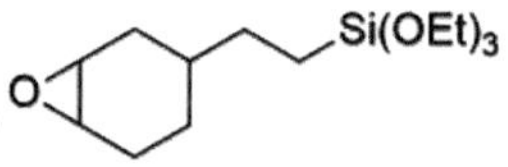 **3l**, 80 °C, 16 h 96% (99%)	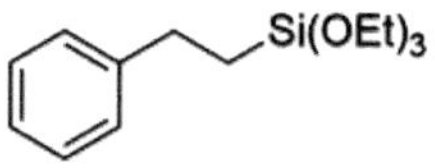 **3m**, 80 °C, 18 h 94% (98%)
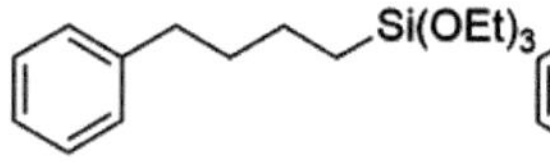 **3n**, 80 °C, 18 h 94% (99%)	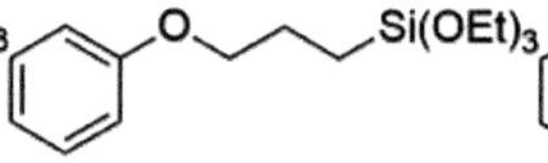 **3o**, 80 °C, 16 h 95% (98%)	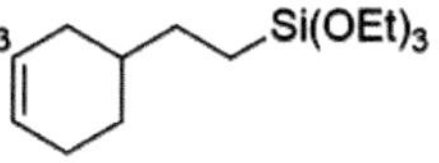 **3p**, 60 °C,16 h 94% (99%)

[a]Reaction conditions: substrate (1.0 mmol) and $HSi(OEt)_3$ (1.3 mmol), Pt_1/NG catalyst, Pt/substrate = 0.25%, p-xylene (3.0 mL) as solvent
[b]Conversions are determined by gas chromatography (GC) analysis with dodecane as internal standard
[c]Selectivities are determined by GC–MS analysis

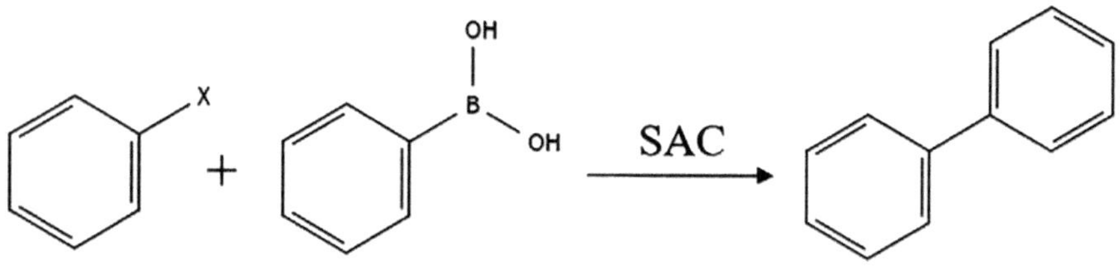

Fig. 4.19 Suzuki–Miyaura coupling reaction

Pd_1/ECN (exfoliated graphitic carbon nitride) is a stable heterogeneous catalyst with 90% selectivity and 56% biphenyl yield, exceeding the most advanced homogeneous catalysts in Suzuki coupling reactions [73]. Pd single atom has an adaptive coordination with the N in ECN, which promotes the adsorption, stabilization, and activation of the substrates and intermediates. The flexible lattice of ECN provides a variable coordination environment for Pd SACs, resulting in high stability without leaching or aggregation. Moreover, Kim et al. used Pd SACs anchored on Zr-based metal–organic polyhedron for Suzuki coupling reactions in aqueous media [74]. The catalyst has a yield of over 90.0% for most substrates, superior to Pd complexes and MOF-restricted Pd atoms.

Pd_1/3DOM-CeO_2 with Pd single atoms supported by 3D ordered porous cerium dioxide was used for Suzuzu-Miyaura coupling reaction, and showed good activity for various substrates [75]. Calculations show that Pd_1/3DOM-CeO_2 can lower the energy barrier of transmetalation as well as reduction elimination steps in Suzuki coupling reaction. Ding and coworkers synthesized Pd_1/ZnO-ZrO_2 for Suzuki reaction under ambient conditions which shows high activity on many substrates [76]. It was found that Pd atom is coordinated with two oxygen atoms.

Conversion over 99% for Suzuki coupling reaction can be obtained over Pt_1/S-MWNT (thiolated multi-walled nanotubes) within 24 h [77]. The introduction of the thiol group in the carbon nanotubes prevents the agglomeration of Pt, and the S atom provides rapid charge compensation for Pt, leading to excellent activity and recyclability.

4.4.5 *Hydroformylation of Olefins*

SACs have exhibited unique catalytic performance for the hydroformylation of olefins to form aldehydes with CO and H_2 (Fig. 4.20).

Lang et al. found that Rh_1/ZnO SACs have comparable or even better activity than the classical homogeneous Wilkinson catalyst $RhCl(PPh_3)_3$ for hydroformylation of various alkenes [78]. The electron transfer from Zn to Rh atoms can stabilize the catalyst and bring good recyclability. Then Amsler and coworkers theoretically and experimentally studied the potential of oxide supported Rh SACs for heterogeneous hydroformylation [79]. Rh_1/CeO_2 SACs exhibit high activity comparable to that of $HRh(PPh_3)_3CO$, and Rh_1/ZnO SACs exhibit excellent stability, and both agree well with theoretical calculations.

Rh_1/CoO exhibits significant activity and selectivity for propylene hydroformylation [80]. DFT calculations show that the Rh SACs undergo structural restructuring

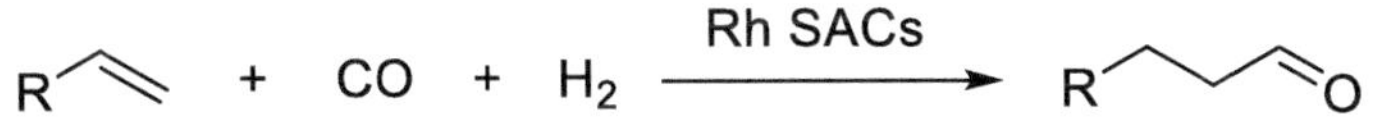

Fig. 4.20 Hydroformylation of olefins with CO and H_2 to aldehydes

under H_2 and CO, which enhances the adsorption and activation of the reactants. The reaction pathway is thought to be limited to Rh_1/CoO, and only certain configurations are stable, effectively altering the selectivity of the product.

Rh_1/nanodiamonds have excellent catalytic properties for hydroformylation of aryl ethylene with wide substrate universality [81]. Rh_1/nanodiamond catalyst has a high aldehyde conversion (>99%), which is comparable to a homogeneous catalyst. Ibuprofen can be synthesized with high efficiency and high yield by using this catalyst, which opens up a new prospect for the application of SACs in drug synthesis. Rh SACs encapsulated in monophosphine polymers also have a linear aldehyde selectivity of 92% for hydroformylation of alkenes, which is much higher than $Rh(CO)_2(acac)$ [82]. The Rh SACs are very stable and can be easily recovered without significant performance degradation.

4.5 Conclusion and Outlook

The research of single atom catalysis has brought the science of catalysis from nanocatalysis to atomic scale. With the help of advanced characterization technologies such as XAFS and HAADF-STEM, we can now understand complex heterogeneous catalytic reactions at atomic scale.

In SACs, the metal is highly dispersed on the support in the form of single atoms, which reaches the limit of metal dispersion and realizes the maximum utilization of atoms. The single atom catalyst has the advantages of both the "isolated site" of homogeneous catalyst and the structural stability and easy separation of heterogeneous catalyst, which closely links heterogeneous catalysis and homogeneous catalysis, and becomes a bridge between heterogeneous catalysis and homogeneous catalysis [83]. The preparation of the single atom catalyst greatly reduces the use of precious metals, reduces the manufacturing cost, and the catalytic performance is obviously better than that of traditional supported catalysts, and the catalytic activity and selectivity have reached an unprecedented high level. The single-atom catalyst has a single dispersed active site, avoiding the occurrence of side reactions and by-products which may bring purification and other subsequent treatment processes and potential environmental problems, so as to save post-treatment costs, make the production process more economic and environmental protection, to achieve the purpose of "green catalysis". Many breakthrough achievements have been made in the study of single atom catalysis, such as the preparation of single Fe center catalysts to solve the scientific problem of selective activation and directed conversion of tetrahedral symmetric methane molecules. Pt /FeO_x SACs and quasi-single atom catalysts selectively reduce nitro groups while preserving C=C and C≡C bonds, which provides a green and efficient way for the selective hydrogenation of aromatic nitrocompounds containing different substituents to prepare aromatic amines.

From the successful preparation of the first single atom catalyst to the present, the research in this field has attracted the attention of many researchers. In the preparation of the single atom catalyst, co-precipitation method and wet impregnation method are simple and universal, and have the development prospect of preparation in large scale. ALD is a new preparation technology, more controllable, but the operating conditions and equipment requirements are high, and it is difficult for large-scale application in a short time.

In terms of research methods and means, emphasis is placed on the combination of theoretical calculation and experimental research to simulate the catalytic reaction mechanism closer to the real one, eliminate the "gap" between the current theoretical calculation and experimental research, and lay an important scientific foundation for the analysis of the structure–activity relationship and catalytic mechanism of SACs, and design of new SACs.

In terms of application fields, the current research focuses on classical chemical reactions such as CO oxidation, selective hydrogenation, and water gas conversion. In the future, on the one hand, the application research of single atom catalysis in traditional fields will be expanded, such as petrochemical industry and the synthesis of fine chemicals. On the other hand, exchanges and cooperation with other disciplines to expand the single atom research system will also be developed. It has been applied to fuel cell, photoelectric catalysis and other emerging fields, so as to give full play to the advantages of SACs. At present, the development and application of SACs are in the laboratory research stage, and there is still a distance from industrial application. However, due to its superior catalytic performance and potential cost advantages, people should pay attention to basic research at the same time, accelerate the industrial application of single atom catalysis research, screen out single atom catalysis system suitable for industrial applications, develop preparation technology with large-scale application prospects, and promote the wide application of single atom catalysis in various fields. It is believed that more and more important research results based on single-atom catalytic systems will emerge in the future.

References

1. Zhang H, Liu G, Shi L, Ye J (2018) Single-atom catalysts: emerging multifunctional materials in heterogeneous catalysis. Adv Energy Mater 8(1):1701343. https://doi.org/10.1002/aenm.201701343
2. Taylor HS (1925) A theory of the catalytic surface. Proc Roy Soc Lond Ser A, Containing Papers Math Phys Char 108(745):105–111
3. Rooney JJ, Webb G (1964) The importance of π-bonded intermediates in hydrocarbon reactions on transition metal catalysts. J Catal 3(6):488–501. https://doi.org/10.1016/0021-9517(64)90049-1
4. Shephard FE, Rooney JJ (1964) Reactions of some C9 aromatics on platinum-alumina catalysts. J Catal 3(2):129–144. https://doi.org/10.1016/0021-9517(64)90121-6
5. Yates JT, Worley SD, Duncan TM, Vaughan RW (1979) Catalytic decomposition of formaldehyde on single rhodium atoms. J Chem Phys 70(3):1225–1230. https://doi.org/10.1063/1.437604
6. Maschmeyer T, Rey F, Sankar G, Thomas JM (1995) Heterogeneous catalysts obtained by grafting metallocene complexes onto mesoporous silica. Nature 378(6553):159–162. https://doi.org/10.1038/378159a0

7. Asakura K, Nagahiro H, Ichikuni N, Iwasawa Y (1999) Structure and catalytic combustion activity of atomically dispersed Pt species at MgO surface. Appl Catal A 188(1):313–324. https://doi.org/10.1016/S0926-860X(99)00247-1
8. Abbet S, Sanchez A, Heiz U, Schneider WD, Ferrari AM, Pacchioni G, Rösch N (2000) Acetylene cyclotrimerization on supported size-selected Pdn clusters ($1 \leq n \leq 30$): one atom is enough! J Am Chem Soc 122(14):3453–3457. https://doi.org/10.1021/ja9922476
9. Fu Q, Saltsburg H, Flytzani-Stephanopoulos M (2003) Active nonmetallic Au and Pt species on ceria-based water-gas shift catalysts. Science 301(5635):935. https://doi.org/10.1126/science.1085721
10. Zhang X, Shi H, Xu B-Q (2005) Catalysis by gold: isolated surface Au^{3+} ions are active sites for selective hydrogenation of 1,3-butadiene over Au/ZrO_2 catalysts. Angew Chem Int Ed 44(43):7132–7135. https://doi.org/10.1002/anie.200502101
11. Hackett SFJ, Brydson RM, Gass MH, Harvey I, Newman AD, Wilson K, Lee AF (2007) High-activity, single-site mesoporous Pd/Al_2O_3 catalysts for selective aerobic oxidation of allylic alcohols. Angew Chem Int Ed 46(45):8593–8596. https://doi.org/10.1002/anie.200702534
12. Qiao B, Wang A, Yang X, Allard LF, Jiang Z, Cui Y, Liu J, Li J, Zhang T (2011) Single-atom catalysis of CO oxidation using Pt1/FeOx. Nat Chem 3(8):634–641. https://doi.org/10.1038/nchem.1095
13. Tierney HL, Baber AE, Sykes ECH (2009) Atomic-scale imaging and electronic structure determination of catalytic sites on Pd/Cu near surface alloys. J Phys Chem C 113(17):7246–7250. https://doi.org/10.1021/jp809766d
14. Tierney HL, Baber AE, Kitchin JR, Sykes ECH (2009) Hydrogen dissociation and spillover on individual isolated palladium atoms. Phys Rev Lett 103(24):246102. https://doi.org/10.1103/PhysRevLett.103.246102
15. Palmer RE, Pratontep S, Boyen HG (2003) Nanostructured surfaces from size-selected clusters. Nat Mater 2(7):443–448. https://doi.org/10.1038/nmat897
16. Li J, Guan Q, Wu H, Liu W, Lin Y, Sun Z, Ye X, Zheng X, Pan H, Zhu J, Chen S, Zhang W, Wei S, Lu J (2019) Highly active and stable metal single-atom catalysts achieved by strong electronic metal-support interactions. J Am Chem Soc 141(37):14515–14519. https://doi.org/10.1021/jacs.9b06482
17. Lin J, Wang A, Qiao B, Liu X, Yang X, Wang X, Liang J, Li J, Liu J, Zhang T (2013) Remarkable performance of $Ir1/FeO_x$ single-atom catalyst in water gas shift reaction. J Am Chem Soc 135(41):15314–15317. https://doi.org/10.1021/ja408574m
18. Qiao B, Liu J, Wang Y-G, Lin Q, Liu X, Wang A, Li J, Zhang T, Liu J (2015) Highly efficient catalysis of preferential oxidation of CO in H_2-rich stream by gold single-atom catalysts. ACS Catal 5(11):6249–6254. https://doi.org/10.1021/acscatal.5b01114
19. Yin P, Yao T, Wu Y, Zheng L, Lin Y, Liu W, Ju H, Zhu J, Hong X, Deng Z, Zhou G, Wei S, Li Y (2016) Single cobalt atoms with precise N-coordination as superior oxygen reduction reaction catalysts. Angew Chem Int Ed 55(36):10800–10805. https://doi.org/10.1002/anie.201604802
20. Gan T, Liu Y, He Q, Zhang H, He X, Ji H (2020) Facile Synthesis of kilogram-scale Co-alloyed Pt single-atom catalysts via ball milling for hydrodeoxygenation of 5-hydroxymethylfurfural. ACS Sustain Chem Eng 8(23):8692–8699. https://doi.org/10.1021/acssuschemeng.0c02065
21. Yang M, Allard LF, Flytzani-Stephanopoulos M (2013) Atomically dispersed $Au–(OH)_x$ species bound on titania catalyze the low-temperature water-gas shift reaction. J Am Chem Soc 135(10):3768–3771. https://doi.org/10.1021/ja312646d
22. Liu P, Zhao Y, Qin R, Mo S, Chen G, Gu L, Chevrier DM, Zhang P, Guo Q, Zang D, Wu B, Fu G, Zheng N (2016) Photochemical route for synthesizing atomically dispersed palladium catalysts. Science 352(6287):797–800. https://doi.org/10.1126/science.aaf5251
23. Krivanek OL, Chisholm MF, Nicolosi V, Pennycook TJ, Corbin GJ, Dellby N, Murfitt MF, Own CS, Szilagyi ZS, Oxley MP, Pantelides ST, Pennycook SJ (2010) Atom-by-atom structural and chemical analysis by annular dark-field electron microscopy. Nature 464(7288):571–574. https://doi.org/10.1038/nature08879

24. Emmrich M, Huber F, Pielmeier F, Welker J, Hofmann T, Schneiderbauer M, Meuer D, Polesya S, Mankovsky S, Ködderitzsch D, Ebert H, Giessibl FJ (2015) Subatomic resolution force microscopy reveals internal structure and adsorption sites of small iron clusters. Science 348(6232):308–311. https://doi.org/10.1126/science.aaa5329
25. Wanjala BN, Fang B, Luo J, Chen Y, Yin J, Engelhard MH, Loukrakpam R, Zhong C-J (2011) Correlation between atomic coordination structure and enhanced electrocatalytic activity for trimetallic alloy catalysts. J Am Chem Soc 133(32):12714–12727. https://doi.org/10.1021/ja2040464
26. Zhang L, Wang A, Miller JT, Liu X, Yang X, Wang W, Li L, Huang Y, Mou C-Y, Zhang T (2014) Efficient and durable Au alloyed Pd single-atom catalyst for the Ullmann reaction of aryl chlorides in water. ACS Catal 4(5):1546–1553. https://doi.org/10.1021/cs500071c
27. Shi Y, Zhou Y, Lou Y, Chen Z, Xiong H, Zhu Y (2022) Homogeneity of supported single-atom active sites boosting the selective catalytic transformations. Adv Sci 9(24):2201520. https://doi.org/10.1002/advs.202201520
28. Kwon Y, Kim TY, Kwon G, Yi J, Lee H (2017) Selective activation of methane on single-atom catalyst of rhodium dispersed on zirconia for direct conversion. J Am Chem Soc 139(48):17694–17699. https://doi.org/10.1021/jacs.7b11010
29. Bai S, Liu F, Huang B, Li F, Lin H, Wu T, Sun M, Wu J, Shao Q, Xu Y, Huang X (2020) High-efficiency direct methane conversion to oxygenates on a cerium dioxide nanowires supported rhodium single-atom catalyst. Nat Commun 11(1):954. https://doi.org/10.1038/s41467-020-14742-x
30. Huang W, Zhang S, Tang Y, Li Y, Nguyen L, Li Y, Shan J, Xiao D, Gagne R, Frenkel AI, Tao F (2016) Low-temperature transformation of methane to methanol on Pd_1O_4 single sites anchored on the internal surface of microporous silicate. Angew Chem Int Ed 55(43):13441–13445. https://doi.org/10.1002/anie.201604708
31. Luo L, Luo J, Li H, Ren F, Zhang Y, Liu A, Li W-X, Zeng J (2021) Water enables mild oxidation of methane to methanol on gold single-atom catalysts. Nat Commun 12(1):1218. https://doi.org/10.1038/s41467-021-21482-z
32. Cui X, Li H, Wang Y, Hu Y, Hua L, Li H, Han X, Liu Q, Yang F, He L, Chen X, Li Q, Xiao J, Deng D, Bao X (2018) Room-temperature methane conversion by graphene-confined single iron atoms. Chem 4(8):1902–1910. https://doi.org/10.1016/j.chempr.2018.05.006
33. Zhu K, Liang S, Cui X, Huang R, Wan N, Hua L, Li H, Chen H, Zhao Z, Hou G, Li M, Jiang Q, Yu L, Deng D (2021) Highly efficient conversion of methane to formic acid under mild conditions at ZSM-5-confined Fe-sites. Nano Energy 82:105718. https://doi.org/10.1016/j.nanoen.2020.105718
34. Tang X, Wang L, Yang B, Fei C, Yao T, Liu W, Lou Y, Dai Q, Cai Y, Cao X-M, Zhan W, Guo Y, Gong X-Q, Guo Y (2021) Direct oxidation of methane to oxygenates on supported single Cu atom catalyst. Appl Catal B 285:119827. https://doi.org/10.1016/j.apcatb.2020.119827
35. Shen Q, Cao C, Huang R, Zhu L, Zhou X, Zhang Q, Gu L, Song W (2020) Single chromium atoms supported on titanium dioxide nanoparticles for synergic catalytic methane conversion under mild conditions. Angew Chem Int Ed 59(3):1216–1219. https://doi.org/10.1002/anie.201913309
36. Ji S, Chen Y, Zhang Z, Cheong W-C, Liu Z, Wang D, Li Y (2019) Single-atomic-site cobalt stabilized on nitrogen and phosphorus co-doped carbon for selective oxidation of primary alcohols. Nanoscale Horizons 4(4):902–906. https://doi.org/10.1039/C9NH00036D
37. Li J, Zhao S, Zhang L, Jiang SP, Yang S-Z, Wang S, Sun H, Johannessen B, Liu S (2021) Cobalt single atoms embedded in nitrogen-doped graphene for selective oxidation of benzyl alcohol by activated peroxymonosulfate. Small 17(16):2004579. https://doi.org/10.1002/smll.202004579
38. Li M, Wu S, Yang X, Hu J, Peng L, Bai L, Huo Q, Guan J (2017) Highly efficient single atom cobalt catalyst for selective oxidation of alcohols. Appl Catal A 543:61–66. https://doi.org/10.1016/j.apcata.2017.06.018
39. Huang K, Fu H, Shi W, Wang H, Cao Y, Yang G, Peng F, Wang Q, Liu Z, Zhang B, Yu H (2019) Competitive adsorption on single-atom catalysts: mechanistic insights into the aerobic

oxidation of alcohols over CoNC. J Catal 377:283–292. https://doi.org/10.1016/j.jcat.2019.06.047
40. Wei Q, Wang J, Shen W (2021) Atomically dispersed $Fe\delta^{+}$ anchored on nitrogen-rich carbon for enhancing benzyl alcohol oxidation through Mott-Schottky effect. Appl Catal B 292:120195. https://doi.org/10.1016/j.apcatb.2021.120195
41. Xin P, Li J, Xiong Y, Wu X, Dong J, Chen W, Wang Y, Gu L, Luo J, Rong H, Chen C, Peng Q, Wang D, Li Y (2018) Revealing the active species for aerobic alcohol oxidation by using uniform supported palladium catalysts. Angew Chem Int Ed 57(17):4642–4646. https://doi.org/10.1002/anie.201801103
42. Lei L, Liu H, Wu Z, Qin Z, Wang G, Ma J, Luo L, Fan W, Wang J (2019) Aerobic oxidation of alcohols over isolated single Au atoms supported on CeO_2 nanorods: catalysis of interfacial [O–Ov–Ce–O–Au] sites. ACS Appl Nano Mater 2(8):5214–5223. https://doi.org/10.1021/acsanm.9b01091
43. Lucci FR, Liu J, Marcinkowski MD, Yang M, Allard LF, Flytzani-Stephanopoulos M, Sykes ECH (2015) Selective hydrogenation of 1,3-butadiene on platinum–copper alloys at the single-atom limit. Nat Commun 6(1):8550. https://doi.org/10.1038/ncomms9550
44. Wei H, Liu X, Wang A, Zhang L, Qiao B, Yang X, Huang Y, Miao S, Liu J, Zhang T (2014) FeO_x-supported platinum single-atom and pseudo-single-atom catalysts for chemoselective hydrogenation of functionalized nitroarenes. Nat Commun 5(1):5634. https://doi.org/10.1038/ncomms6634
45. Peng Y, Geng Z, Zhao S, Wang L, Li H, Wang X, Zheng X, Zhu J, Li Z, Si R, Zeng J (2018) Pt single atoms embedded in the surface of Ni nanocrystals as highly active catalysts for selective hydrogenation of nitro compounds. Nano Lett 18(6):3785–3791. https://doi.org/10.1021/acs.nanolett.8b01059
46. Wang H, Wang Y, Li Y, Lan X, Ali B, Wang T (2020) Highly efficient hydrogenation of nitroarenes by N-doped carbon-supported cobalt single-atom catalyst in ethanol/water mixed solvent. ACS Appl Mater Interfaces 12(30):34021–34031. https://doi.org/10.1021/acsami.0c06632
47. Fujita S-I, Watanabe H, Katagiri A, Yoshida H, Arai M (2014) Nitrogen and oxygen-doped metal-free carbon catalysts for chemoselective transfer hydrogenation of nitrobenzene, styrene, and 3-nitrostyrene with hydrazine. J Mol Catal A: Chem 393:257–262. https://doi.org/10.1016/j.molcata.2014.06.021
48. He J, Li N, Li Z-G, Zhong M, Fu Z-X, Liu M, Yin J-C, Shen Z, Li W, Zhang J, Chang Z, Bu X-H (2021) Strategic defect engineering of metal-organic frameworks for optimizing the fabrication of single-atom catalysts. Adv Func Mater 31(41):2103597. https://doi.org/10.1002/adfm.202103597
49. Cheong W-C, Yang W, Zhang J, Li Y, Zhao D, Liu S, Wu K, Liu Q, Zhang C, Wang D, Peng Q, Chen C, Li Y (2019) Isolated iron single-atomic site-catalyzed chemoselective transfer hydrogenation of nitroarenes to arylamines. ACS Appl Mater Interfaces 11(37):33819–33824. https://doi.org/10.1021/acsami.9b09125
50. Mitsudome T, Mikami Y, Matoba M, Mizugaki T, Jitsukawa K, Kaneda K (2012) Design of a silver-cerium dioxide core-shell nanocomposite catalyst for chemoselective reduction reactions. Angew Chem Int Ed 51(1):136–139. https://doi.org/10.1002/anie.201106244
51. Takale BS, Wang S, Zhang X, Feng X, Yu X, Jin T, Bao M, Yamamoto Y (2014) Chemoselective reduction of α, β-unsaturated aldehydes using an unsupported nanoporous gold catalyst. Chem Commun 50(92):14401–14404. https://doi.org/10.1039/C4CC07068B
52. Lou Y, Zheng Y, Li X, Ta N, Xu J, Nie Y, Cho K, Liu J (2019) Pocketlike active site of Rh1/MoS2 single-atom catalyst for selective crotonaldehyde hydrogenation. J Am Chem Soc 141(49):19289–19295. https://doi.org/10.1021/jacs.9b06628
53. Lou Y, Zhao Y, Liu H, Gu Q, Yang B, Shi Y, Yao T, Yang B (2021) Edge-confined Pt1/MoS2 single-atom catalyst promoting the selective activation of carbon-oxygen bond. ChemCatChem 13(12):2783–2793. https://doi.org/10.1002/cctc.202100325
54. Hu F, Leng L, Zhang M, Chen W, Yu Y, Wang J, Horton JH, Li Z (2020) Direct synthesis of atomically dispersed palladium atoms supported on graphitic carbon nitride for efficient

selective hydrogenation reactions. ACS Appl Mater Interfaces 12(48):54146–54154. https://doi.org/10.1021/acsami.0c13881
55. Fan Y, Zhuang C, Li S, Wang Y, Zou X, Liu X, Huang W, Zhu G (2021) Efficient single-atom Ni for catalytic transfer hydrogenation of furfural to furfuryl alcohol. J Mater Chem A 9(2):1110–1118. https://doi.org/10.1039/D0TA10838C
56. Tian S, Wang Z, Gong W, Chen W, Feng Q, Xu Q, Chen C, Chen C, Peng Q, Gu L, Zhao H, Hu P, Wang D, Li Y (2018) Temperature-controlled selectivity of hydrogenation and hydrodeoxygenation in the conversion of biomass molecule by the Ru1/mpg-C_3N_4 Catalyst. J Am Chem Soc 140(36):11161–11164. https://doi.org/10.1021/jacs.8b06029
57. Huang F, Deng Y, Chen Y, Cai X, Peng M, Jia Z, Ren P, Xiao D, Wen X, Wang N, Liu H, Ma D (2018) Atomically dispersed Pd on nanodiamond/graphene hybrid for selective hydrogenation of acetylene. J Am Chem Soc 140(41):13142–13146. https://doi.org/10.1021/jacs.8b07476
58. Feng Q, Zhao S, Xu Q, Chen W, Tian S, Wang Y, Yan W, Luo J, Wang D, Li Y (2019) Mesoporous nitrogen-doped carbon-nanosphere-supported isolated single-atom Pd catalyst for highly efficient semihydrogenation of acetylene. Adv Mater 31(36):1901024. https://doi.org/10.1002/adma.201901024
59. Wei S, Li A, Liu J-C, Li Z, Chen W, Gong Y, Zhang Q, Cheong W-C, Wang Y, Zheng L, Xiao H, Chen C, Wang D, Peng Q, Gu L, Han X, Li J, Li Y (2018) Direct observation of noble metal nanoparticles transforming to thermally stable single atoms. Nat Nanotechnol 13(9):856–861. https://doi.org/10.1038/s41565-018-0197-9
60. Segura Y, López N, Pérez-Ramírez J (2007) Origin of the superior hydrogenation selectivity of gold nanoparticles in alkyne + alkene mixtures: triple-versus double-bond activation. J Catal 247(2):383–386. https://doi.org/10.1016/j.jcat.2007.02.019
61. Liu Z, Huang F, Peng M, Chen Y, Cai X, Wang L, Hu Z, Wen X, Wang N, Xiao D, Jiang H, Sun H, Liu H, Ma D (2021) Tuning the selectivity of catalytic nitriles hydrogenation by structure regulation in atomically dispersed Pd catalysts. Nat Commun 12(1):6194. https://doi.org/10.1038/s41467-021-26542-y
62. Kaiser SK, Surin I, Amorós-Pérez A, Büchele S, Krumeich F, Clark AH, Román-Martínez MC, Lillo-Ródenas MA, Pérez-Ramírez J (2021) Design of carbon supports for metal-catalyzed acetylene hydrochlorination. Nat Commun 12(1):4016. https://doi.org/10.1038/s41467-021-24330-2
63. Sun X, Dawson SR, Parmentier TE, Malta G, Davies TE, He Q, Lu L, Morgan DJ, Carthey N, Johnston P, Kondrat SA, Freakley SJ, Kiely CJ, Hutchings GJ (2020) Facile synthesis of precious-metal single-site catalysts using organic solvents. Nat Chem 12(6):560–567. https://doi.org/10.1038/s41557-020-0446-z
64. Hai X, Xi S, Mitchell S, Harrath K, Xu H, Akl DF, Kong D, Li J, Li Z, Sun T, Yang H, Cui Y, Su C, Zhao X, Li J, Pérez-Ramírez J, Lu J (2022) Scalable two-step annealing method for preparing ultra-high-density single-atom catalyst libraries. Nat Nanotechnol 17(2):174–181. https://doi.org/10.1038/s41565-021-01022-y
65. Chen Z, Chen Y, Chao S, Dong X, Chen W, Luo J, Liu C, Wang D, Chen C, Li W, Li J, Li Y (2020) Single-atom AuI–N_3 site for acetylene hydrochlorination reaction. ACS Catal 10(3):1865–1870. https://doi.org/10.1021/acscatal.9b05212
66. Kaiser SK, Fako E, Manzocchi G, Krumeich F, Hauert R, Clark AH, Safonova OV, López N, Pérez-Ramírez J (2020) Nanostructuring unlocks high performance of platinum single-atom catalysts for stable vinyl chloride production. Nat Catal 3(4):376–385. https://doi.org/10.1038/s41929-020-0431-3
67. Li W-H, Yang J, Jing H, Zhang J, Wang Y, Li J, Zhao J, Wang D, Li Y (2021) Creating high regioselectivity by electronic metal-support interaction of a single-atomic-site catalyst. J Am Chem Soc 143(37):15453–15461. https://doi.org/10.1021/jacs.1c08088
68. Zhang J, Wu X, Cheong W-C, Chen W, Lin R, Li J, Zheng L, Yan W, Gu L, Chen C, Peng Q, Wang D, Li Y (2018) Cation vacancy stabilization of single-atomic-site Pt1/Ni$(OH)_x$ catalyst for diboration of alkynes and alkenes. Nat Commun 9(1):1002. https://doi.org/10.1038/s41467-018-03380-z

69. Huang W-Y, Wang G-Q, Li W-H, Li T-T, Ji G-J, Ren S-C, Jiang M, Yan L, Tang H-T, Pan Y-M, Ding Y-J (2020) Porous ligand creates new reaction route: bifunctional single-atom palladium catalyst for selective distannylation of terminal alkynes. Chem 6(9):2300–2313. https://doi.org/10.1016/j.chempr.2020.06.020
70. Chen Y, Ji S, Sun W, Chen W, Dong J, Wen J, Zhang J, Li Z, Zheng L, Chen C, Peng Q, Wang D, Li Y (2018) Discovering partially charged single-atom Pt for enhanced anti-markovnikov alkene hydrosilylation. J Am Chem Soc 140(24):7407–7410. https://doi.org/10.1021/jacs.8b03121
71. Zhu Y, Cao T, Cao C, Luo J, Chen W, Zheng L, Dong J, Zhang J, Han Y, Li Z, Chen C, Peng Q, Wang D, Li Y (2018) One-pot pyrolysis to N-doped graphene with high-density Pt single atomic sites as heterogeneous catalyst for alkene hydrosilylation. ACS Catal 8(11):10004–10011. https://doi.org/10.1021/acscatal.8b02624
72. Liu K, Badamdorj B, Yang F, Janik MJ, Antonietti M (2021) Accelerated anti-markovnikov alkene hydrosilylation with humic-acid-supported electron-deficient platinum single atoms. Angew Chem Int Ed 60(45):24220–24226. https://doi.org/10.1002/anie.202109689
73. Chen Z, Vorobyeva E, Mitchell S, Fako E, Ortuño MA, López N, Collins SM, Midgley PA, Richard S, Vilé G, Pérez-Ramírez J (2018) A heterogeneous single-atom palladium catalyst surpassing homogeneous systems for Suzuki coupling. Nat Nanotechnol 13(8):702–707. https://doi.org/10.1038/s41565-018-0167-2
74. Kim S, Jee S, Choi KM, Shin D-S (2021) Single-atom Pd catalyst anchored on Zr-based metal-organic polyhedra for Suzuki-Miyaura cross coupling reactions in aqueous media. Nano Res 14(2):486–492. https://doi.org/10.1007/s12274-020-2885-7
75. Tao X, Long R, Wu D, Hu Y, Qiu G, Qi Z, Li B, Jiang R, Xiong Y (2020) Anchoring positively charged Pd single atoms in ordered porous ceria to boost catalytic activity and stability in Suzuki coupling reactions. Small 16(43):2001782. https://doi.org/10.1002/smll.202001782
76. Ding G, Hao L, Xu H, Wang L, Chen J, Li T, Tu X, Zhang Q (2020) Atomically dispersed palladium catalyses Suzuki-Miyaura reactions under phosphine-free conditions. Commun Chem 3(1):43. https://doi.org/10.1038/s42004-020-0289-y
77. Lee E-K, Park S-A, Woo H, Hyun Park K, Kang DW, Lim H, Kim Y-T (2017) Platinum single atoms dispersed on carbon nanotubes as reusable catalyst for Suzuki coupling reaction. J Catal 352:388–393. https://doi.org/10.1016/j.jcat.2017.05.005
78. Lang R, Li T, Matsumura D, Miao S, Ren Y, Cui Y-T, Tan Y, Qiao B, Li L, Wang A, Wang X, Zhang T (2016) Hydroformylation of olefins by a rhodium single-atom catalyst with activity comparable to $RhCl(PPh_3)_3$. Angew Chem Int Ed 55(52):16054–16058. https://doi.org/10.1002/anie.201607885
79. Amsler J, Sarma BB, Agostini G, Prieto G, Plessow PN, Studt F (2020) Prospects of heterogeneous hydroformylation with supported single atom catalysts. J Am Chem Soc 142(11):5087–5096. https://doi.org/10.1021/jacs.9b12171
80. Wang L, Zhang W, Wang S, Gao Z, Luo Z, Wang X, Zeng R, Li A, Li H, Wang M, Zheng X, Zhu J, Zhang W, Ma C, Si R, Zeng J (2016) Atomic-level insights in optimizing reaction paths for hydroformylation reaction over Rh/CoO single-atom catalyst. Nat Commun 7(1):14036. https://doi.org/10.1038/ncomms14036
81. Gao P, Liang G, Ru T, Liu X, Qi H, Wang A, Chen F-E (2021) Phosphorus coordinated Rh single-atom sites on nanodiamond as highly regioselective catalyst for hydroformylation of olefins. Nat Commun 12(1):4698. https://doi.org/10.1038/s41467-021-25061-0
82. Zhao K, Wang H, Wang X, Li T, Dai X, Zhang L, Cui X, Shi F (2021) Confinement of atomically dispersed Rh catalysts within porous monophosphine polymers for regioselective hydroformylation of alkenes. J Catal 401:321–330. https://doi.org/10.1016/j.jcat.2021.08.004
83. Zhou Y, Jiang Y, Ji Y, Lang R, Fang Y, Wu C-D (2023) The opportunities and challenges in single-atom catalysis. Chem Cat Chem 15(5):e202201176. https://doi.org/10.1002/cctc.202201176

Chapter 5
Organic Synthesis Catalyzed by Nanoporous Metals

Abstract Nanoporous metal (MNPore) skeleton catalysts have attracted much interest as a new class of heterogeneous catalysts in the field of green and sustainable heterogeneous catalysis. Due to their interesting features such as peculiar three dimensional (3D) bio-continuous network structures, high surface area, non-toxic nature, rather simple recovery, MNPore catalysts have been widely applied as active catalysts in organic molecule transformation. In the present chapter, we summarize the preparation, characterization, and application of MNPore catalysts in organic synthesis based on current research works and studies. This chapter will be discussed in terms of the catalytic performance of nanoporous Au, Pd, Ag, Ru, Ni, Cu, AuPd, PtRu, PtFe, and PtCo/Co_3O_4 with respect to the achievements on catalytic applications in various molecular transformations. Finally, AuNPore functionalized with Ce–TiO_x or Pr–TiO_x catalyst will also be discussed.

Nanoporous metals (MNPore), are amazingly active and selective green heterogeneous catalysts, have three-dimensional bicontinuous structures, controlled feature sizes, low density, high surface area, fine electrical and thermal conductivity, and reduced stiffness. The ease of simple preparation in combination with high reactivity and selectivity, rather simple recovery, reusability and long-term stability suggested that nanoporous metals are a green and environment-friendly heterogeneous catalyst. In contrast to their supported metal nanoparticle counterparts, nanoporous metal catalysts have the unique unsupported nanoporous 3D network structure composed of ligaments, which is helpful to understand the insight of the reaction mechanism and catalytic origins without considering the support-effect in experimental and theoretical studies. Therefore, nanoporous metal catalysts are recently used in many organic synthesis fields.

In this chapter, we primarily describe the heterogeneous catalytic performances of nanoporous Au, Pd, Ag, Ru, Ni, Cu, AuPd, PtRu, PtFe, PtCo/Co_3O_4, and AuNPore functionalized with Ce–TiO_x or Pr–TiO_x catalysts in a variety of organic molecular transformations.

M. Bao et al., *Nanometal Catalysis in Organic Synthesis*, Molecular Catalysis 3,
https://doi.org/10.1007/978-981-97-4573-9_5

5.1 Nanoporous Au Catalyst

As early as 1979, Forty developed a AuNPore catalyst by the selective leaching of Ag from a monolithic AuAg alloy (30% Au and 70% Ag) in 70 wt.% nitric acid at ambient temperature for 18 h [1]. Subsequently, Erlebacher and co-workers proposed a continuum model to demonstrate that nanoporosity in AuNPore catalyst is due to an intrinsic dynamical pattern formation process [2]. In this work, they gave the evolution of nanoporosity in dealloying. In Au–Ag alloys, the more noble Au atoms are chemically driven to aggregate into two-dimensional clusters by a phase separation process at the solid-electrolyte interface, and the surface area continuously increases due to etching.

5.1.1 Oxidation Reactions

As early as in 2006, Bäumer and co-workers demonstrated that AuNPore has an unexpectedly high catalytic activity for CO oxidation at ambient pressures and temperatures down to 20 °C [3]. They found that the AuNPore catalyst exhibited a superiority in thermal stability compared to gold nanoparticles. The morphology of AuNPore can be easily reproduced. Subsequently, Bäumer et al. investigated AuNPore with respect to its morphology, surface composition, and catalytic properties. The results revealed that Ag in the AuNPore played a significant role in facilitating the adsorption and dissociation of molecular oxygen [4, 5]. And, the catalytic activity of AuNPore for CO oxidation will be strongly enhanced when increasing the content of Ag [6].

Basically at the same time as Bäumer, Ding and co-workers reported their research work about AuNPore-catalyzed CO oxidation at low temperature [7]. They employed a modified dealloying method, by etching Ag/Au in concentrated nitric acid in the presence of applied anodic potential, to fabricate AuNPore with ligament sizes less than 6 nm. This catalyst exhibited significant activities for CO oxidation reaction at low temperatures, such as-30 °C (>85% conversion). The author mentioned that the structure coarsening of AuNPore was significantly retarded at lower temperatures, and the pore size of the catalyst increased from the original 5 to 8 nm when the catalytic reaction was conducted at −30 °C for over 30 h. This property offered the AuNPore catalyst a unique advantage of performing CO oxidation reaction at low temperatures with sustained stability. Ding et al. then systematically studied the catalytic properties of AuNPore toward CO oxidation, including the active sites of AuNPore, the reaction kinetics, and activity dependence as functions of space velocity and temperature [8]. The results indicated that the metallic gold on AuNPore is the intrinsic active species. The kinetic study found that the reaction rate of CO oxidation on the AuNPore catalyst depended significantly on CO concentration but only slightly on O_2 concentration. They also suggest that CO adsorption was the rate-limiting step in CO oxidation. In 2008, Ding et al. investigated the catalytic performance of AuNPore for aerobic oxidation of D-glucose to D-gluconic acid [9].

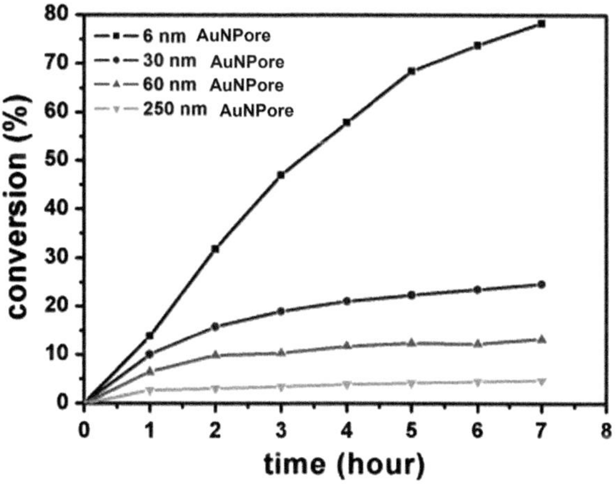

Fig. 5.1 Catalytic activity of AuNPore catalysts with different ligament sizes (6, 30, 60, and 250 nm). Experiments were carried out at 323 K and pH 9 using 20 mg catalyst [9]. Copyright 2008, with permission from ACS Publications

They found that AuNPore catalyst with a ligament size of 6 nm exhibited the highest catalytic activity and was more resistant to deactivation (Fig. 5.1). In addition, they found that the 30 nm sample had better structure stability during the whole catalytic process, while maintaining a decent catalytic activity. And the presence of residual Ag atoms didn't contribute to the activity of AuNPore catalyst for the glucose oxidation reaction. On the contrary, Au atoms on the corners and step edges were suggested to be the active sites.

Besides the ability of AuNPore to catalyze oxidation reactions already at very low temperatures, AuNPore, as a highly selective catalyst, was used for the first time to catalyze the oxidative coupling of methanol to methyl formate in the gas phase [10, 11]. The high selectivity (above 97%) and high turnover frequencies (0.26 s^{-1}) at temperatures below 80 °C were achieved. The experiments result confirmed that Au surface sites were the reactive sites of the coupling reaction. Bäumer and their workers deduced that the selective surface chemistry of Au was unaltered but that O_2 could be readily activated with AuNPore. Residual Ag in AuNPore regulated the availability of reactive oxygen on the surface. When increasing Ag content, the selectivity of methyl formate decreased (partial oxidation), and the selectivity of CO_2 increased (total oxidation) [11].

Wittstock et al. also studied AuNPore-catalyzed gas-phase aerobic oxidation and coupling of primary alcohols (methanol, ethanol, and *n*-butanol) at ambient pressures and low temperatures (Scheme 5.1 and Fig. 5.2) [12]. As shown in Scheme 5.1 and Fig. 5.2, the two predominant products, acetaldehyde and ethyl acetate, were observed for AuNPore-catalyzed ethanol oxidation at reaction temperatures as low as 40 ºC. In this reaction, acetaldehyde and ethyl acetate were produced in a molar ratio of about 2:1, corresponding to an equal conversion of ethanol into two products. The total oxidation of ethanol to CO_2 was found to be close to the detection limit (ca. 10^{-3} vol.%). However, the oxidation of *n*-butanol led exclusively to the product *n*-butanal and the generation of CO_2 was close to the detection limit of about 10^{-3} vol.% (Scheme 5.1c and Fig. 5.2). The pattern of reactivity for oxidation and coupling of alcohols on AuNPore was shown in Scheme 5.2. The three main reaction pathways for the oxidation and coupling of primary alcohols were as follows: (1) coupling to form esters, (2) formation of aldehydes, and (3) complete oxidation (combustion). As shown in Scheme 5.2, one of the main intermediates was the surface-bonded carboxylate, such as formate and acetate. These products stemmed from the oxidation of surface-bonded aldehyde. After reaction, the carboxylate surface species either desorbed with water to form the corresponding carboxylic acid, or were further oxidized to the ketene and eventually CO_2. Therefore, the amount of this carboxylate species increased with increasing chain length and increasing tendency to form a surface-bonded aldehyde (methanol < ethanol < n-butanol).

Kong et al. also studied the gas-phase selective oxidation of ethanol using AuNPore catalyst with Ag content of 26.56% (AuNPore 16 min) and 0.72% (AuNPore 12 h). They examined the influence of Ag content, partial pressure of oxygen and temperature on selectivity. Through adjusting Ag content (Fig. 5.3), temperature and O_2 partial pressure, acetaldehyde, ethyl acetate and acetic acid were obtained respectively. Under optimal conditions, selectivity to acetaldehyde, ethyl acetate and acetic acid reached 90%, 85% and 48% respectively [13].

Chen and co-workers characterized AuNPore using spherical-aberration-corrected transmission electron microscopy (TEM) and environmental TEM in order to obtain the atomic insights about the AuNPore catalyst [14]. They found that the average diameter of both gold ligaments and nanopores is ~30 nm. And the curved surfaces of AuNPore possessed a high density of atomic steps and kinks, which were active for chemical reactions (Fig. 5.4).

$$2\,CH_3OH + O_2 \xrightarrow{\text{AuNPore}} HCOOCH_3 + 2\,H_2O \qquad \text{(a)}$$

$$3\,C_2H_5OH + 3/2\,O_2 \xrightarrow{\text{AuNPore}} C_2H_5O + CH_3COOC_2H_5 + 3\,H_2O \qquad \text{(b)}$$

$$C_4H_9OH + 1/2\,O_2 \xrightarrow{\text{AuNPore}} C_4H_8O + H_2O \qquad \text{(c)}$$

Scheme 5.1 Oxidation and self-coupling of alcohols over AuNPore [12]

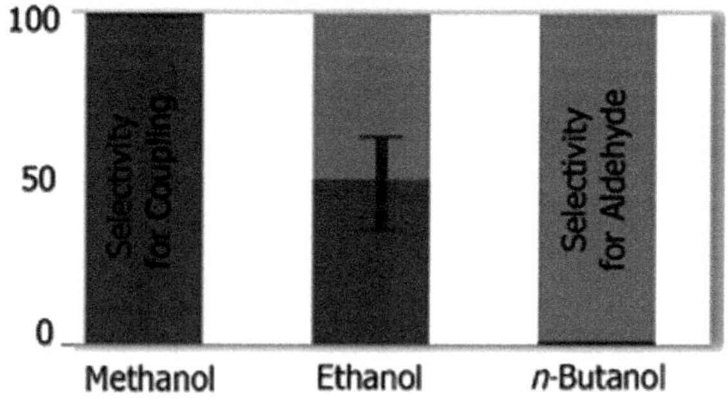

Fig. 5.2 Selectivity towards the aldehyde or self-coupling product (ester) depends on the chain length of the alcohol. The greater facility for β-hydride elimination from longer-chain alkoxy groups favors aldehyde production [12]. Copyright 2012, with permission from John Wiley and Sons

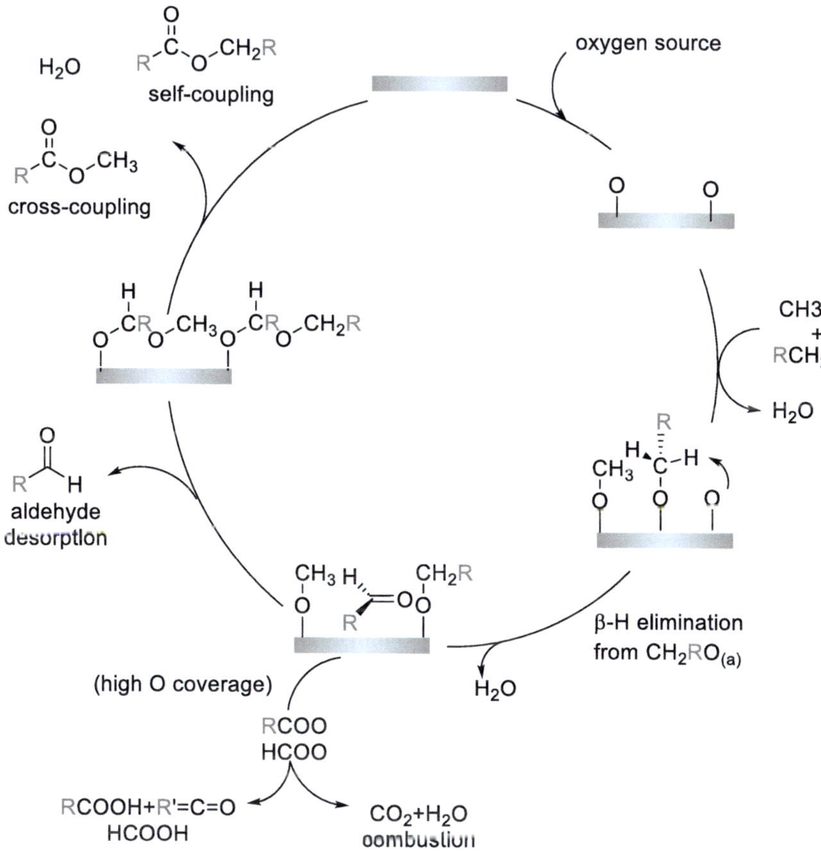

Scheme 5.2 Pattern of reactivity for oxidation and coupling of alcohols on AuNPore. Surface oxygen initiates scission of the alcoholic O–H bond, forming an alkoxy unit. Subsequent β-H elimination from the alkoxy by reaction with surface oxygen or an adjacent alkoxy leads to adsorbed aldehyde which is then attacked by unreacted alkoxy, forming a hemiacetal species which eliminates a hydrogen to yield the ester [12]. Copyright 2012, with permission from John Wiley and Sons

In 2010, Yamamoto and Asao reported AuNPore catalyzed oxidation of organosilanes with water for the first time [15]. The nanoporous gold catalyst was prepared by selective leaching of Ag from an $Au_{30}Ag_{70}$ (in atom%) alloy with thickness of 40 mm in 70 wt.% HNO_3 for 18 h at room temperature [1, 2]. Figure 5.5a showed the microstructure of the dealloyed AuNPore material; the nanopore size was measured to be around 30 nm.

As shown in Table 5.1, AuNPore catalyst exhibited a remarkable catalytic activity in the oxidation of a wide range of organosilanes with water. The corresponding silanols were produced in high yields under mild conditions together with the evolution of hydrogen gas. The catalyst was easily recoverable and could be used at least five times without leaching and loss of activity.

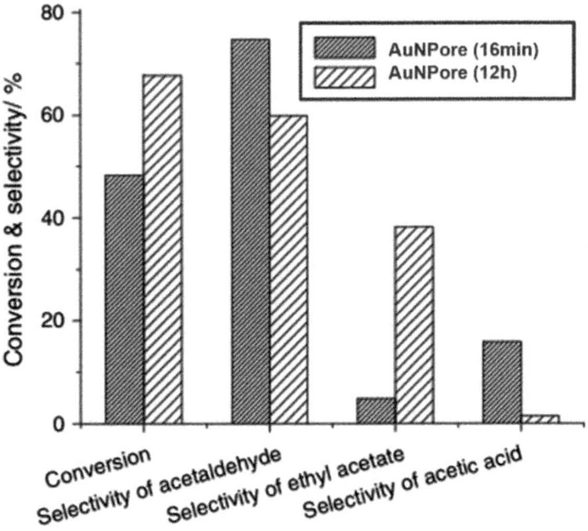

Fig. 5.3 Influence of Ag content on oxidation of ethanol comparing reactions over AuNPore (16 min) and AuNPore (12 h) at 240 °C with total gas flow of 14 cm^3/min (ethanol 4 cm^3/min, $O_2 + N_2$ 10 cm^3/min) [13]. Copyright 2012, with permission from Elsevier

Subsequently, they studied AuNPore-catalyzed aerobic oxidation of alcohols in the liquid phase in batch and flow systems (Table 5.2) [16]. This oxidation reaction didn't need any additives, such as bases, stabilizers, and ligands as well as any cumbersome work-up procedure like filtration or centrifugation. And adsorptions of 1-phenylethanol and O_2 into the AuNPore catalyst were confirmed by thermal desorption spectroscopy (TDS) analysis.

Direct *N*-formylation of amines from methanol and amines still remained difficult since methanol is far less reactive than ordinary alcohols under mild conditions. Kaneda et al. reported that AuNPore-catalyzed oxidative amidation of methanol with the oxidizable amines under mild conditions (Table 5.3) [17]. The AuNPore-1 catalyst containing 3 at.% Ag exhibited excellent catalytic activity for the chemoselective aerobic oxidation of methanol in the coexistence of various oxidizable primary and secondary amines. For example, the reaction of octylamine with MeOH proceeded smoothly in the presence of AuNPore-1, to produce the desired product *N*-octylforamide (**2a**) in 97% yield under an O_2 atmosphere at 60 ºC. Interestingly, the Ag-free AuNPore-2 catalyst prepared from an $Au_{20}Al_{80}$ alloy in 5 wt.% HCl was almost inactive. The AuNPore-4 prepared by dealloying of an $Au_{19}Ag_1Al_{80}$ alloy in 20% NaOH, dramatically increased catalytic activity by adding a small amount of Ag. These results clearly showed that the residual Ag enhanced the catalytic activity of gold through a synergistic effect with gold.

Subsequently, Wittstock and co-workers studied the key parameters for the AuNPore-catalyzed oxidative coupling of alcohols and amines to understand the mechanism [18]. Firstly, Wittstock et al. investigated the oxidative coupling of the two most simple reactants, methanol and dimethylamine, to the corresponding product dimethylformamide (DMF) at oxygen pressures of up to 8 bar. The results showed that the activation of molecular oxygen was a key parameter for this reaction. No catalytic activity was detected in the absence of molecular oxygen in the reaction

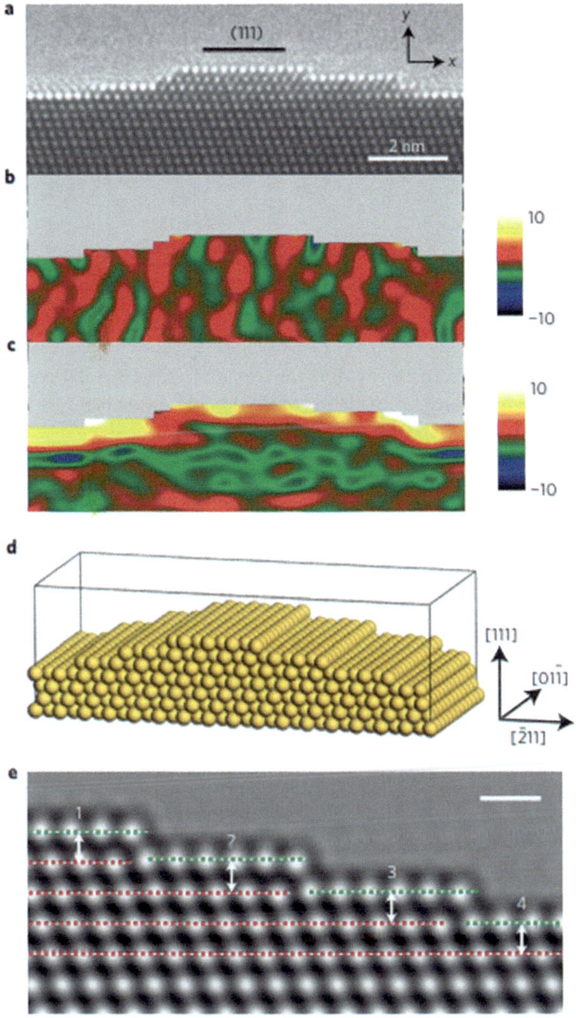

Fig. 5.1 HRTEM characterization and theoretical modelling of AuNPore surface strain. **a** HRTEM micrograph of a convex (111) surface plane viewed along [01ī]. Predefined directions (x and y) were chosen to be parallel and perpendicular to the surface plane, respectively. **b**, **c** Strain maps of εxx (**b**) and εyy (**c**) with a spatial resolution of 0.6 nm obtained by AuNPore. Color represents variation of strain from −10.0 to 10.0%. **d** Structural model with a (111) surface plane, based on **a**, for near-surface strain estimation. **e** Simulated HRTEM image of the surface configuration after structural relaxation to evaluate the surface strain. Average strains relative to the (111) lattice spacing of the bulk sample are calculated for the top four layers (1–4). Dotted lines represent upward displacement. Parameters for the image simulation were as follows: accelerating voltage, 200 kV; spherical aberration coefficient, 1 μm; focus spread, 80 Å; beam convergence, 0.1 Å−1; defocus, 10 nm (overfocus). Scale bar, 5 Å [14]. Copyright 2012, with permission from Springer Nature

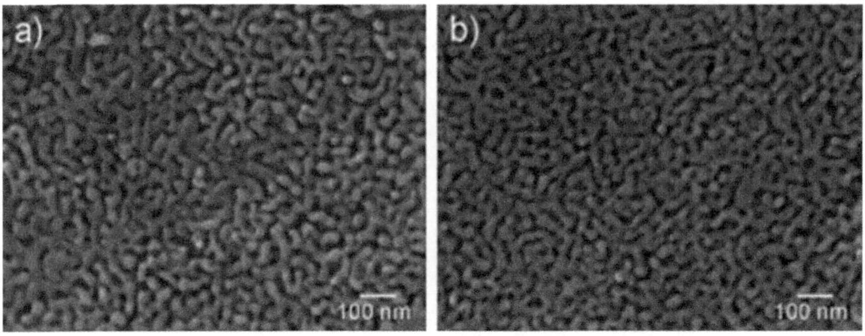

Fig. 5.5 Scanning electron microscopy (SEM) images of dealloyed AuNPore leaf: **a** before reaction, **b** after being used five times for oxidation of $PhMe_2SiH$. Scale bars: 100 nm [15]. Copyright 2010, with permission from John Wiley and Sons

Table 5.1 Scope of the oxidation of organosilanes[a] [16]. Copyright 2010, with permission from John Wiley and Sons

$$\underset{\mathbf{1}}{R_{(4\text{-}n)}SiH_n} + n\,H_2O \xrightarrow[\text{acetone, RT}]{\text{AuNPore cat.}} \underset{\mathbf{2}}{R_{(4\text{-}n)}Si(OH)_n} + n\,H_2$$

Entry	1	$R_{(4\text{-}n)}SiH_n$	AuNPore [mol%]	t [h]	2	Yield [%][b]
1	**1b**	Et_3SiH	1	2	**2b**	94
2	**1c**	Bu_3SiH	3	3	**2c**	95
3	**1d**	iPr_3SiH	3	5	**2d**	88
4	**1e**	Ph_3SiH	1	5	**2e**	99
5	**1f**	Ph_2SiH_2	1	9	**2f**	90
6	**1g**	$PhSiH_3$	5	6	**2g**	80
7	**1h**	$(H_2C{=}CH)MePh\text{-}SiH$	1	1	**2h**	98
8	**1i**	$(PhC{\equiv}C)Me_2SiH$	3	1.5	**2i**	92

[a] Reactions were performed using **1** (1.0 mmol), H_2O (0.1 mL), and AuNPore (n mol %) in 1.5 mL of acetone at room temperature. [b] Yield of isolated product.

chamber (nitrogen, atmospheric pressure). The coupling product DMF was obtained at 40 °C with 100% selectivity after introducing oxygen (≥1 bar). Then, they tuned the surface chemistry of the AuNPore catalyst by adding small amount of metal Ag or Ru, to explore whether such mechanistic insight could be used for generating catalysts with improved catalytic performance (Fig. 5.6). The results indicated that doping an admetal (Ru, Ag) in the AuNPore catalyst increased the activity of the catalyst considerably for this reaction. The conversion to DMF was increased by

Table 5.2 Aerobic oxidation of alcohols with AuNPore catalyst[a,b] [16]. Copyright 2012, with permission from Royal Society of Chemistry

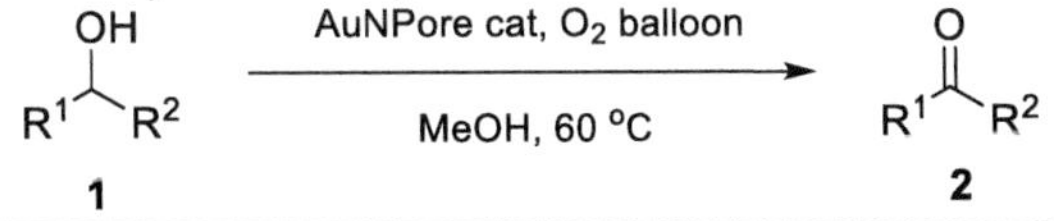

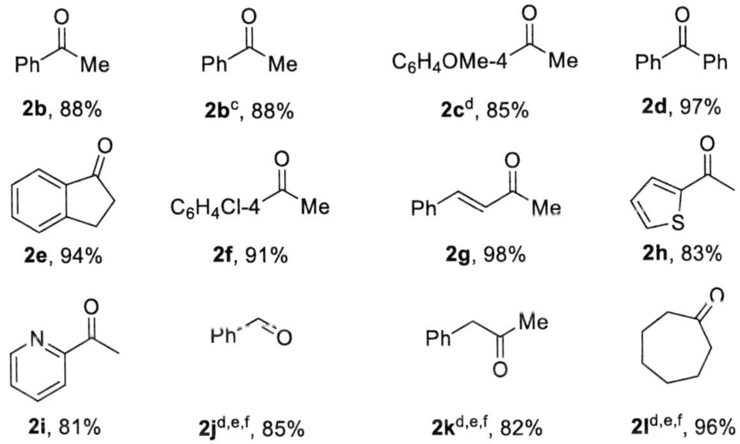

[a] Reactions were carried out with **1** (0.30 mmol) and O_2 balloon in the presence of 10 mol% of AuNPore catalyst at 60 °C unless otherwise noted. [b] Determined by ^{1}H NMR analysis with *p*-xylene as an internal standard. [c] Air balloon was used instead of O_2 balloon. [d] 1,4-Dioxane was used as asolvent. [e] Reaction temperature was 80 °C. [f] 20 mol% of catalyst was used.

approximately 33% (turnover frequency = 52 h^{-1}) by doping the AuNPore with Ag. With the Ru-doped AuNPore catalyst, the activity was increased by a factor of approximately 2.5 (turnover frequency = 97 h^{-1}).

5.1.2 Reduction Reactions

For a long time, AuNPore has not been employed in a heterogeneous catalytic hydrogenation reaction due to the limited ability of H_2 dissociation of gold as compared to its counterpart of the Pd group metals [19, 20]. In 2012, Jin and Yamamoto reported for the first time the highly selective semihydrogenation of alkynes using the unsupported AuNPore and organosilanes/or water as the catalyst and hydrogen source, respectively [21]. They prepared AuNPore catalyst by the selective leaching of Ag from a homogeneous $Au_{30}Ag_{70}$ alloy with thickness of 40 μm in 70 wt.% nitric acid

Table 5.3 AuNPore-catalyzed *N*-formylation of various amines[a] [17]. Copyright 2013, with permission from John Wiley and Sons

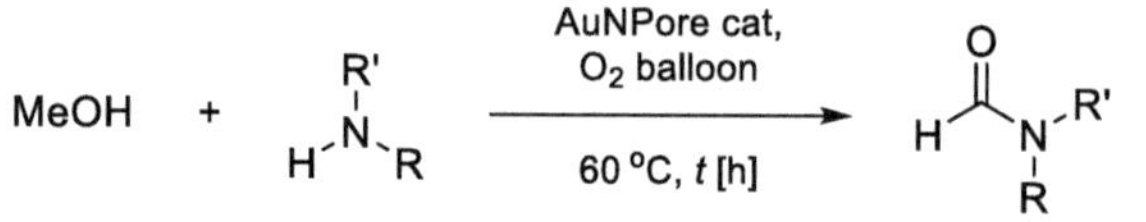

Entry	1		*t* [h]	Product, yield [%][b]
1	**1b**		20	**2b**, 99
2	**1c**	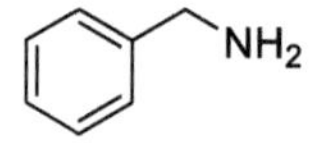	72	**2c**, 67
3	**1d**	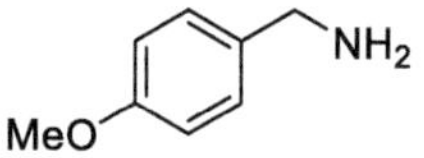	72	**2d**, 77
4	**1e**	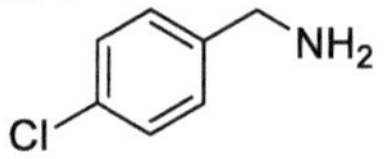	72	**2e**, 72
5	**1f**	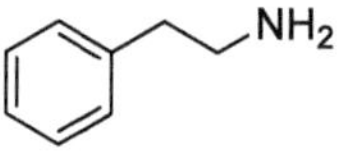	72	**2f**, 70
6	**1g**	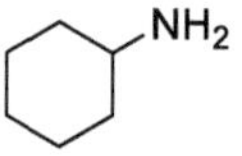	40	**2g**, 73
7	**1h**	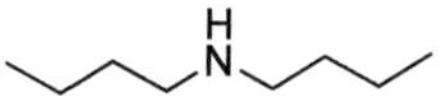	30	**2h**, 95
8	**1i**	Me_2NH	20	**2i**, 40
9	**1j**	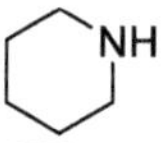	20	**2j**, 98
10	**1k**	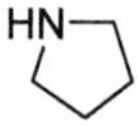	40	**2k**, 93
11	**1l**	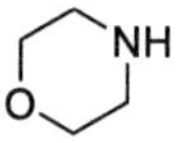	48	**2l**, 91

[a] Reaction conditions: **1** (0.30 mmol), AuNPore-1 (0.03 mmol), MeOH (3.0 mL). [b] Yield was determined by 1H NMR spectroscopy using CH_2Br_2 as an internal standard.

at an ambient temperature for 18 h [1, 2]. The scanning electron microscope (SEM) image showed that AuNPore consisted of a clear 3D network of Au ligaments with an average diameter around 30 nm across the entire nanopore channels (Fig. 5.7a). The transmission scanning microscope (TEM) image clearly revealed the multiple layers of Au ligaments with convex and concave columnar curvatures. The characterization results of high-resolution TEM (HRTEM) and scanning TEM (STEM) mode with

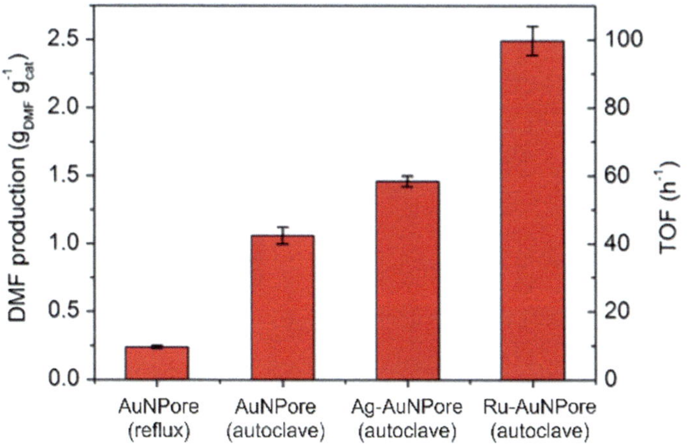

Fig. 5.6 DMF production catalyzed by different npAu samples. Experiments were performed at 40 °C over 4 h with either AuNPore with 50 mL min^{-1} constant oxygen flow or AuNPore in comparison with silver or ruthenium-modified AuNPore in an autoclave (4 bar oxygen, 30 mmol dimethylamine and excess methanol) [18]. Copyright 2015, with permission from John Wiley and Sons

high-angle annular dark-field (HAADF) detector (Fig. 5.7c, d) showed a very high density of atomic steps and kinks at curved surfaces, implying the presence of a high concentration of low-coordination surface atoms.

Initially, in the process of screening of various solvents for the hydrogenation of phenylacetylene (**1a**) with $PhMe_2SiH$ and H_2O, they found that DMF was the only effective solvent for affording styrene (**2a**) in 96% yield (Scheme 5.3a), and other solvents were almost ineffective. The powerful solvent effect prompted them to study the role of DMF in this reaction. The results indicated that amines generated in situ from DMF might play an important role in the semihydrogenation of alkynes, which could suppress the formation of H_2 gas. Therefore, several amines were examined using CH_3CN as solvent instead of DMF. It was found that pyridine was the most effective amine in the semihydrogenation of alkynes **1b**, to produce the corresponding alkene **2b** with 98% yield (Scheme 5.3b). Under the optimized reaction conditions, a wide range of terminal and internal alkynes was reduced to the Z-alkenes in high yields with excellent chemo and stereoselectivity in DMF or pyridine-additive methods. The AuNPore catalyst was easily recovered and reused several times without leaching and any loss of catalytic activity.

Subsequently, Jin and Yamamoto reported AuNPore-catalyzed highly regioselective hydrogenation of quinolines using organosilane and water as a hydrogen source [22]. When using toluene as the reaction solvent, the hydrogenation of quinoline (**1a**) proceeded smoothly to produce the corresponding 1,2,3,4-tetrahydroquinoline (py-THQ) in high yields with excellent regioselectivity (Table 5.4). The deuterium labeling experiments indicated that both $PhMe_2SiH$ and water were the hydrogen source in this reaction. The mechanism study showed that the formation of a H-bond

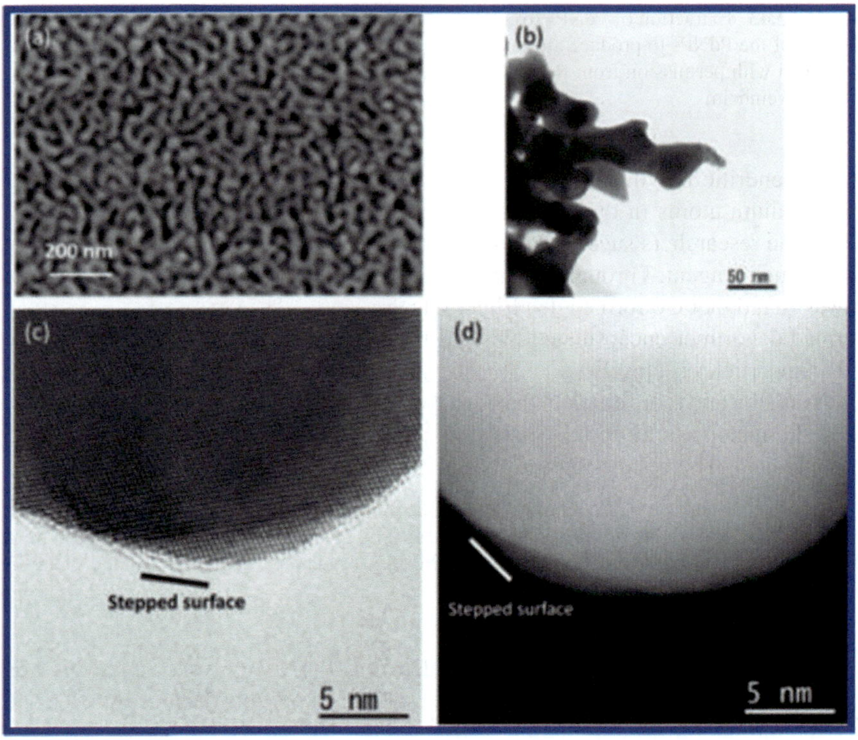

Fig. 5.7 **a** SEM image of AuNPore, **b** representative bright-field TEM image, **c** HRTEM image with stepped surface, and **d** HAADF-STEM image with stepped surface [21]. Copyright 2012, with permission from Springer Nature

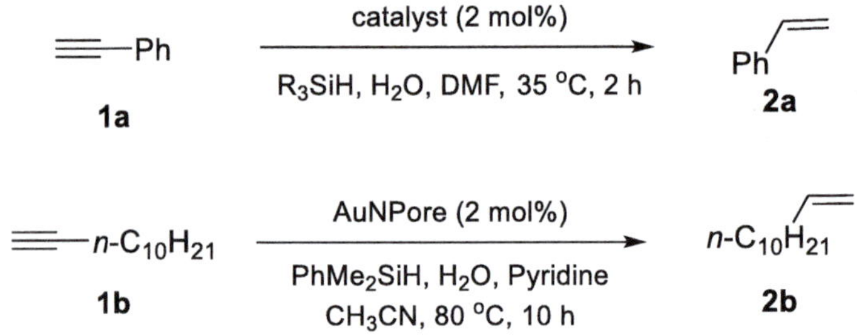

Scheme 5.3 AuNPore-catalyzed semihydrogenation of alkynes. **a** The use of DMF as an amine source. **b** The use of pyridine as an amine source [21]. Copyright 2012, with permission from Springer Nature

between quinoline and water is crucial. The gold in AuNPore catalyst was a real catalytic species and the residual Ag component in AuNPore was inactive. However, the following hydride addition into [H-quinoline]$^+$, such as 1,4- or 1,2-addition, remained unclear at this stage. The AuNPore catalyst can be easily recovered and reused several times without any loss of catalytic activity.

After successfully achieving the selective reduction of alkynes to alkenes and quinolines to tetrahydroquinolines, Bao and Yamamoto et al. shifted their interest to the AuNPore-catalyzed selective reduction reaction of carbon−heteroatom multiple bonds, such as C=O and C=N. They accidentally discovered that the AuNPore/$PhMe_2SiH/H_2O$ system (condition A) had an exclusive chemoselectivity for the reduction of amines in the presence of aldehydes (Table 5.5) [23]. The reactions with $NaBH_4$, $LiAlH_4$, and homogeneous Pd catalyst $Pd(OAc)_2$, under conditions B−D, favored reduction of the C=O bond rather than the C=N bond.

Table 5.4 AuNPore-catalyzed hydrogenation of various quinolines[a] [22]. Copyright 2013, with permission from Springer Nature

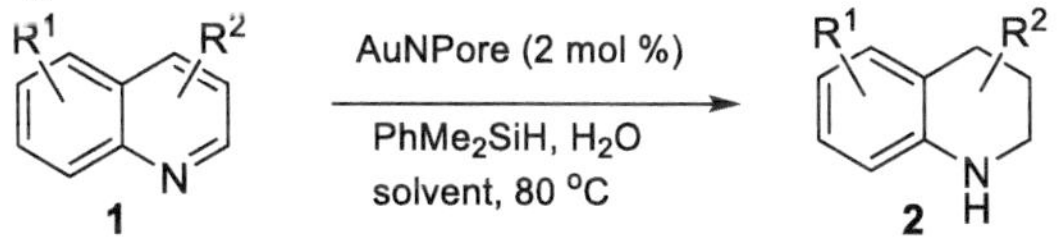

entry	substrate	1	time (h)	2	yield (%)[b]
1	R^1 = H, R^2 = H	**1a**	24	**2a**	86[c]
2	R^1 = H, R^2 = 2-Me	**1b**	7	**2b**	91
3	R^1 = H, R^2 = 3-Me	**1c**	18	**2c**	96
4	R^1 = 8-Me, R^2 = H	**1d**	6	**2d**	83
5	R^1 = 3-Me, R^2 = 6-Me	**1e**	7	**2e**	98
6	R^1 = H, R^2 = 2-*n*-Pr	**1f**	3	**2f**	97
7	R^1 = H, R^2 = 2-Ph	**1g**	10	**2g**	08
8	R^1 = 6-Cl, R^2 = H	**1h**	39	**2h**	97
9	R^1 = 2-Me, R^2 = 3-Me	**1i**	24	**2i**	78[d]
10	R^1 = H, R^2 = 2-CHCHPh	**1j**	18	**2j**	85[e]

[a] Reaction conditions: Quinolines (**1**, 0.5 mmol), $PhMe_2SiH$ (2.0 mmol), H_2O (2.0 mmol), toluene (0.5 mL), AuNPore (2 mol %), 80 °C. [b] Isolated yield. [c] $PhMe_2SiH$ (1.25 mmol) and H_2O (1.5 mmol) were used. [d] A 2.5:1 mixture of *syn* and *anti* isomers. [e] The product is 2-phenethyl-1,2,3,4-tetrahydroquinoline.

Table 5.5 Chemoselectivities in the reduction of imines and aldehydes with representative reducing reagents/catalysts[a] [23]. Copyright 2014, with permission from ACS Publications

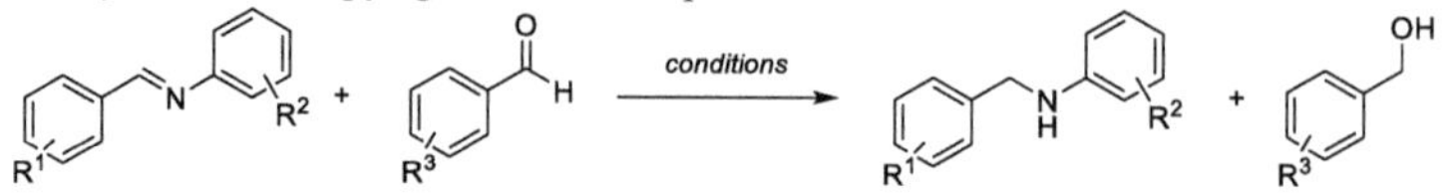

imine **1** aldehyde **2** amine **3** alcohol **4**

conditions:
- A: 5 mol % AuNPore/1.2 equiv $PhMe_2SiH$/1.2 equiv H_2O, acetonitrile, rt, 5 h
- B: 1.2 equiv $NaBH_4$, MeOH, rt, 5 h
- C: 1.2 equiv $LiAlH_4$, diethyl ether, rt, 5 h
- D: 5 mol % $Pd(OAc)_2$, H_2 10 bar, MeOH, rt,5 h

entry	imine **1**	aldehyde **2**		product	
				3 (%)[b]	**4** (%)[c]
1	**1a**,	**2a**,	A	**3a**, 89	**4a**, 0
	$R^1 = H$	$R^3 = H$	B	**3a**, 58[d]	**4a**, 60[d]
	$R^2 = H$		C	**3a**, 34[e]	**4a**, 83[e]
			D	**3a**, 29	**4a**, 93
2	**1b**,	**2a**	A	**3b**, 98	**4a**, 0
	$R^1 = H$		B	**3b**, 91	**4a**, 89
	$R^2 = OMe$		C	**3b**, 63	**4a**, 87
			D	**3b**, trace[f]	**4a**, 72
3	**1c**,	**2a**	A	**3c**, 99	**4a**, 0
	$R^1 = H$		B	**3c**, 85	**4a**, 85
	$R^2 = OMe$		C	**3c**, 80	**4a**, 89
			D	**3c**, 23	**4a**, 73

Subsequently, the authors extended this exclusive imine reduction property of AuNPore to a variety of imines and enimines (Tables 5.6 and 5.7). The reduction of the imines bearing electron-donating or weak electron-withdrawing groups at the aromatic ring proceeded smoothly to produce the desired amines in good to excellent yields, while strong electron-withdrawing groups such as –COOH and –COOMe led to moderate yields of the desired products, respectively (entries 6 and 7, Table 5.6). Encouraged by these results, the authors utilized the AuNPore/$PhMe_2SiH$/H_2O system for the selective 1,2-reduction of enimines to allylamines (Table 5.7). It should be noted that the bulky substitutions at R^2 resulted in lower yields of the desired products (entry 6), and tosyl-protected enimine **7g** also underwent 1,2 reduction uneventfully to produce the corresponding allylamine in good yield (entry 7).

Interestingly, Bao and Yamamoto et al. then investigated the reductive amination of aldehydes and ketone using using AuNPore/$PhMe_2SiH$ (Table 5.8 and Scheme 5.4) [24]. As shown in Table 5.8, not only aliphatic aldehydes but also aromatic and heteroaromatic aldehydes could undergo the reductive amination to produce the corresponding amines in moderate to good yields. In this work, water was not necessarily needed because it was produced in situ during the reaction. And in all cases,

Table 5.6 Reduction of simple imines using AuNPore/$PhMe_2SiH^a$ [23]. Copyright 2014, with permission from ACS Publications

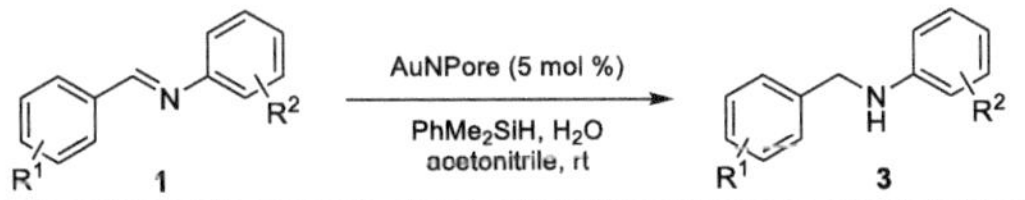

entry	substrate **1**	time (h)	product **3**	yield of **3** (%)[b]
1	**1a**, R^1 = H; R^2 = H	5	**3a**	90
2	**1b**, R^1 = H; R^2 = 4-OMe	3	**3b**	99
3	**1d**, R^1 = 4-Cl; R^2 = H	5	**3d**	92
4	**1e**, R^1 = 4-Br; R^2 = H	4	**3e**	99
5	**1g**, R^1 = H; R^2 = 4-Br	3	**3g**	95
6[c]	**1h**, R^1 = H; R^2 = 4-COOH	7	**3h**	76
7[c]	**1i**, R^1 = H; R^2 = 4-COOMe	7	**3i**	73
8[c]	**1j**, R^1 = 4-OH; R^2 = H	5	**3j**	98
9	**1k**, R^1 = 4-NMe_2; R^2 = H	4	**3k**	98

[a]Reaction conditions: **1** (0.5 mmol), AuNPore (5 mol %), $PhMe_2SiH$ (0.6 mmol), H_2O (0.6 mmol), acetonitrile (2.0 mL), at rt. [b]Isolated yield. [c]Reactions were carried out at 80 °C.

Table 5.7 Allylamines from enimines via reduction using AuNPore/$PhMe_2SiH^a$ [23]. Copyright 2014, with permission from ACS Publications

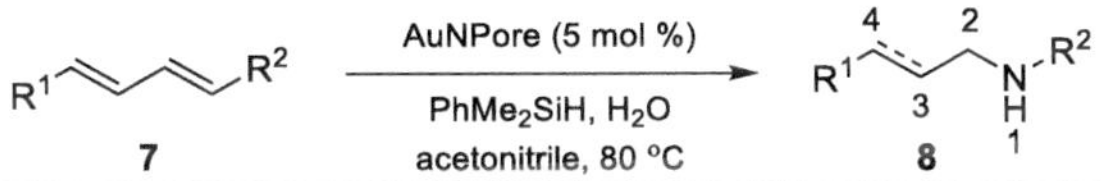

entry	substrate **7**	yield of 8 (%)[b]	selectivity (1,2 reduction/complete reduction)
1	**7a**, R^1 = C_6H_5; R^2 = C_6H_5	75	**8a-1/8a-2 = 90/10**
2	**7b**, R^1 = C_6H_5; R^2 = Me-C_6H_4	85	**8b-1/8b-2 = 77/23**
3	**7c**, R^1 = C_6H_5; R^2 = 3-Me-C_6H_4	83	**8c-1/8c-2 = ~100/-**
4[c]	**7d**, R^1 = C_6H_5; R^2 = 3,4-$(Me)_2$-C_6H_3	75	**8d-1/8d-2 = 89/11**
5[c]	**7e**, R^1 = C_6H_5; R^2 = 4-MeO-C_6H_4	75	**8e-1/8e-2 = ~100/-**
6	**7f**, R^1 = C_6H_5; R^2 = 1-naphthyl	67	**8f-1/8f-2 = 99/1**
7	**7g**, R^1 = C_6H_5; R^2 = Ts	72	**8g-1/8g-2 = 83/17**

[a]Reaction conditions: **7** (0.5 mmol), AuNPore (5 mol %), $PhMe_2SiH$(0.6 mmol), H_2O (0.6 mmol), acetonitrile (2.0 mL), 80 °C, 5 h. [b]Isolated yield. [c]Reaction time was increased to 7 h. [d]Reaction time was increased to 10 h.

alcohols as the products of the reduction of aldehydes were not detected. It is well-known that ketimine formation from ketones is much slower than aldimine formation from aldehydes, so liberation of H_2 from Au surface is much quicker than the reduction of ketimines, which resulted in poor yield in this case. Performing this reaction in two steps effectively solved this problem, and after stirring for 24 h in the first step followed by 5 h in the second step, an overall yield of 62% could be obtained (Scheme 5.4).

Table 5.8 Allylamines from enimines via reduction using AuNPore/$PhMe_2SiH^a$ [24]. Copyright 2015, with permission from Elsevier

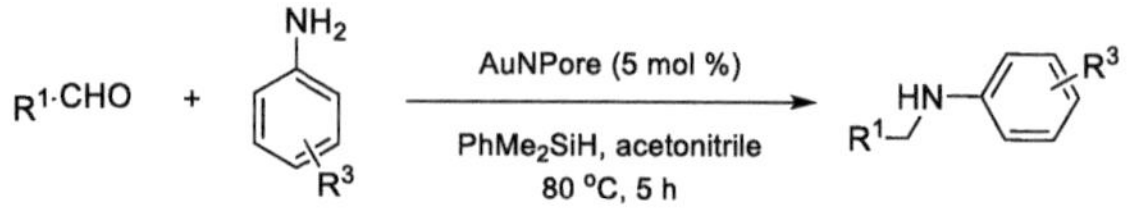

4a-4f **9a-9c** **6a-6j**

Entry	Substrate **4**	Substrate **9**	Product **6**	Yield of **6** (%)[b]
1	**4a**; R^1 = Ph	**9a**; R^3 =H	**6a**	69
2	**4a**; R^1 = Ph	**9b**; R^3 =4-OMe	**6b**	85
3	**4a**; R^1 = Ph	**9c**; R^3 =4-Me	**6c**	79
4	**4b**; R^1 = 4-ClC_6H_4	**9a**; R^3 =H	**6d**	65
5	**4c**; R^1 = 4-BrC_6H_4	**9a**; R^3 =H	**6e**	88
6	**4d**; R^1 = $(Ch_2)_2Ph$	**9a**; R^3 =H	**6f**	79
7	**4d**; R^1 = $(Ch_2)_2Ph$	**9c**; R^3 =4-Me	**6g**	63
8	**4e**; R^1 = *n*-pr	**9a**; R^3 =H	**6h**	69
9	**4e**; R^1 = *n*-pr	**9b**; R^3 =4-OMe	**6i**	65
10	**4f**; R^1 = furyl-2-	**9a**; R^3 =H	**6j**	64

[a] Reaction conditions: a mixture of **4** (0.5 mmol), **9** (0.5 mmol), $PhMe_2SiH$ (0.6 mmol), and acetonitrile (2.0 mL) was stirred in the presence of AuNPore (5 mol %) at 80 °C for 5 h.
[b] Isolated yield.

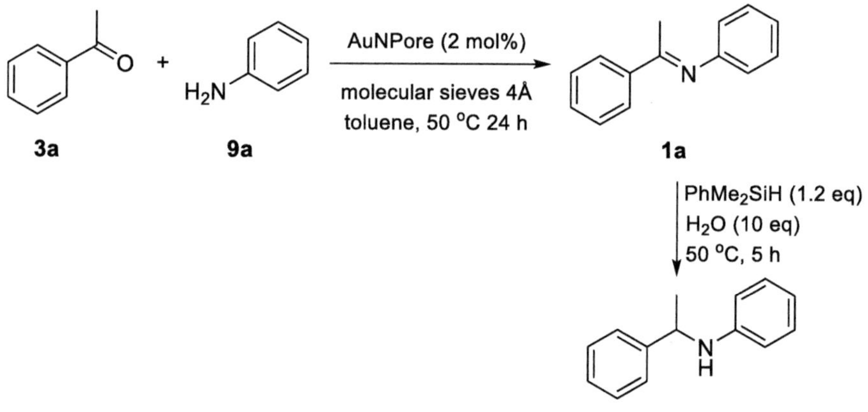

2a; 62%

Scheme 5.4 One pot reductive amination of a ketone using AuNPore/silane [24]. Copyright 2015, with permission from Elsevier

In general, the C=C bond is more easily reduced than the C=O bond by approximately 35 kJ mol^{-1} from a thermodynamic perspective [25], the chemoselective catalytic hydrogenation of α,β-unsaturated aldehydes to allyl alcohols still remained greatly challenging. In 2014, Bao and Yamamoto et al. reported that AuNPore-catalyzed chemoselective reduction of α,β-unsaturated aldehydes to produce the corresponding allylic alcohols using silane and water as a hydrogen source [26]. A variety of α,β-unsaturated aldehydes gave the corresponding allylic alcohols in satisfactory chemical yields with excellent chemoselectivity (Table 5.9). However, aldehydes bearing electron withdrawing substituents such as $-NO_2$ could not furnish the desired product. The deuterium labelling experiment showed that 50% yield of the desired deuterated product with 83% D content as determined by NMR was obtained when treatment of cinnamaldehyde **1a** with 3 equiv. of $PhMe_2SiD$, 6 equiv. of H_2O and 0.5 equiv. of Et_3N (Scheme 5.5).

In 2019, Bao and Jin et al. demonstrated that the unsupported AuNPore was a green and highly efficient heterogeneous catalyst for the reduction of amides to amines using hydrosilanes as a hydrogen source (Tables 5.10 and 5.11) [27]. Under

Table 5.9 Reduction of α,β-unsaturated aldehydes using AuNPore/Et_3SiH^a [26]. Copyright 2014, with permission from Royal Society of Chemistry

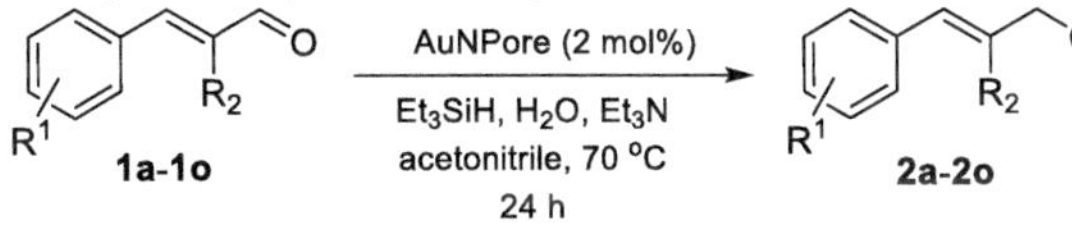

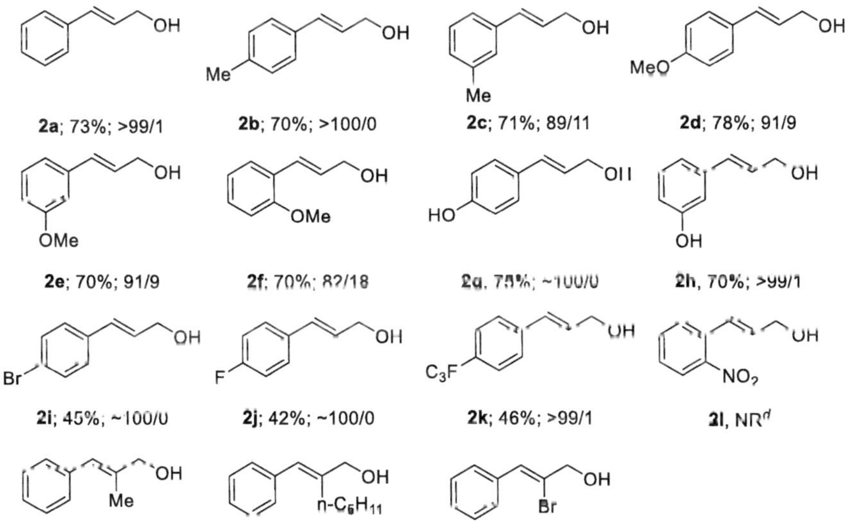

2a; 73%; >99/1
2b; 70%; >100/0
2c; 71%; 89/11
2d; 78%; 91/9
2e; 70%; 91/9
2f; 70%; 82/18
2g; 75%; ~100/0
2h; 70%; >99/1
2i; 45%; ~100/0
2j; 42%; ~100/0
2k; 46%; >99/1
2l, NR[d]
2m; 69%; >99/1
2n; 62%; 97/3
2o, NR

[a] Reaction conditions: a mixture of aldehyde **1** (0.5 mmol), Et_3SiH (1.5 mmol), H_2O (3.0 mmol), Et_3N (0.25 mmol), and solvent (1.5 mL) was stirred in the presence of AuNPore (2 mol%) at 70 °C for 24 h. [b] Isolated yield. [c] % selectivity for unsaturated alcohol/saturated alcohol. [d] NR: no reaction.

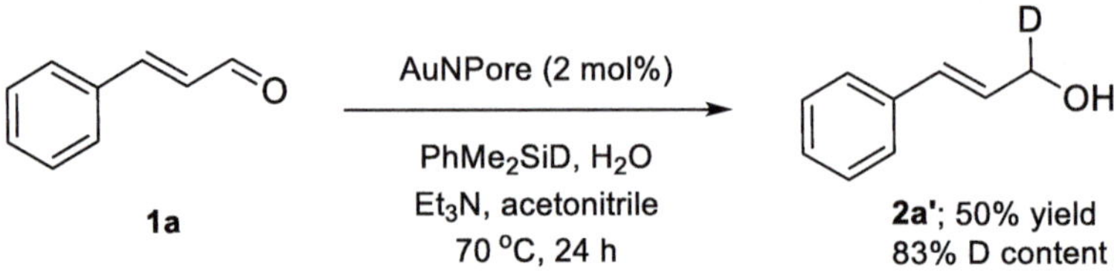

Scheme 5.5 Deuterium labelling in the reduction of cinnamaldehyde [26]. Copyright 2014, with permission from Royal Society of Chemistry

optimized condtions, the reduction of various amides proceeded smoothly in the presence of AuNPore to afford the corresponding amines in 55−94% yields. The substrates with electron-withdrawing groups on the nitrogen atom were unfavorable for this reaction, and the influence of the electronic properties of substituents around the carbonyl group was not obvious. AuNPore catalyst can be simply recovered by filtration and reused for eleven times without any loss of the catalytic activity. The system was also successfully applied for the reduction of sulfoxide and *N*-oxide compounds.

Although researchers have made a great progress in AuNPore-catalyzed reduction reaction in the presence of organosilanes as a hydrogen source, the use of excess amounts of costly organosilanes inevitably resulted in the formation of stoichiometric amounts of siloxane, which rendered this protocol less green. Bio-renewable formic

Table 5.10 Reduction of various amides using $PhMe_2SiH$[a,b] [27]. Copyright 2019, with permission from John Wiley and Sons

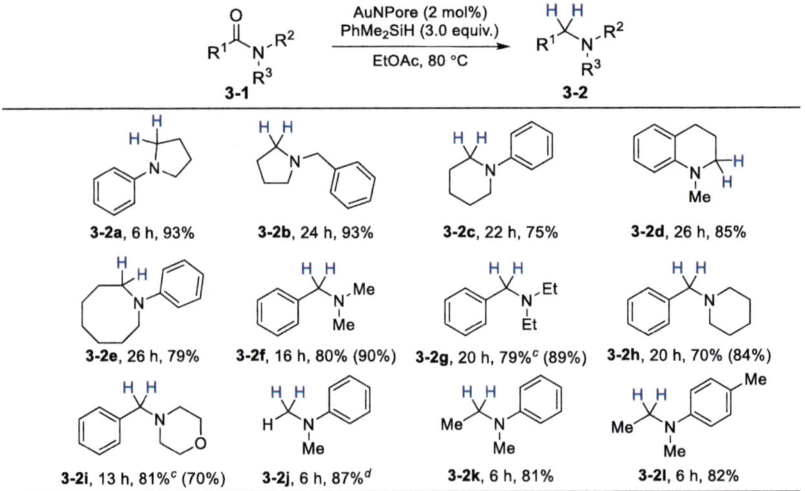

[a]Reaction conditions: amides **3-1** (0.5 mmol), AuNPore (2 mol%), $PhMe_2SiH$ (3.0 equivalents) and EtOAc (0.5 mL) at 80 °C. [b]Isolated yields are shown. The yields obtained by using $(Me_2SiH)_2O$ (1.5 equivalents) as a reductant are shown in parentheses. [c]5 mol% of AuNPore was used. [d]Reaction was conducted at 60 °C.

Table 5.11 Reduction of various amides using $(Me_2SiH)_2O$[a, b] [27]. Copyright 2019, with permission from John Wiley and Sons

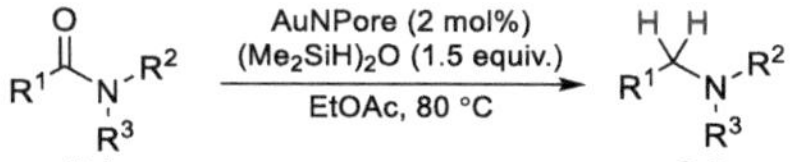

3-1 → 3-2

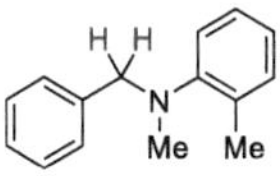

3-2m, 10 h, 92% (18%)

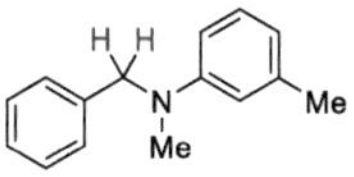

3-2n, 10 h, 90% (25%)

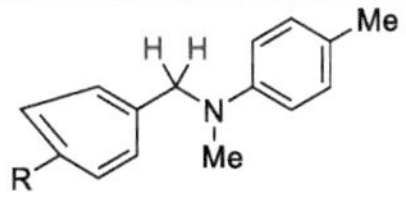

3-2o, R = H, 10 h, 93% (20%)
3-2p, R = Me, 12 h, 90% (30%)
3-2q, R = OMe, 16 h, 89% (40%)
3-2r, R = CO_2Me, 12 h, 72% (32%)

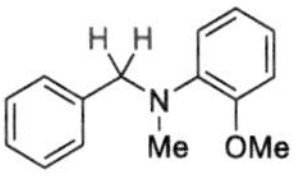

3-2s, 12 h, 82% (23%)

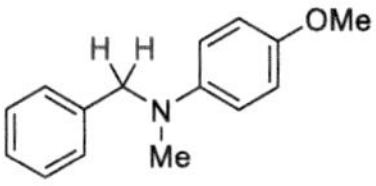

3-2t, 11 h, 67% (9%)

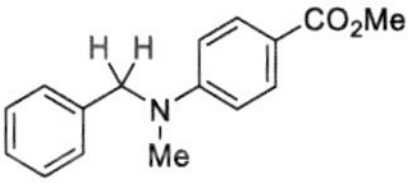

3-2u, 13 h, 55% (0%)

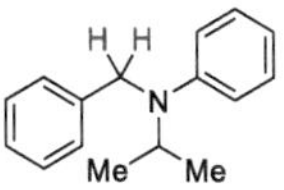

3-2v, 16 h, 84% (9%)

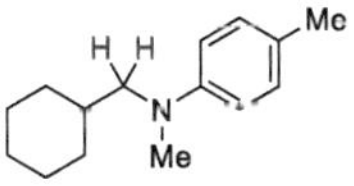

3-2w, 48 h, 83% (14%)

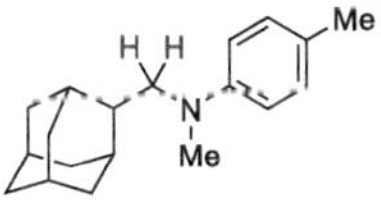

3-2x, 25 h, 94% (0%)

[a]Reaction conditions: amides **3-1** (0.5 mmol), AuNPore (2 mol%), $(Me_2SiH)_2O$ (1.5 equivalents) and EtOAc (0.5 mL) at 80 °C. [b]Isolated yields are shown. The yields obtained by using $PhMe_2SiH$ (3.0 equivalents) as a reductant are shown in parentheses.

acid (HCO_2H) is a safe and convenient hydrogen source that possesses inherent strong hydrogen-donating capabilities [28–30].

Asao et al. reported a AuNPore-catalyzed semihydrogenation of alkynes using HCO_2H as a hydrogen donor for the first time [31]. They demonstrated that HCO_2H was a cost-effective and easily available hydrogen source in the presence of an amine additive for the AuNPore-catalyzed semihydrogenation of alkynes. With this method, a wide range of terminal and internal aromatic/aliphatic alkynes were reduced to the corresponding Z-alkenes in high yields with excellent chemo and stereoselectivity (Scheme 5.6).

Scheme 5.6 AuNPore-catalyzed semihydrogenation of alkynes using formic acid as hydrogen source [31]. Copyright 2015, with permission from ACS Publications

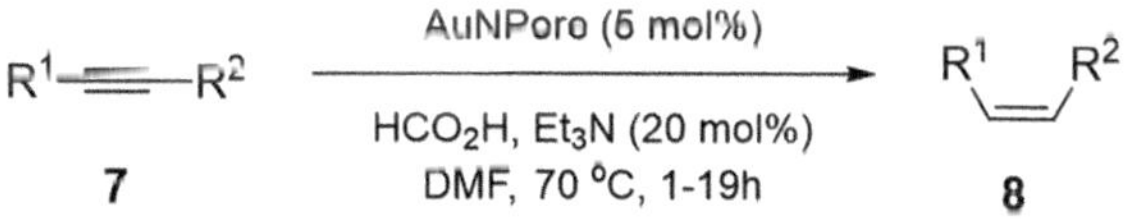

R^1 = Ph, R^2 = H	91%
R^1 = CH_2Ph, R^2 = H	90%
R^1 = CH_2OTHP, R^2 = H	98%
R^1 = CH_2NHTs, R^2 = H	99%
R^1 = Ph, R^2 = Ph	99%
R^1 = n-C_5H_{11}, R^2 = n-C_5H_{11}	99%

In 2017, Bao and Feng et al. reported AuNPore-catalyzed selective reduction of α,β-unsaturated aldehydes using formic acid (HCO_2H) as a hydrogen donor [32]. Under the optimized conditions, various α,β-unsaturated aldehydes were reduced to their corresponding allylic alcohols in 76–98% yields with high selectivity (Table 5.12). This modified protocol exhibits good functional groups tolerance, such as methyl, methoxy, trifluoromethyl, carbonyl, and halides (F). However, no reaction was observed when α,β-unsaturated aldehydes bearing a stronger electron-withdrawing group (nitro, cyano or formyl) were examined; this could be due to their weak coordination ability to the AuNPore catalyst. No reaction was observed when the α-bromosubstituted substrate was tested, and the reason behind this behavior remains unclear. No Au leached from AuNPore during the reaction, and the catalyst was easily recycled and reused without losing the catalytic activity. The chemoselective hydrogenation of α,β-unsaturated aldehydes was scaled up to the gram scale, and the yield was only slightly decreased, specifying that this protocol has good practicability.

Table 5.12 Reduction of α,β-unsaturated aldehydes using AuNPore/HCO_2H[a] [32]. Copyright 2017, with permission from John Wiley and Sons

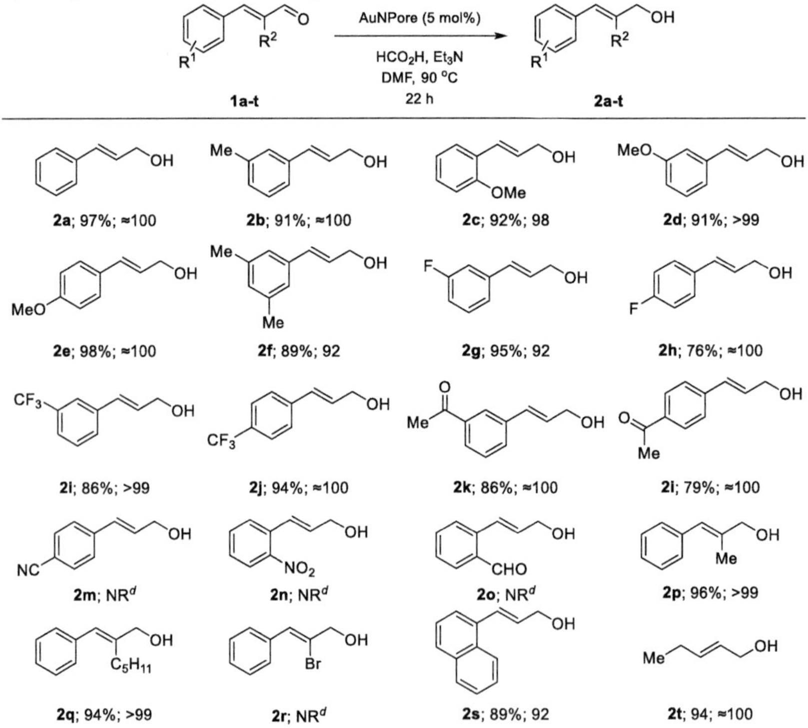

[a]Reaction conditions: cinnamaldehyde (**1**, 0.3 mmol), AuNPore (5 mol%), HCO_2H (0.9 mmol), and Et_3N (1.5 mmol) in DMF (2.0 mL) at 90 °C for 22 h. [b] Isolated yield. [c] The selectivity was determined by ^1H NMR. [d] No reaction; the starting material **1** was recovered.

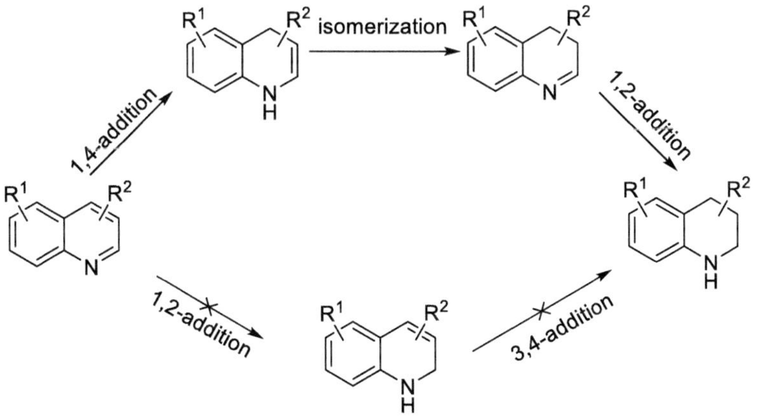

Scheme 5.7 Possible pathways for quinoline reduction [33]. Copyright 2019, with permission from John Wiley and Sons

Subsequently, they studied AuNPore-catalyzed chemoselective reduction of quinolines with formic acid as a hydrogen source [33]. This AuNPore-catalyzed transfer hydrogenation of quinolines also proceeded very smoothly in the presence of low amounts of HCO_2H at a relatively low temperature to produce the desired 1,2,3,4-tetrahydroquinolines (py-THQs) in 74–94% yields. This modified protocol also showed good tolerance for various functional groups, such as methyl, methoxy, halides (F, Cl, and Br), and acyl. Mechanistic studies revealed that the AuNPore-catalyzed transfer hydrogenation of quinolines proceeded via a sequence of 1,4-hydride addition, isomerization, and 1,2-hydride addition (Scheme 5.7). This operationally simple experimental protocol offers a practical and environmentally friendly alternative to the currently known methods for the chemoselective reduction of quinolines.

In the previous work, researchers reported that AuNPore was an effective heterogeneous catalyst for the reduction of various π-bonds using excess amounts of formic acid or costly organosilanes as a hydrogen source. Therefore, the formation of stoichiometric amounts of the byproduct is inevitable. The use of hydrogen gas has well-accepted advantages over H transfer reagents because no waste is generated after the reaction. In 2016, Bao and Yamamoto et al. reported for the first time that AuNPore was an efficient and highly chemoselective catalyst for the hydrogenation of C≡C, C=C, C=N, and C=O bonds under mild conditions (8 atm H_2 pressure, 90 °C) [34]. As described earlier, the residual Ag which was left from the preparation of AuNPore played a promising effect in O_2 activation residual, where they tested $Au_{90}Ag_{10}$NPore and $Au_{>99}Ag_1$NPore by performing two oxidation reactions. And their experimental results agreed with this phenomenon, $Au_{90}Ag_{10}$NPore showed enhanced reactivity compared to $Au_{>99}Ag_1$NPore in oxidation reactions (Scheme 5.8).

On the other hand, they applied two catalysts ($Au_{>99}Ag_1$NPore and $Au_{90}Ag_{10}$NPore) for reduction of aldehydes to check if such Ag effect could be

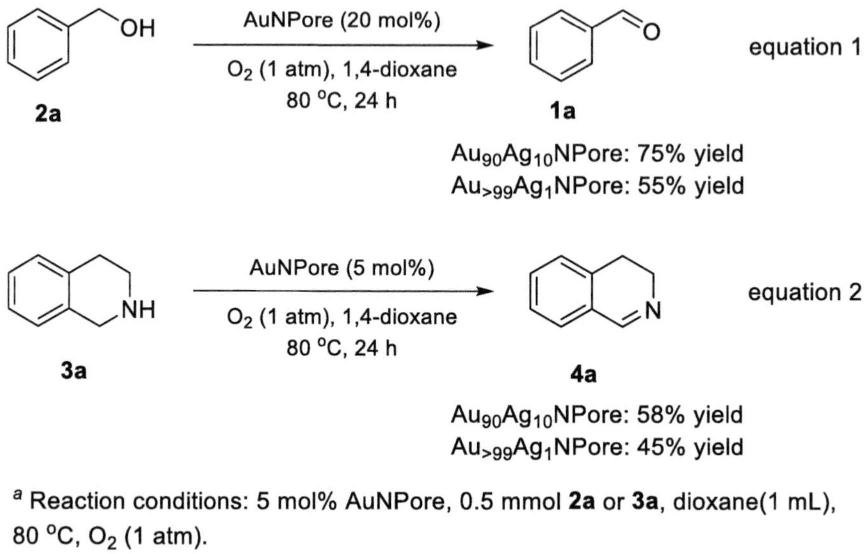

Scheme 5.8 Effect of residual silver on oxidation of benzyl alcohol and tetrahydroisoquinoline[a] [34]. Copyright 2016, with permission from ACS Publications

observed. They found that almost pure nanoporous gold catalyst $Au_{>99}Ag_1$NPore showed much higher catalytic reactivity in hydrogenation than $Au_{90}Ag_{10}$NPore; this means that a lesser amount of residual Ag is more desirable in reduction reactions (Table 5.13).

Then, the wide applicability of $Au_{>99}Ag_1$NPore was examined for hydrogenation of a variety of functional groups (Scheme 5.9). The results showed that $Au_{>99}Ag_1$NPore had a high chemoselectivity toward hydrogenation of terminal alkynes in the presence of internal alkynes which was not achieved using commercially available supported AuNPs. In a similar manner, the $Au_{>99}Ag_1$NPore/H_2

Table 5.13 Residual silver effect on reduction of aldehydes using AuNPore/H_2[a] [34]. Copyright 2016, with permission from ACS Publications

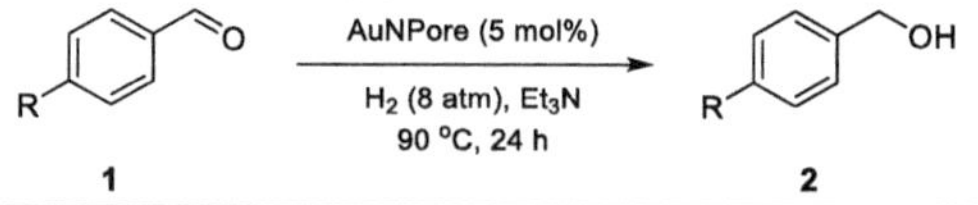

entry	AuNPore	R; 1	Yield[b] (%) of 2
1	$Au_{99}Ag_1$NPore	H; **1a**	99
2		CH_3; **1b**	98
3		Cl; **1c**	96
4	$Au_{90}Ag_{10}$NPore	H; **1a**	51
5		CH_3; **1b**	32
6		Cl; **1c**	73

[a] Reaction conditions: 5 mol% AuNPore, 0.5 mmol **1**, and H_2 (8 atm) in triethylamine at 90 °C for 24 h. [b] Yield determined by NMR using CH_2Br_2, as an internal standard. In the cases where the yields of **2** were low (for example, entries 4–6), **1** was recovered in ca. 30–60% yields.

Scheme 5.9 $Au_{>99}Ag_1$NPore-catalyzed hydrogenation reaction [34]. Copyright 2016, with permission from ACS Publications

system showed a preference for the hydrogenation of aldehydes including unsaturated aldehydes in the presence of imines, to produce the corresponding alcohols in high yields with excellent chemoselectivities. Further, reductive amination, which has great significance in the synthesis of amines due to its atom-economical nature, had also been realized using $Au_{>99}Ag_1$NPore. Apart from this, less reactive functional groups such as nitro, quinoline, ketone, and alkene could also be hydrogenated to the corresponding products in lower yields. High chemoselectivity, easy recovery and high reusability of AuNPore made it a promising heterogeneous catalyst for hydrogenation reactions.

5.1.3 Other Reactions

In 2012, Asao and Yamamoto reported AuNPore-catalyzed [4 + 2] benzannulation between ortho-alkynylbenzaldehydes and alkynes, leading to naphthalene derivatives (Table 5.14) [35]. The AuNPore catalyst exhibited a high catalytic activity in this transformation. They examined the effect of the porous size on the catalytic activity. The results indicated that the catalytic activity of AuNPore catalyst was highly dependent on the pore size, and the size should be less than 30 nm for this reaction. The AuNPore catalyst with 100 nm pore size did not have any catalytic activity for this reaction. The AuNPore catalyst was easily recovered and reused without significant loss in activity.

Asao and co-workers reported the AuNPore-catalyzed selective hydrosilylation of alkynes without any additives, such as bases, stabilizers, and ligands (Scheme 5.10) [36]. Because this hydrosilylation of terminal alkynes could possibly produce three isomeric products, α-, β-(E)-, and β-(Z)-isomers, the control of regio- and stereoselectivity in this reaction was highly desirable for selective preparation of these vinylsilane compounds. Asao et al. fabricated the AuNPore catalyst from Au–Al alloy by dealloying in 20% NaOH. The experimental results indicate that AuNPore was an effective green catalyst for hydrosilylation of terminal alkynes, leading to β-(E)-products with high regio- and stereoselectivities (Table 5.15). They also prepared

Table 5.14 AuNPore-Catalyzed Benzannulation of **1** with **2** [35]. Copyright 2012, with permission from Georg Thieme Verlag KG

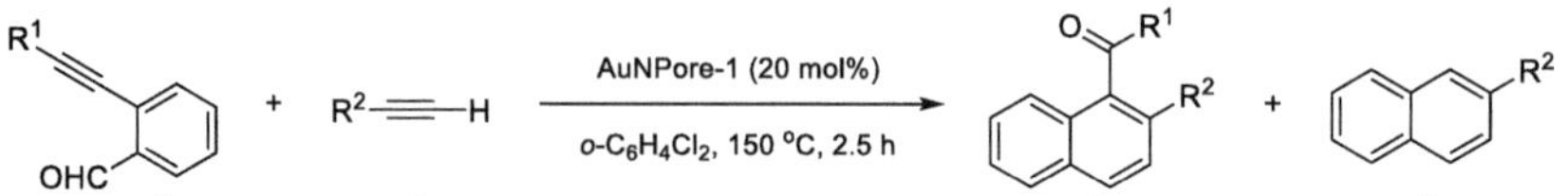

1 **2** **3** **4**

Entry	1	R^1	2	R^2	3[a]	Yield (%)	4[a]	Yield (%)
1	**1b**	4-$F_2CC_6H_4$	**2a**	Ph	**3b**	76	**4a**	8
2	**1c**	4-$MeOC_6H_4$	**2a**	Ph	**3c**	51[b]	**4a**	32
3	**1d**	Bu	**2a**	Ph	**3d**	12[c]	**4a**	47
4	**1a**	Ph	**2b**	Pr	**3e**	68[d]	**4b**	22

[a] Isolated yield.
[b] Ratio between **3c** and its regioisomer is 5:1.
[d] Ratio between **3d** and its regioisomer is 3:1.
[c] Ratio between **3e** and its regioisomer is 6:1.

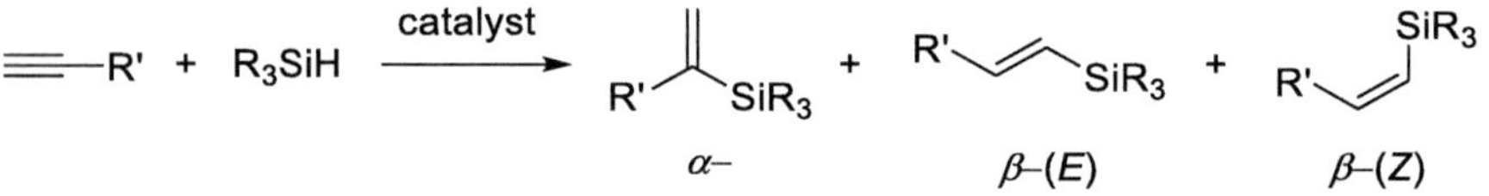

Scheme 5.10 Hydrosilylation of alkynes giving three possible isomeric vinylsilane compounds [36]

other AuNPore catalysts from Au–Ag alloy or Au–Ag–Al alloy. They found that aluminum had no special activation effect on this reaction and the residual metal Ag had the deactivation effect on this reaction.

In 2018, Ding and co-workers disclosed that AuNPore could activate Si–H bonds through a homolytic pathway, forming hydrogen radicals and [Au]–[Si] intermediates. They utilized this new reactivity to achieve three new silylation reactions: controlled monoalcoholysis of dihydrosilanes (Scheme 5.11a), one-pot formation of unsymmetrical silaketals (Scheme 5.11b) and ring-opening bis-silylation of cyclic ethers (Scheme 5.11c) [37]. Specifically, this new activation mode allowed efficient monoalcoholysis of dihydrosilanes in a controlled (only one Si–H is activated) and chemoselective (hydrogen radicals tolerate many functional groups) manner. Then by tuning the reaction conditions, the second Si–H bond activation/alcoholysis could proceed to produce unsymmetrical silaketals in a one-pot reaction. Finally, the [Au]–[Si] intermediates could serve as both as an electrophile (forming Si–O bonds) and nucleophile (forming C–Si bonds), leading to the ring opening/bis-silylation reaction of a range of cyclic ethers.

In this work, the authors proposed a plausible mechanism of this cyclic ether ring opening/bis-silylation reaction (Fig. 5.8). First, the Si–H bond of silane is activated on the surface of AuNPore, to give the intermediate A. and species B was formed

Table 5.15 AuNPore-catalyzed hydrosilylation with a variety of alkynes and hydrosilanes[a] [36]. Copyright 2013, with permission from Royal Society of Chemistry

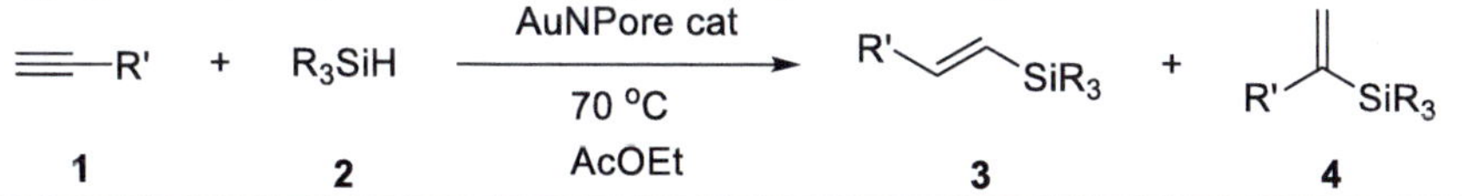

Entry	R'		R_3SiH	Cat (mol%)		Time (h)	Yield[b] (%)	
							3	4
1	C_6H_5	**1a**	Et_3SiH	**2a**	2	3	98	1
2	*p*-FC_6H_4	**1b**	Et_3SiH	**2a**	2	2	97	2
3	*p*-$MeOC_6H_4$	**1c**	Et_3SiH	**2a**	2	6	97	2
4	C_6H_{13}	**1d**	Et_3SiH	**2a**	2	3	99	2
5	$PHCH_2$	**1e**	Et_3SiH	**2a**	2	3	97	1
6	*c*-C_6H_{11}	**1f**	Et_3SiH	**2a**	2	3	97	1
7	C_6H_5	**1a**	$PhMe_2SiH$	**2b**	2	6	96	3
8	C_6H_5	**1a**	$(EtO)_3SiH$	**2c**	5	12	85	3
9	C_6H_5	**1a**	Bu_3SiH	**2d**	10	12	98	2
10	C_6H_5	**1a**	$(i\text{-}Pr)_3SiH$	**2e**	20	18	80	2

[a] Reactions were carried out with **1** (1 mmol) and **2** (1.5 mmol) in the presence of AuNPore catalyst at 70 °C. [b] Determined by ^1H NMR analysis with *p*-xylene as an internal standard.

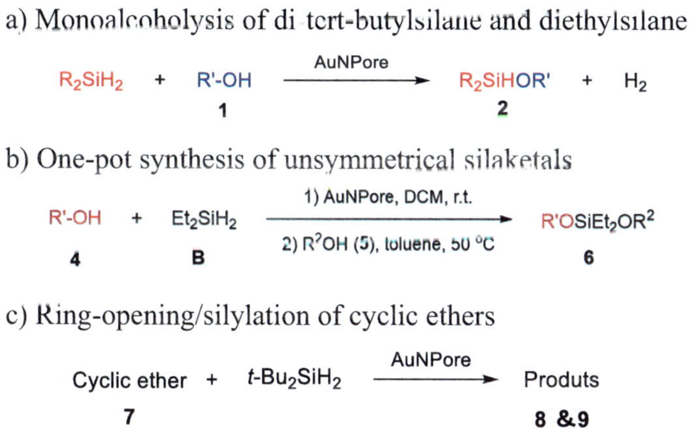

Scheme 5.11 AuNPore-catalyzed three new silylation reactions [37]

from species A after hydrogen gas formation in the reaction. Then a radical type C–O bond cleavage of the cyclic ether was triggered either by the intermediate A or by the intermediate B, leaving the radical on the more substituted carbon (i.e., C). This radical was instantly trapped by the AuNPore surface, thus forming the intermediate D. The intermediate D underwent elimination to give alkene products when it is a highly stable benzyl radical. The intermediate D also could form the hydrosilylation products by a formal hydrogen shift on the surface of A, or form the bis-silylation product through a formal Si shift on the surface of B. This work showed that the surface reactivity of AuNPore may provide an exciting opportunity for new reaction discovery.

In 2013, Jin and co-workers reported for the first time that AuNPore could activate a B–B bond of bis(pinacolato)diboron (B_2pin_2) in direct diboration of alkynes without using any additives (Scheme 5.12) [38]. The diboration of aryl terminal alkynes bearing an electron-donating group (methyl or methoxy group) or an electron-withdrawing group (F) on the benzene ring proceeded smoothly to produce the corresponding 1,2-diborylalkenes in good to high yields with high cis-selectivities. Aryl and aliphatic terminal alkynes as well as internal alkynes could also undergo this reaction, and in each case a small amount of trans-adducts was observed. This was the first demonstration that Au could cleave a B–B bond directly without using any additives.

Although the detailed driving force for the cleavage of a B–B bond on the AuNPore surface remains uncertain, on the basis of the experimental observations, the authors proposed a reaction mechanism (Scheme 5.13). After the adsorption of B_2pin_2 onto the low coordinated Au atoms on the stepped surface of AuNPore, the B–B bond was

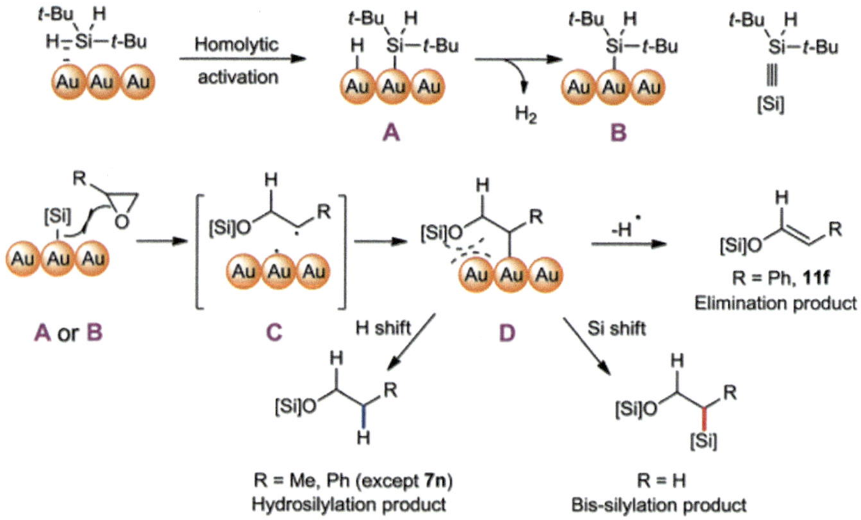

Fig. 5.8 Proposed mechanism of the ring-opening reaction [37]. Copyright 2018, with permission from Royal Society of Chemistry

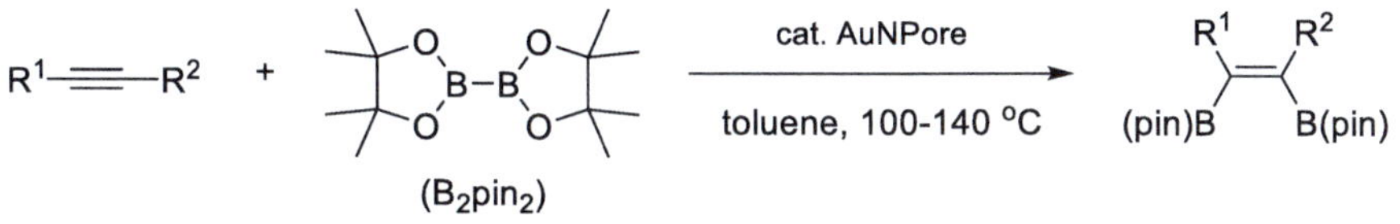

Scheme 5.12 AuNPore-Catalyzed Diboration of Various Alkynes [38]

cleaved to give [Au–Bpin] species under the reaction conditions. Then, the adsorbed alkyne reacted rapidly with two [Au–Bpin] species either through the simultaneous addition path to form the corresponding cis-adduct 3 (path a) or through the stepwise addition (path b). In the case of terminal alkyne (R^2 = H), the trans-adducts 3' would be obtained through the mode B due to the less sterical hindrance between the AuNPore surface and the hydrogen substituent compared to that in the mode A, especially for the more stable vinyl cation (path c).

Subsequently, Jin and co-workers studied the AuNPore-catalyzed diboration of methylenecyclopropanes (MCPs) with $B_2(pin)_2$ through the cleavage of a distal C−C bond of the cyclopropane ring in a heterogeneous manner (Scheme 5.14) [39]. The diaryl-substituted MCPs having electron-rich or electron-poor groups (such as Me, F, and Cl) on the phenyl ring produced the corresponding 1,3-diboron products **2b**−**e** in good to high yields, in which the electronic effects on the catalytic efficiency could be ignored under the method A (2:1 mol ratio of 1 to $B_2(pin)_2$). However, much lower yields of **2b**−**e** were obtained under the method B (1:2 mol ratio of 1 to $B_2(pin)_2$).

The experimental results and computational studies of the mechanism showed that the present diboration proceeds through the formation of a relatively stable trimethylenemethane (TMM) intermediate on the AuNPore surface (Scheme 5.15). The increased negative charges on the trimethylene carbons in the TMM intermediate gave rise to the nucleophilic attack to $B_2(pin)_2$, producing the corresponding 1,3-diboron products selectively.

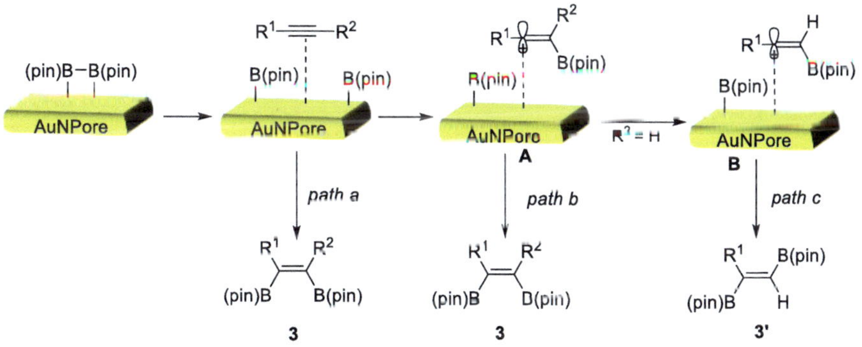

Scheme 5.13 Plausible reaction pathway of the AuNPore-catalyzed diboration of alkynes [38]. Copyright 2013, with permission from ACS Publications

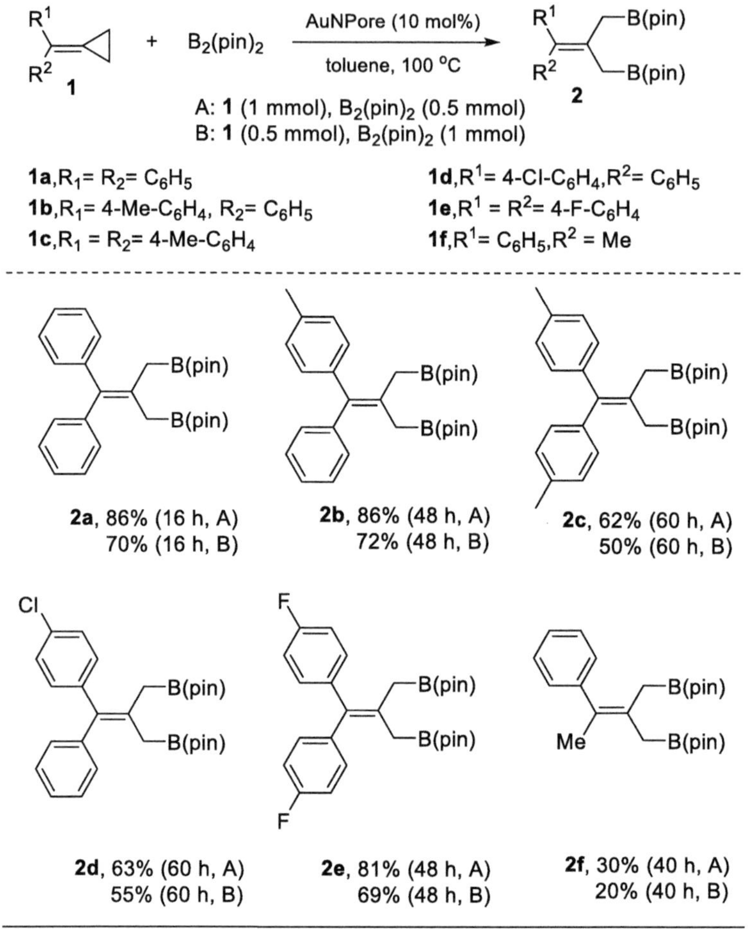

[a]Isolated yields are shown.

Scheme 5.14 Substrate Scope for Diboration of MCPs[a] [39]. Copyright 2018, with permission from ACS Publications

In 2015, Jin and co-workers reported for the first time that AuNPore showed a highly efficient and reusable heterogeneous catalytic activity for the C–H functionalization of tertiary amines with nucleophiles under aerobic conditions, either by using molecular oxygen at 80 °C or tert-butyl hydrogen peroxide (TBHP) at room temperature (Scheme 5.16) [40]. In this study, the α-C(sp3)–H bond of tertiary amines could be cleaved by the AuNPore catalyst under mild conditions in the presence of molecular oxygen or TBHP. A wide range of tertiary amines and nucleophiles underwent the present heterogeneous cross-dehydrogenative coupling (CDC) reaction efficiently to afford the corresponding C–C and C-heteroatom (C–P and C–O) coupling products in good to excellent yields with excellent reusability.

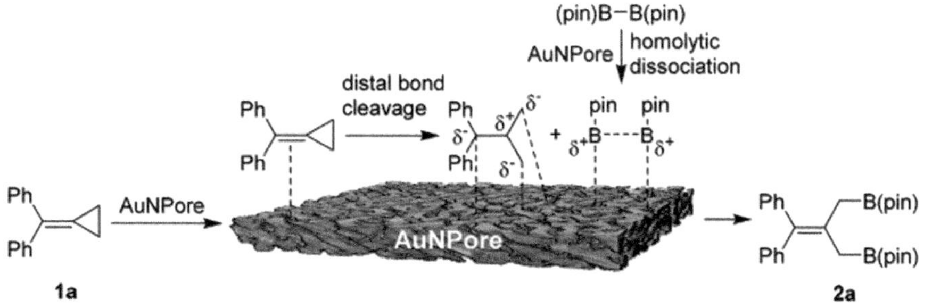

Scheme 5.15 Plausible reaction mechanism [39]. Copyright 2018, with permission from ACS Publications

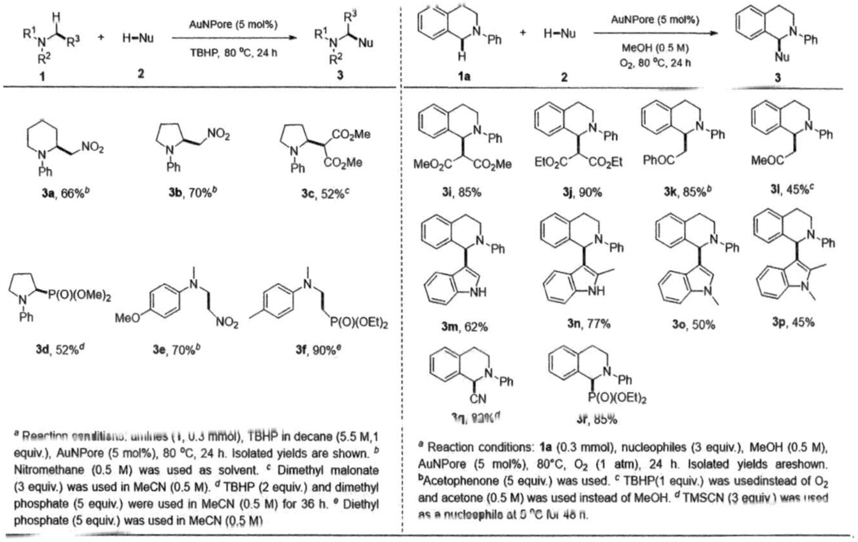

Scheme 5.16 AuNPore-catalyzed CDC reaction [40]. Copyright 2015, with permission from Royal Society of Chemistry

On the basis of the abovementioned experimental observations, the authors proposed a conceivable mechanism (Scheme 5.17). In the O_2-based method, the adsorption of both O_2 and 1,2,3,4-tetrahydroisoquinoline (THIQ) **1a** onto the active sites of AuNPore promoted the cleavage of the α-C–H bond of **1a**, resulting in the formation of an iminium intermediate, which was stabilized by AuNPore or through the interconversion with the methoxy-substituted **4a** in MeOH. In the presence of a strong nucleophile ($MeNO_2$), the iminium intermediate or **4a** underwent the forward reaction to produce the corresponding product **3a**. Similarly, in the TBHP-based method, the *t*-BuO species that dissociated from TBHP on AuNPore initiated the cleavage of the α-C–H bond of **1a** to form the iminium intermediate, which was stabilized by the formation of tert-butylperoxyl-substituted **4b**. Likewise, the iminium

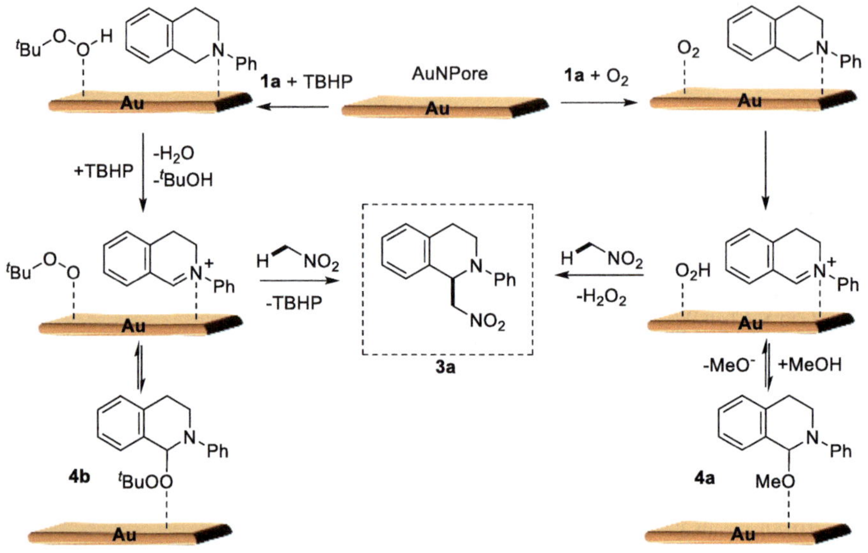

Scheme 5.17 Proposed reaction mechanism [40]. Copyright 2015, with permission from Royal Society of Chemistry

intermediate and **4b** adsorbed on AuNPore further reacted with nitromethane to produce the desired product **3a**.

In 2020, Bao and co-workers reported the AuNPore-catalyzed selective hydrodebromination of aryl bromines at relatively low hydrogen pressure and reaction temperature (Scheme 5.18) [41]. The hydrodebromination of various aromatic and some special aromatic bromides proceeded smoothly in the presence of AuNPore to produce the corresponding products in 59–93% yields. This protocol showed good functional group tolerance, such as acyl, carboxyl and a protected aldehyde, and the selective hydrodebromination reaction occurred exclusively in the coexistence of chlorine atom. Due to these advantages, this AuNPore catalyst could be practically applied for the degradation of bromine-containing herbicide, insecticide, and fire retardant. More importantly, the developed method can also be used as a promising deprotection strategy in organic synthesis. Leaching experiment indicated there was no gold atom leaching during the reaction process and the AuNPore catalyst could be easily recovered and reused several times.

The heterolytic cleavage of the H–H bonds that occurred on the AuNPore surface was verified through the deuterium-hydrogen exchange reaction. Control experiment

Scheme 5.18 AuNPore-catalyzed hydrodebromination of aryl bromides [41]

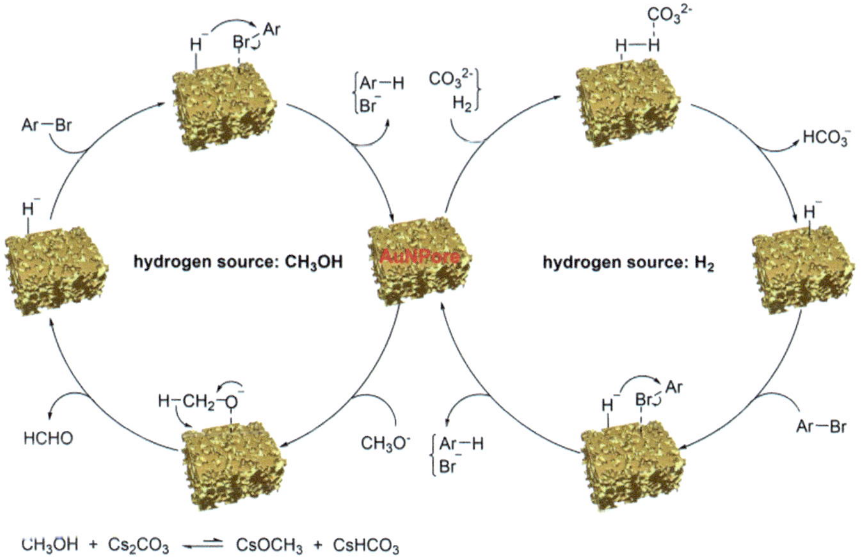

Scheme 5.19 Proposed mechanism for hydrodebromination reaction [41]. Copyright 2020, with permission from John Wiley and Sons

indicated that the solvent MeOH also acted as a hydrogen donor in the hydrodebromination reaction. Therefore, the authors proposed a plausible catalytic cycle. As shown in Scheme 5.19, a H_2 molecule was adsorbed onto the surface of the AuNPore catalyst, and then heterolytic cleavage of the H–H bond occurred through the interaction of carbonate anion (CO_3^{2-}) to produce Au–H species and HCO_3^- (right cycle). The coordination of Ar–Br to the surface of the AuNPore catalyst and the nucleophilic substitution of Ar–Br with Au–H species successively occurred to give the debromination product and Br^-. In addition, a CH_3O^- anion generated in situ from the reaction of CH_3OH with Cs_2CO_3 coordinated to the surface of AuNPore catalyst, and then the Au–H species was produced via the dehydrogenation reaction (left cycle); the formation of formaldehyde was verified by LC–MS determination. The coordination of Ar–Br to surface of AuNPore catalyst and the nucleophilic substitution of Ar–Br with Au–H species successively took place to provide the debromination product and Br^-.

5.2 Nanoporous Pd Catalyst

Chemical dealloying and electrochemical dealloying are the two most advantageous methods for fabricating nanoporous metals with bicontinuous open nanoporosity extending in three dimension. In the dealloying process, it is required that the alloy system must be a monolithic phase because the nanoporosity is made by a self-assembly process through surface diffusion, not by the simple excavation of one phase from a preseparated multiphase system. Therefore, PdNPore had not been prepared for a long time. Until 2008, Chen and co-workers prepared PdNPore catalyst by electrochemical dealloying of a metallic glass $Pd_{30}Ni_{50}P_{20}$ ribbon (about 15 μm thickness) in 1 M H_2SO_4 using a classical three-electrode setup [42, 43]. The nanopore size was measured using SEM to be around of 30–60 nm. Subsequently, Zhang et al. prepared the other PdNPore catalyst through chemical dealloying of a $Pd_{20}Al_{80}$ alloy ribbon in 20% NaOH aqueous solution [44]. The PdNPore prepared by chemical dealloying method also exhibited an open, three-dimensional bicontinuous interpenetrating ligament-channel structure with nanometer length scales, and was used in the in the following hydrogenation reactions.

5.2.1 Cross-Coupling Reactions

Asao and Yamamoto et al. reported that PdNPore prepared by electrochemical dealloying of a glassy metallic alloy $Pd_{30}Ni_{50}P_{20}$ showed high catalytic activity and reusability for the Suzuki coupling of iodoarenes with arylboronic acids [45].

They examined the substrate generality under the reaction conditions (Table 5.16). As shown in Table 5.16, various aryl iodides bearing electron-donating and electron-withdrawing groups on the phenyl ring underwent the coupling reaction to give the corresponding products in excellent yields. The reaction of iodobenzene (**1a**) 4-iodoanisole (**1b**) with *p*-tolylboronic acid proceeded smoothly in the presence of PdNPore and KOH in MeOH at 50 °C to produce the desired coupling product **3a** and **3b** in 99 and 95% yields. The reaction took place smoothly even with sterically hindered 2-iodoanisole (**1d**) (entry 4). Aryl iodides possessing electron-withdrawing groups (entries 6–8) underwent the coupling reaction faster than those having electron-donating groups (entries 2 and 5). Moreover, the PdNPore catalyst was reused four times without a significant loss of catalytic activity. And no leached Pd atoms were detected by inductivity coupled plasma (ICP-AES) analysis (<0.0005%).

The success in the preparation of PdNPore and their application in Suzuki coupling reactions prompted the authors to explore the scope and limitation of PdNPore towards the Heck reaction. Asao and Yamamoto et al. reported that PdNPore prepared from a glassy metallic alloy $Pd_{30}Ni_{50}P_{20}$ exhibited a remarkable catalytic activity for the Heck reaction of various aryl halides with olefins (Scheme 5.20) [46].

Various aryl iodides underwent the Heck reaction with acrylic acid in the presence of PdNPore catalyst using KOH and tetrabutylammonium iodide (TBAI) as additives in MeOH at 80 °C, producing the corresponding products in high yields (Scheme 5.21a). Significantly, the Heck reaction of various arylbromides and terminal alkenes also proceeded smoothly in the presence of PdNPore and NaOAc as catalyst and additive, in *N*,*N*-dimethylacetamide (DMAc) at 140 °C (Scheme 5.21b), and the corresponding products were obtained in good-to-excellent yields. Moreover,

Table 5.16 Scope and limitations of the Suzuki-coupling reaction catalyzed by PdNPore[a] [45]. Copyright 2011, with permission from Royal Society of Chemistry

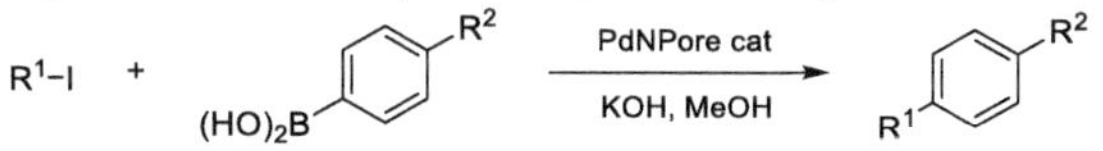

1 **2a**: R^2 = Me **3**

Entry	1	R^1-I	2	Condition	Product % yield[b]
1	**1a**	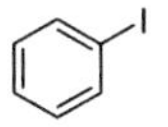	**2a**	50 °C, 3 h	**3a**, 99
2	**1b**	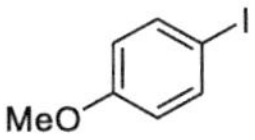	**2a**	50 °C, 4 h	**3b**, 95
3	**1c**	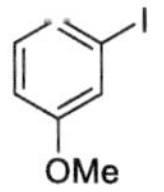	**2a**	50 °C, 5 h	**3c**, 95
4	**1d**	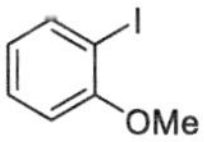	**2a**	50 °C, 6 h	**3d**, 79
5	**1e**	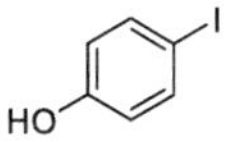	**2a**	80 °C, 12 h	**3e**, 89
6	**1f**	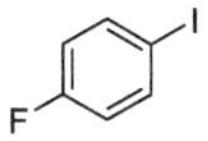	**2a**	50 °C, 3 h	**3f**, 99
7	**1g**	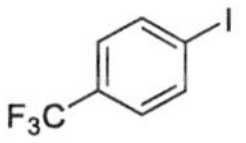	**2a**	50 °C, 0.5 h	**3g**,99
8	**1h**	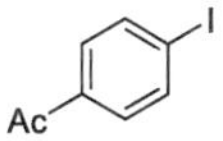	**2a**	50 °C, 2 h	**3h**, 95
9	**1i**	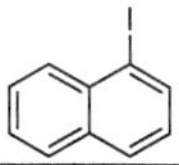	**2a**	50 °C, 3 h	**3i**, 99

[a] Reaction conditions: 1 (0.5 mmol), 2 (0.75 mmol), KOH (2.5 mmol), PdNPore (2 mol%), MeOH (2 mL). [b] Isolated yield. [c] The reaction was conducted in a mixture of MeOH (1.2 mL) and H_2O (0.6 mL). [d] The reaction was conducted in the absence of KOH.

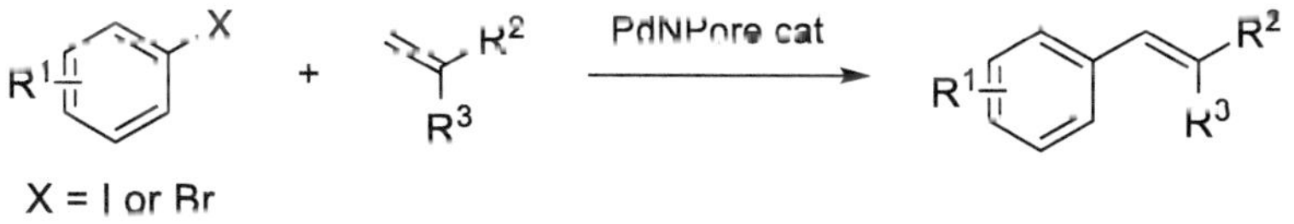

Scheme 5.20 PdNPore-catalyzed Heck reaction [46]. Copyright 2011, with permission from John Wiley and Sons

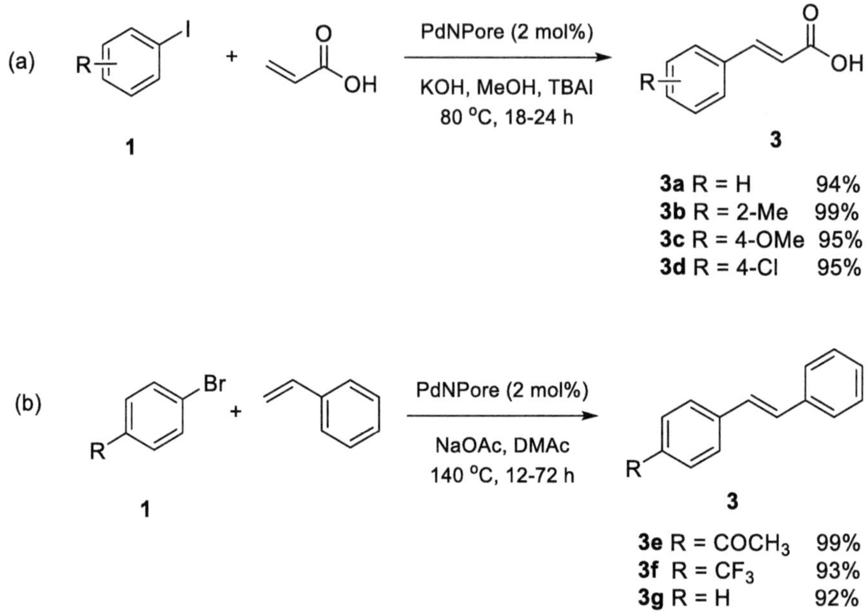

Scheme 5.21 PdNPore-catalyzed Heck reaction of **a** aryl iodides and **b** aryl bromides

the PdNPore was reused several times without a significant loss of catalytic activity, and the PdNPore exhibited a high resistance to leaching than Pd/C and Pd black.

Subsequently, Zhang and Ding prepared the PdNPore catalyst by selective leaching of aluminium from an $Al_{77}Pd_{23}$ (at.%) alloy foil (chemical dealloying) using 10 wt.% NaOH solution for 6 h at room temperature [47]. The ligaments and pores were formed uniformly across the entire PdNPore and their diameter was about 5–10 nm (Fig. 5.9). Compared with the size of PdNPore (30–60 nm) made by electrochemical dealloying of $Pd_{30}Ni_{50}P_{20}$ which has been used in Suzuki and Heck reactions [43, 44], the size of the nanopores obtained here was much smaller, which indicated that they might possess higher catalytic activity.

The authors developed a one-pot synthesis of aryl–aryl products by combining organosilane oxidation with ROH (water or alcohols) and silicon-based cross-coupling reaction in the presence of PdNPore catalyst (Scheme 5.22). The reactions were completed under mild conditions and the PdNPore catalyst could be reused several times without evident loss of its catalytic activity.

5.2.2 Reduction Reactions

The use of H_2 gas as hydrogen source had well-accepted advantages over H transfer reagents (such as organosilanes and formic acid, etc.) because no waste was

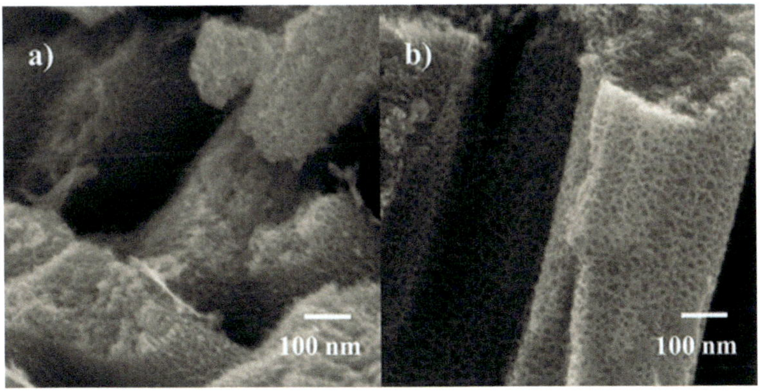

Fig. 5.9 SEM images of dealloyed PdNPore **a** before catalytic reaction and **b** after being used five times [47]. Copyright 2014, with permission from Royal Society of Chemistry

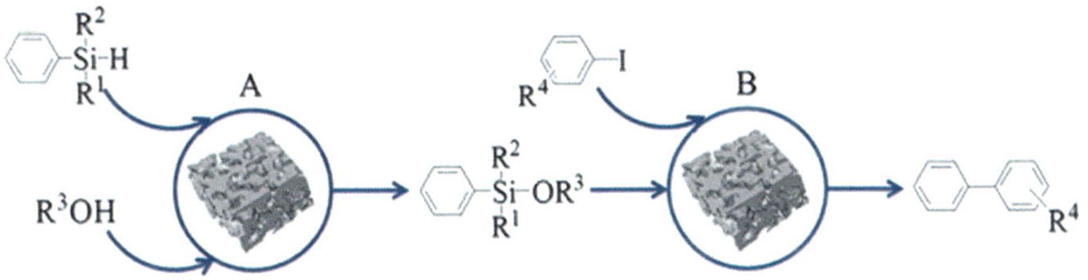

Scheme 5.22 Schematic representation of silicon-based cross-coupling reaction. Cycle A: oxidation of organosilanes. Cycle B: C–C cross-coupling reaction [47]. Copyright 2014, with permission from Royal Society of Chemistry

generated after the hydrogenation reaction. In 2017, Bao and co-workers reported the highly chemoselective and stereoselective semihydrogenation of alkynes to Z-internal and terminal alkenes in the presence of PdNPore and hydrogen gas as catalyst and hydrogen source under mild conditions (room temperature and 1 atm of H_2, Scheme 5.23) [48]. In this work, PdNPore catalyst was prepared by the chemical dealloying method described previously [42]. The nanopore size was determined to be 3.6 nm from the isotherm adsorption branch based on DFT. The EDX analysis exhibited that the composition of this PdNPore was $Pd_{88}Al_{12}$. The XPS analysis indicated that Pd was electronically more positive in PdNPore compared to metallic Pd, indicating that the electron transfer took place from Pd to Al oxide derived from the residual Al.

Under optimized conditions, the semihydrogenation of various terminal/internal and aromatic/aliphatic alkynes afforded the corresponding alkenes in 70–95% yields with high selectivities. This protocol showed good functional-group tolerance, and various synthetically useful functional groups, such as methyl, methoxyl, *tert*-butyl, trifluoromethyl, formyl, carboxyl and halide (F, Cl and Br) groups that remained intact

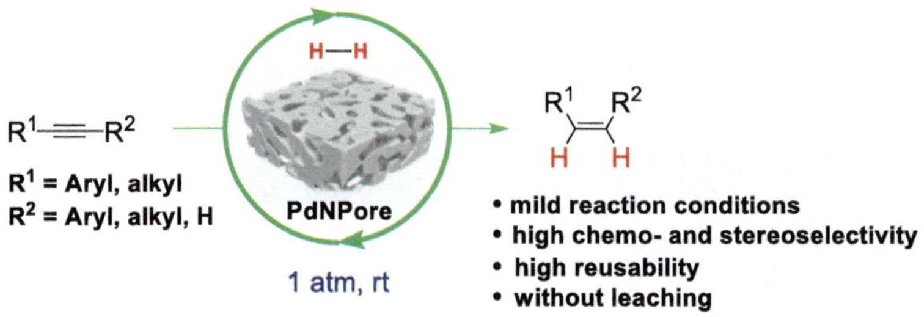

Scheme 5.23 Schematic representation of PdNPore-catalyzed selective hydrogenation of alkynes [48]. Copyright 2017, with permission from ACS Publications

during the semihydrogenation of alkynes. PdNPore further showed high chemoselectivity toward terminal alkynes in the presence of internal alkynes. No Palladium leached from PdNPore during the reaction, and the catalyst was easily recovered and reused without a loss of activity. For comparison, the use of Lindlar Pd and Pd/C as catalysts under standard conditions led to poor chemoselectivity, comparable amounts of over-reduced products, and large amounts of palladium leaching.

A deuterium-labelling experiment was conducted to understand the excellent selectivity of PdNPore in the hydrogenation of alkynes (Scheme 5.24). The semihydrogenation of **1k** using D_2 gas proceeded smoothly to produce a mixture of deuterium-labeled products **2k-d**$_\alpha$ and **2k-d**$_\beta$ (1:1) in 30% yield under standard conditions, while double deuterated alkene **2k-d**$_\gamma$ was not formed at all. This result indicated that a heterolytic cleavage of the H–H bond might occur through the interaction of hydroxide ion (OH^-) to one of H atoms to produce Pd–H species. The excellent chemoselectivity between alkynes and alkenes might be attributed to the weak nucleophilic characteristics of the Pd–H species.

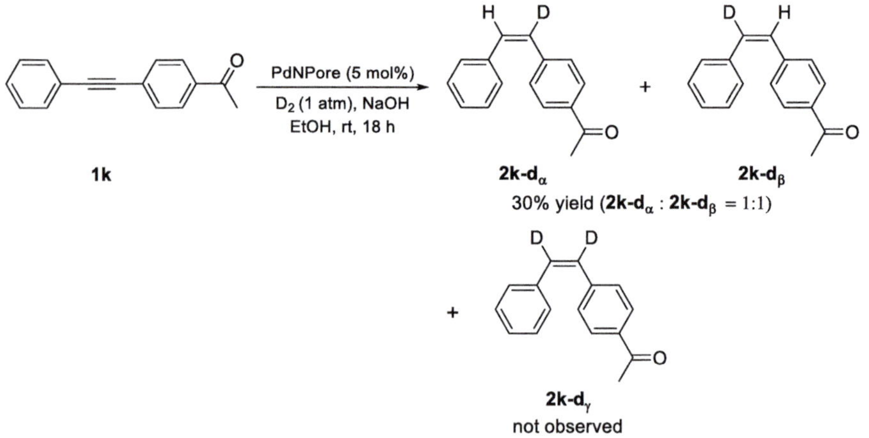

Scheme 5.24 Deuterium labelling experiment using D_2 gas for the semihydrogenation of alkyne **1k** [48]. Copyright 2017, with permission from ACS Publications

Scheme 5.25 PdNPore-catalyzed hydrogenation of nitrogen-containing heteroaromatic compounds [49]. Copyright 2018, with permission from Elsevier

Subsequently, Bao and co-workers studied PdNPore-catalyzed chemoselective hydrogenation of quinolones (Scheme 5.25) [49]. Under optimized conditions, the PdNPore-catalyzed chemoselective hydrogenation of quinoline took place smoothly under mild reaction conditions (low temperature and H_2 pressure) to produce 1,2,3,4-tetrahydroquinolines (py-THQs) in 62–95% yields. This protocol showed good functional-group tolerance, and various synthetically useful functional groups, such as methyl, methoxyl, hydroxyl, ester, formyl, amide and halide (F, Cl and Br) groups which remained intact during the hydrogenation of quinolones.

The chemoselective hydrogenation of quinoline was also scaled up to the gram scale, the yield was only slightly decreased, indicating that this method had good practicability. The result of the kinetic experiment indicated that CH_3CN solvent, ethylamine and py-THQ produced during the reaction, would act as polar and Lewis-basic additives in the hydrogenation of quinolines. Deuterium-hydrogen exchange reaction between D_2O and H_2 was conducted under standard conditions (Scheme 5.26). As shown in Scheme 5.26, the generation of D_2 gas was observed by gas chromatography-mass spectrometry (GC–MS); it was considered that D_2 was produced through the formation of DH. This result indicated that the present hydrogenation involved heterolytic H_2 splitting on the surface of PdNPore to produce Pd–H species, and the in situ-generated Pd–H species played an important role in the high chemoselectivity achieved with the PdNPore catalyst.

In 2019, Li and co-workers reported the PdNPore-catalyzed reduction of aromatic nitro compounds to prepare aromatic amines with silane/alcohol system as the hydrogen source (Scheme 5.27) [50]. In this process, alcohol was used as the oxidant for hydrosilanes and the solvent for the system. $PhMe_2SiH$ and MeOH were found to be the suitable hydrogen source combination. PdNPore catalyst was fabricated by

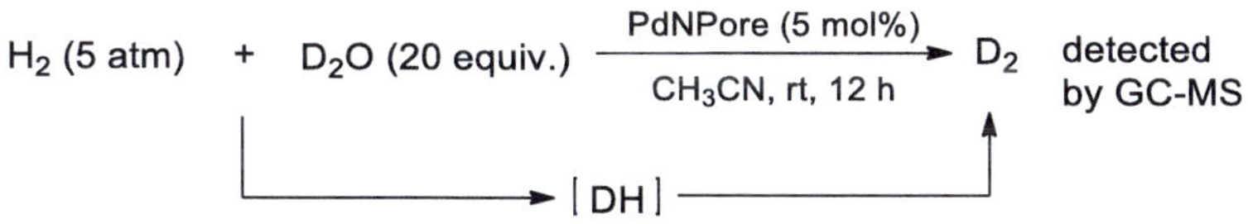

Scheme 5.26 Deuterium-hydrogen exchange reaction [49]. Copyright 2018, with permission from Elsevier

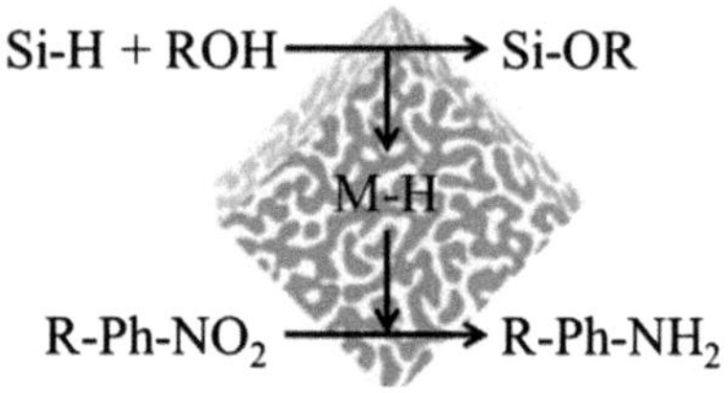

Scheme 5.27 PdNPore-catalyzed reduction of aromatic nitro compounds to aromatic amines [50]

chemical dealloying of aluminum-palladium alloy ($Al_{75}Pd_{25}$, at.%) in 2.5 M NaOH at room temperature for 24 h.

Under the optimized conditions, a series of aromatic nitro compounds underwent the reduction using $PhMe_2SiH$ and MeOH as hydrogen sources. Aromatic nitro compounds bearing electron-withdrawing group or electron-donating group could be reduced with high yields and selectivities. Notably, the most reduction of nitro was focused on aromatic nitro compounds, and the reduction of aliphatic nitro compounds was rarely studied due to its low activity in the reduction. But aliphatic nitro compound nitrocyclohexane was successfully reduced under the PdNPore/hydrosilane/alcohol system to produce the corresponding products in good yields (85%).

The two accepted routes (direct route and condensation route) for the reduction of aromatic nitro compounds were proposed (Scheme 5.28). As shown in Scheme 5.28, nitrobenzene and aniline were detected by gas chromatography-mass spectrometer (GC–MS), while azobenzene oxide was also observed. The result showed that this reduction reaction follows the condensation route.

Subsequently, Li and co-workers studied the PdNPore-catalyzed reductive *N*-alkylation of nitroarenes using silane/alcohol as the hydrogen source in the presence of aromatic aldehydes (Scheme 5.29) [51]. In this process, nitroarenes and aromatic aldehydes were used as the starting material, and the oxidation of hydrosilane produced active hydrogen, which would reduce nitroarenes into aromatic amines. Aromatic amines were coupled with aromatic aldehydes to afford imines, which would be reduced to yield the corresponding product *N*-alkyl amines (Scheme 5.30). Under mild reaction conditions (room temperature and atmospheric pressure), various nitroarenes and aromatic aldehydes underwent the reductive *N*-alkylation smoothly, to produce the corresponding product *N*-alkyl amines in high yields. PdNPore catalyst could be reused atleast eight times.

Feng and Bao et al. reported the selective hydrogenation of nitro compounds catalyzed by PdNPore (Scheme 5.31) [52]. They prepared six types of PdNPore catalysts with different nanoporous structures by chemical dealloying of Pd_xAl_{100-x} ($Pd_{15}Al_{85}$, $Pd_{20}Al_{80}$, and $Pd_{30}Al_{70}$) ribbons in a 20 wt.% NaOH or 5 wt.% HCl aqueous solution. The obtained catalysts were referred to as PdNPore-1, PdNPore-2, PdNPore-3, PdNPore-4, PdNPore-5, and PdNPore-6, respectively. Energy-dispersive X-ray spectroscopy (EDS) indicated that the components of PdNPore-1, -2, -3, -4, -5, and -6 were $Pd_{90}Al_{10}$, $Pd_{82}Al_{18}$, $Pd_{86}Al_{14}$, $Pd_{94}Al_6$, $Pd_{53}Al_{47}$, and $Pd_{57}Al_{43}$, respectively.

Scheme 5.28 A proposed reaction pathway for the PdNPore-catalyzed reduction of aromatic nitro compounds [50]. Copyright 2019, with permission from Elscvier

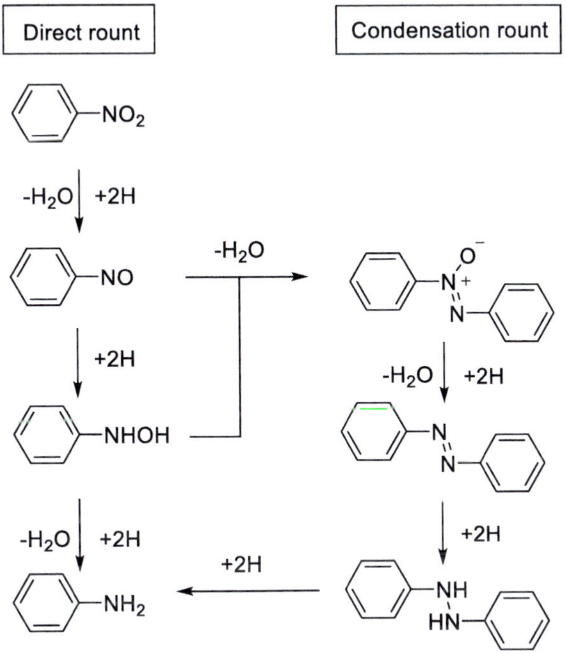

Scheme 5.29 PdNPore-catalyzed reductive *N*-alkylation of nitroarenes with silane/alcohol [51]. Copyright 2020, with permission from Elsevier

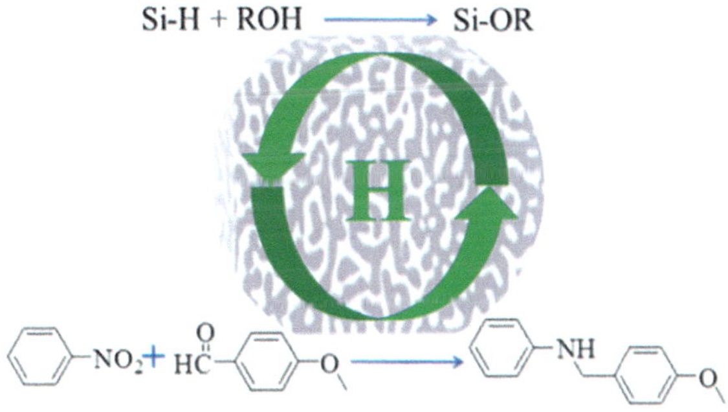

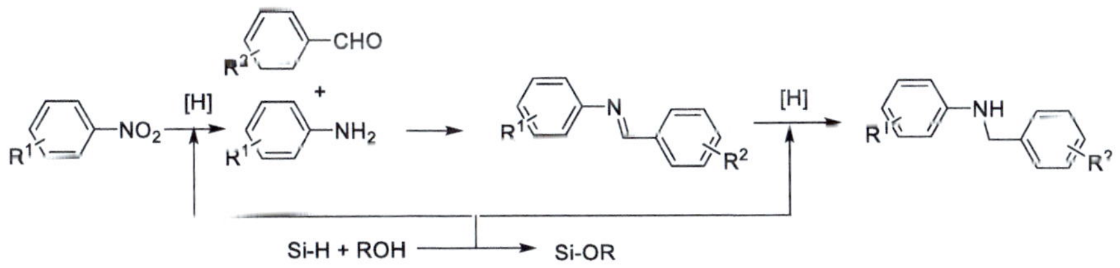

Scheme 5.30 Schematic representation of one-pot synthesis of *N*-alkyl amines by reductive *N*-alkylation of nitroarenes with silane/alcohol as the hydrogen source [51]. Copyright 2020, with permission from Elsevier

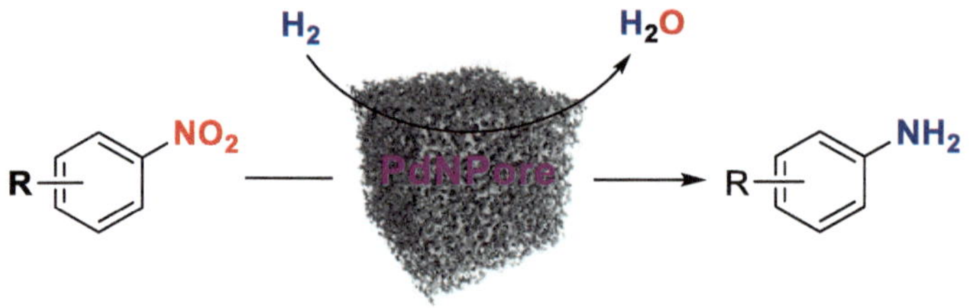

Scheme 5.31 Selective hydrogenation of aromatic nitro compounds catalyzed by PdNPore [52]. Copyright 2020, with permission from John Wiley and Sons

Then, the relevance between catalytic activity and nanoporous structure was investigated. The results indicated that the order of the catalytic activity of these PdNPore catalysts nearly coincided with the order of their BET area. And the PdNPore-3 catalyst with a specific surface area of 40.8 m^2/g showed relatively high activity (Fig. 5.10). The hydrogenation reactions of various nitro compounds took place smoothly with good functional group tolerance (acyl, ester, aminde, cyano and halogen) in the presence of PdNPore-3 catalyst. Moreover, the PdNPore-3 catalyst could be simply recovered and reused several times without significant loss of catalytic activity. Compared with the commercially available Pd/C and Raney-Ni catalysts, PdNPore-3 was more useful for industrial applications due to its unique features, including high stability and no palladium atom leaching.

CO_2 is an abundant, non-toxic, inexpensive and renewable C1 building block. Converting CO_2 into high-value-added chemicals in organic synthesis has attracted broad interests. Direct CO_2 hydrogenation was one of the most promising strategies for converting CO_2 into useful feedstock chemicals or liquid fuels. Compared with homogeneous catalytic hydrogenation protocols, the more economical and environmentally friendly heterogeneous hydrogenation protocols were highly desirable for CO_2 hydrogenation.

Feng and Bao et al. reported the high yield and selectivity for CO_2 hydrogenations using the PdNPore as a heterogeneous catalyst under mild reaction conditions (Scheme 5.32) [53].

They prepared four types of PdNPore catalysts from different prealloyed ingots, and systematic characterization indicated that the Al content in the ingot would affect the structure and catalytic activity of PdNPore. The PdNPore-1 catalyst prepared from $Pd_{10}Al_{90}$ (atomic ratio), showed high catalytic activity for CO_2 hydrogenation to formate. Under the optimized conditions, 86% yield of formate was produced with excellent chemoselectivities in MeCN at 80 °C under H_2 (1.0 MPa) and CO_2 (1.5 MPa) using PdNPore-1 and 1,8-diazabicyclo[5.4.0]undec-7-ene (DBU) as the catalyst and base, respectively. And 87% yield of format was produced at 80 °C for 24 h under 1.5 MPa of H_2 using sodium bicarbonate and PdNPore-1 as the starting material and catalyst, respectively.

The deuterium–hydrogen exchange reaction between deuterium (D_2) and H_2O was conducted under standard conditions to understand the excellent performance of PdNPore-1 in the hydrogenation of CO_2 (Scheme 5.33 and Fig. 5.11). The results

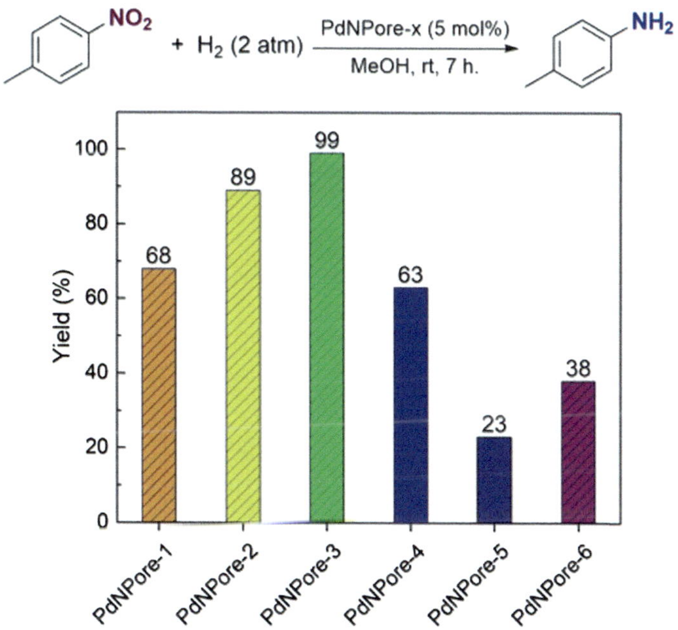

Fig. 5.10 Investigation of catalytic activity of PdNPore catalysts 1–6 in hydrogenation of 4-nitrotoluene [52]. Copyright 2020, with permission from John Wiley and Sons

Scheme 5.32 PdNPore-catalyzed hydrogenation of CO_2 and sodium bicarbonate to FA [53]. Copyright 2021, with permission from John Wiley and Sons

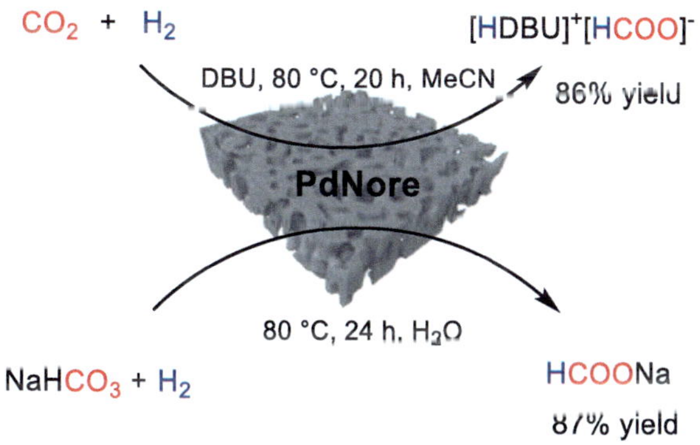

revealed that the H–H bond split in a heterolysis manner on the surface of the PdNPore-1 catalyst, and the Pd–H species generated in situ played an important role in CO_2 hydrogenation. The PdNPore-1 was easily recovered and reused ten times without any obvious loss in activity, which was better than commercial palladium catalyst (Pd/C and lindlar catalyst).

Subsequently, they studied the PdNPore-catalyzed *N*-formylation of amines and CO_2 with $PhSiH_3$ as hydrogen source under mild reaction conditions (Scheme 5.34) [54]. In this work, PdNPore showed a remarkably high reusability and functional

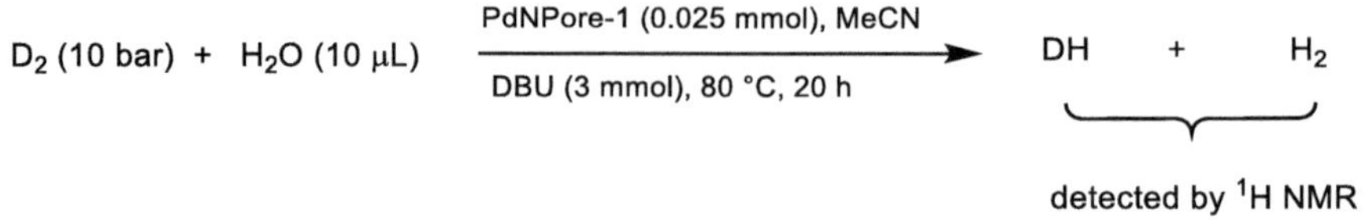

Scheme 5.33 Deuterium–hydrogen exchange reaction [53]. Copyright 2021, with permission from John Wiley and Sons

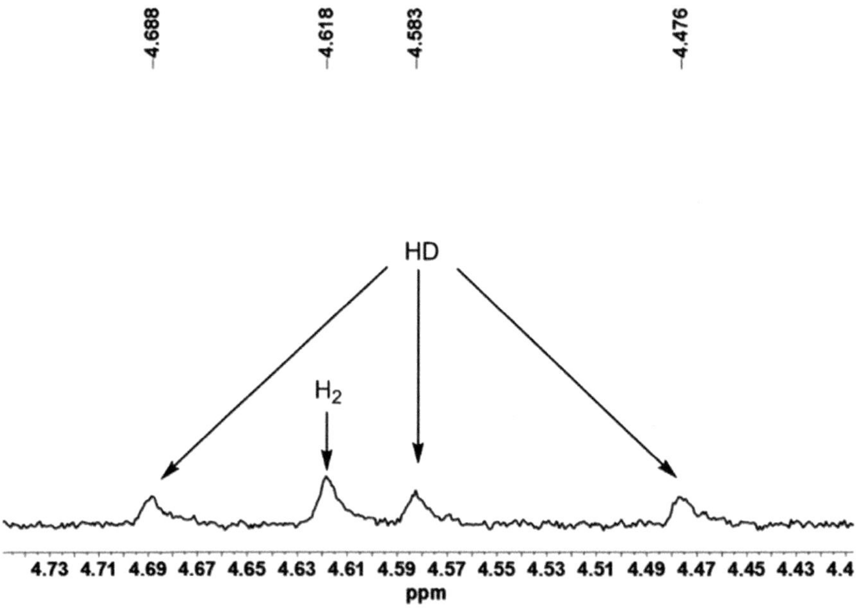

Fig. 5.11 ^{1}H NMR spectrum of deuterium–hydrogen exchange reaction [53]. Copyright 2021, with permission from John Wiley and Sons

group compatibility. Under the optimized conditions, various secondary amines and primary amines underwent the *N*-formylation reaction smoothly to afford the corresponding formamides bearing different functional groups in 32−91% yields. And a variety of synthetically useful functional groups, such as hydroxyl, allyl, halogen, and cyano groups, remained intact during the *N*-formylation of amines.

Leaching experiments indicated that no palladium leached from PdNPore into the reaction solution, and the current *N*-formylation of *N*-methylaniline (**1a**) with CO_2 was conducted in a heterogeneous catalytic manner. Under the standard conditions, the *N*-formylation of **1a** with CO_2 and $PhSiH_3$ was proceeded in deuterated acetonitrile and the mixture at different reaction times was analyzed using ^{1}H NMR spectroscopy. As shown in Fig. 5.12. A new proton signal intensity at 8.09 ppm appeared after 1 h assigned to the proton of formic acid, and the proton signal intensity at 8.51 ppm appeared after 4 h assigned to N-CHO of **2a**. Moreover, $HCOO^-$

Scheme 5.34 PdNPore-catalyzed *N*-formylation of amines with CO_2 [54]. Copyright 2022, with permission from John Wiley and Sons

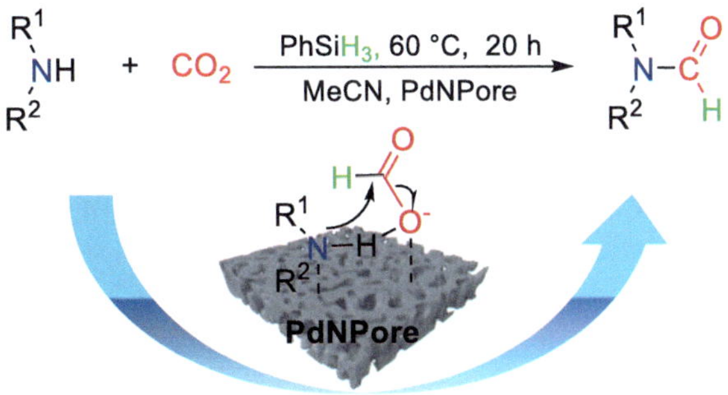

species was confirmed in the reaction solution after 1 h by ion chromatography (IC). Therefore, the authors proposed that the formic acid generated from $PhSiH_3$ and CO_2 was an important intermediate in the *N*-formylation reactions. The catalyst was also easily recovered from the catalytic system and reused ten times without any obvious loss in activity.

Recently, Feng and Bao et al. reported the PdNPore-catalyzed selective hydrogenation of nitriles without using any additive (Scheme 5.35) [55]. This reaction was conducted smoothly without any additives under mild reaction conditions (low H_2 pressure and low temperature), and exhibited good functional group tolerance. Under

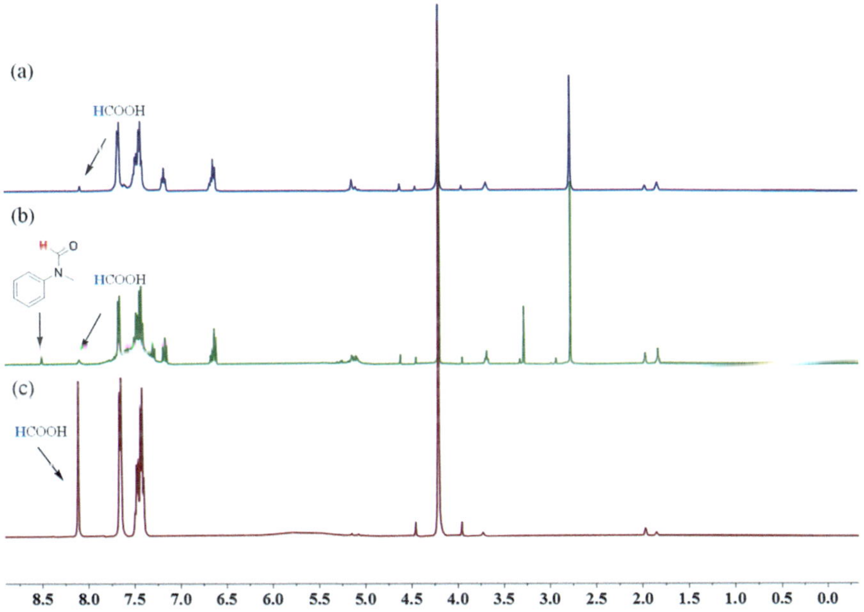

Fig. 5.12 ^{1}H NMR spectra of the reaction mixture of PdNPore/**1a**/CO_2 for 1 h (**a**) and 4 h (**b**) and formic acid/$PhSiH_3$ in CD_3CN (**c**) [54]. Copyright 2022, with permission from John Wiley and Sons

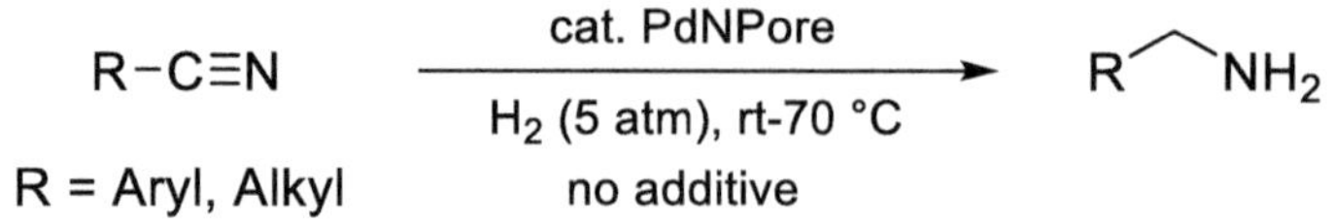

Scheme 5.35 Hydrogenation of nitriles catalyzed by PdNPore catalyst [55]. Copyright 2022, with permission from Royal Society of Chemistry

the optimized conditions, various aromatic/aliphatic nitriles underwent the hydrogenation smoothly to afford the corresponding primary amines in 60–95% yields with excellent selectivities. The reactivity of nitriles was influenced by the electronic properties of substituents and electron-withdrawing groups that were more favorable for this reaction.

Systematic studies showed that the good Lewis acidity and porous structure of the PdNPore led to the high activity and excellent selectivity. The PdNPore catalyst was easily recovered and reused without obvious loss of catalytic activity. A deuterium-hydrogen exchange experiment clearly showed that the hydrogenation of nitriles involved heterolytic H_2 splitting on the surface of the PdNPore catalyst.

5.3 Nanoporous Ag Catalyst

In 2011, Zhang et al. prepared the AgNPore catalyst by chemical dealloying of Mg–Ag based alloys in acid media under free corrosion conditions [56]. Subsequently, Ding and co-workers studied AgNPore-catalyzed aerobic oxidation of alcohols in gas-phase systems under relatively mild reaction conditions [57]. In this work, AgNPore catalyst was fabricated by chemical leaching of $Ag_{77}Mg_{23}$ alloy foil in aqueous solution of tartaric acid (1 wt.%) for 1 h at room temperature. The average pore/ligament diameter was measured using SEM to be about 25 nm (Fig. 5.13a).

The catalytic activity of AgNPore was examined by using a continuous-flow fixed-bed microreactor. The results showed that the AgNPore catalyst possessed

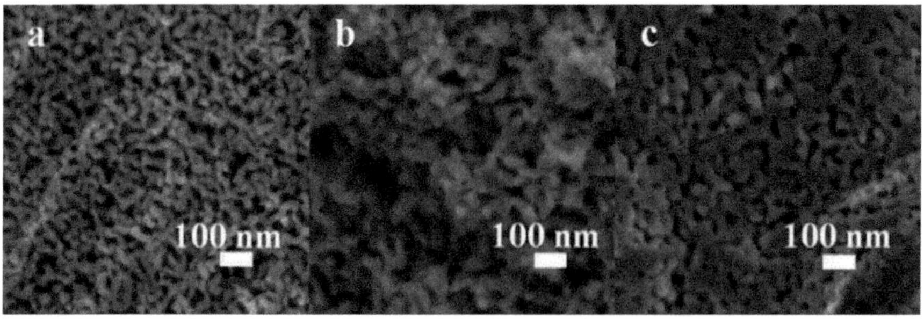

Fig. 5.13 SEM images of dealloyed AgNPore, **a** before the reaction, **b** after one use for the oxidation of MeOH, and **c** after four uses [57]. Copyright 2013, with permission from John Wiley and Sons

remarkable activity, durability, and selectivity in the oxidation of alcohols. Not only MeOH could be oxidized into methyl formate with high selectivity, but also many other primary and secondary aliphatic alcohols could be oxidized into the desired product aldehydes or ketones with high selectivity (Scheme 5.36).

Then, Zhang and co-workers applied the AgNPore catalyst to selective oxidation of organosilanes to organosilanols and organosilyl ethers at room temperature (Scheme 5.37) [58]. By using water or alcohols as the oxidant, various aromatic and aliphatic silanes underwent the oxidation reaction smoothly to produce the corresponding silanols and silyl ethers in high yields with high selectivity.

In this selective oxidation of organosilanes, H_2 gas was the only by-product and the AgNPore catalyst could be recycled for several times without any loss of activity and selectivity.

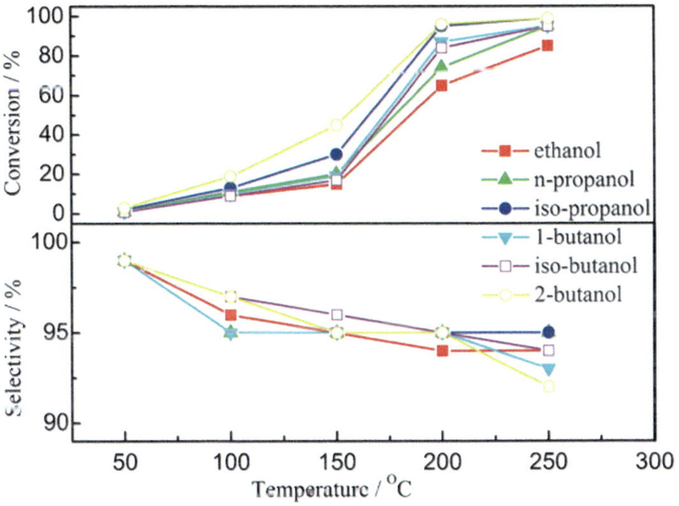

Scheme 5.36 Temperature dependence of the conversion of various alcohols and the selectivity for aldehydes or ketones over AgNPore (alcohol/O_2, 2:1; total flow rate: 100 mL min^{-1} (5% O_2 v/v); catalyst: 45 mg) [57]. Copyright 2013, with permission from John Wiley and Sons

$$R_{(4\text{-}n)}SiH_n + n\,H_2O \xrightarrow{\text{AgNPore}} R(4\text{-}n)Si(OH)n + n\,H_2$$

$$R_{(4\text{-}n)}SiH_n + n\,R^4OH \xrightarrow{\text{AgNPore}} R(4\text{-}n)Si(OR^4)n + n\,H_2$$

Scheme 5.37 AgNPore-catalyzed oxidation of organosilanes to organosilanols and organosilyl ethers [58]. Copyright 2014, with permission from Elsevier

5.4 Nanoporous Ru Catalyst

In 2014, Xu and co-workers prepared the RuNPore catalyst by dealloying $Ru_{20}Al_{80}$ alloy in 1 M HCl solution for 48 h at 50 °C. The nanopore size for RuNPore (50 °C) is mainly distributed around 10 nm with the specific surface area around 52.5 $m^2\ g^{-1}$. Then, they applied the RuNPore catalyst to hydrolysis of ammonia borane (Eq. 5.1) [59]. The results indicated that the RuNPore catalyst had superior catalytic activity for the hydrolysis of ammonia borane at room temperature under an ambient atmosphere.

$$H_3NBH_3(aq) + 2H_2O \xrightarrow{\text{catalyst}} NH_4BO_2(aq) + 3H_2(g) \quad (5.1)$$

The recyclability of RuNPore was further explored in the hydrolysis of ammonia borane (Fig. 5.14). As shown in Fig. 5.14, RuNPore catalyst showed high reusability, maintaining 67% of its initial catalytic activity even after five runs. Moreover, the activation energy of the hydrolysis of ammonia borane over RuNPore was examined based on the kinetic studies depending on the reaction temperature, and the activation energy (Ea) was estimated to be 66.5 kJ mol^{-1} from the slope of the straight line. The effect of catalystconcentration was also explored on the reaction rate of hydrolysis of ammonia borane, indicating that the hydrolysis reaction is the first-order reaction with respect to the concentration of RuNPore.

Because of clean catalyst surface, simple preparation, and high structure stability, the RuNPore material had promising application potential as the hydrogen storage material.

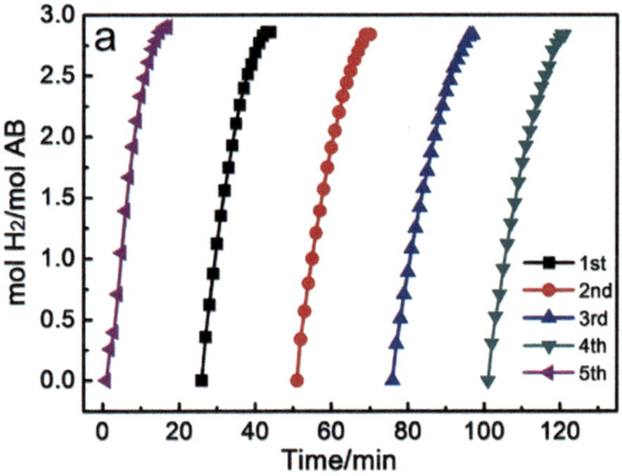

Fig. 5.14 The plot of mol H_2 per mol AB versus time toward catalytic dehydrogenation of AB in the presence of RuNPore from 1st to 5th cycles at 25 ± 1 °C (AB = 100 mM, [Ru] = 1.0 mM) [59]. Copyright 2016, with permission from Elsevier

5.5 Nanoporous Ni Catalyst

As early as 1927, commercial Raney Ni was prepared by the alkali leaching of Al from NiAl alloy, and has been used as an industry catalyst for more than 90 years [60]. However, because of the multiple intermetallic phases in the precursor Ni–Al alloy, the porous structure in Raney Ni was heterogeneous with a large variation in pore size and porosity. Moreover, due to the brittle nature of the precursor Ni–Al alloy, the Raney Ni was always in the form of powders [61]. Recently, NiNPore was fabricated by electrochemical dealloying of a single-phase alloy foil of $Ni_{30}Mn_{70}$ in various $(NH_4)_2SO_4$ aqueous solutions at different temperatures and applied potentials [62, 63]. Subsequently, Huo and Jin et al. achieved the hydrogenation of carbonates (as CO_2 sources) to formic acid (FA) using NiNPore catalyst for the first time (Scheme 5.38) [64].

They prepared NiNPore catalyst by dealloying in 1.0 M $(NH_4)_2SO_4$ aqueous solution at 50 °C according to the method of Chen [62, 63]. The structure of NiNPore with bright nano-pores and dark-contrast ligaments was observed clearly (Fig. 5.15a) and the pore size is around 7 nm (Fig. 5.15b).

The NiNPore catalyst exhibited remarkable catalytic activity in the hydrogenation of a wide range of carbonates to afford formic acid in excellent yields with high selectivity. Under the optimized conditions (H_2 6 MPa, catalyst 22 mol%, H_2O 10 mL, 200 °C, 2 h), NiNPore gave the highest yields of 86.6% or 92.1% when using $NaHCO_3$ or $KHCO_3$ as the starting material, respectively. The hydrogen pressure and pKa of the carbonates had a significant influence on the formation of FA. The NiNPore

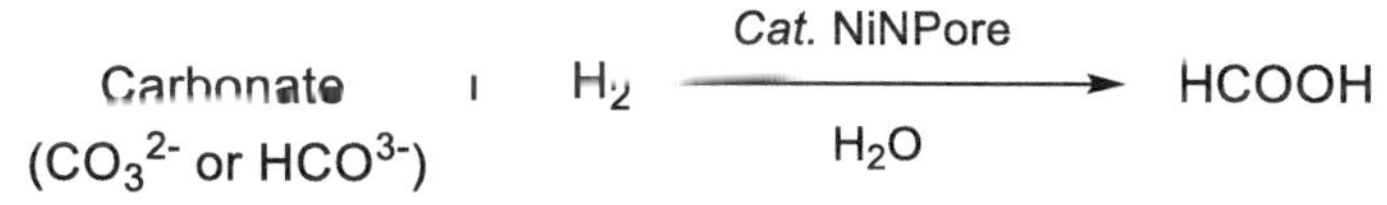

Scheme 5.38 Hydrogenation of carbonates to FA over NiNPore in water

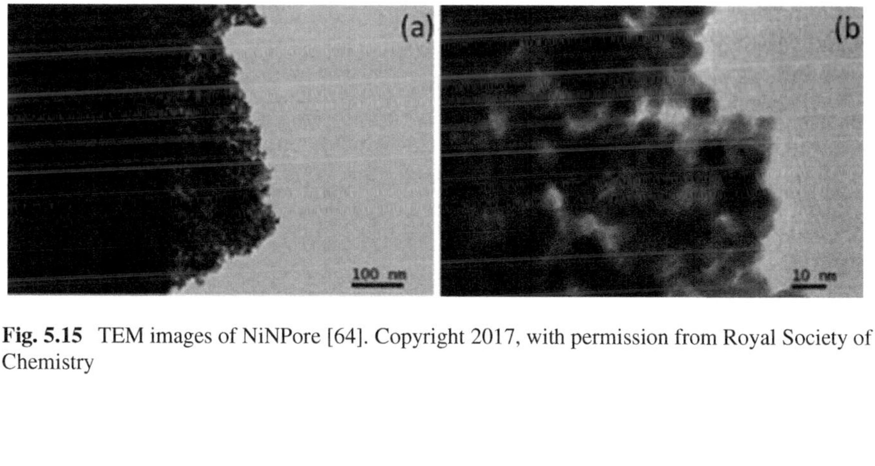

Fig. 5.15 TEM images of NiNPore [64]. Copyright 2017, with permission from Royal Society of Chemistry

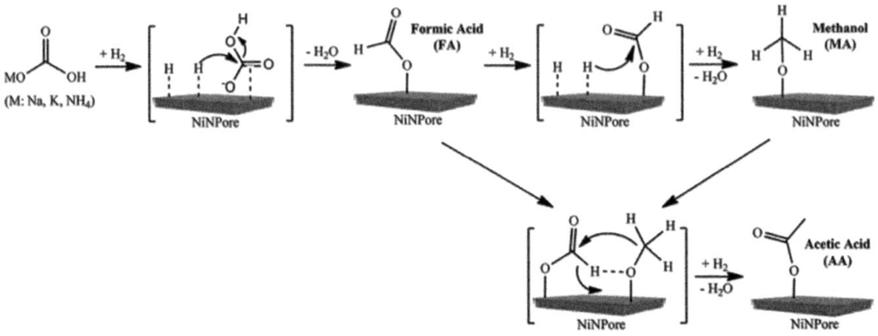

Scheme 5.39 Plausible pathway for the hydrogenation of HCO_3^- into FA over NiNPore [64]. Copyright 2017, with permission from Royal Society of Chemistry

catalyst was easily recovered and recycled at least five times without leaching and loss of activity.

A plausible mechanism for the hydrogenation of bicarbonates to FA over NiNPore was proposed according to the present findings and previous literature (Scheme 5.39). Firstly, hydrogen was absorbed on the surface of the NiNPore catalyst by the formation of hydrogen bonds between NiNPore and hydrogen. At the same time, bicarbonate was adsorbed on the surface of NiNPore through the combination of NiNPore with carbonylic C and O atoms. Then, the hydrogen reacted with a bicarbonate molecule to produce the formate (FA) and water. And methanol was produced by further hydrogenation of the formed formate. Moreover, the formed formic acid (FA) adsorbed on the surface of NiNPore reacted with methanol adsorbed on the surface of NiNPore to produce acetic acid.

5.6 Nanoporous Cu Catalyst

As early as 1980, Pryor and co-workers prepared monolithic CuNPore by electrochemical de-alloying of binary $Mn_{0.5}Cu_{0.5}$ and $Mn_{0.25}Cu_{0.75}$ alloys in 0.5 M NaCl at 25 °C, and subsequently studied the mechanism of dealloying of copper base binary alloys [65, 66]. Then, Min and Li also synthesized monolithic CuNPore by electrolytic dealloying of a Cu–Mn alloy (Incramute: Cu, 40–48 wt.% Mn, 1.4–2.3 wt.% Al) in 0.3 M NaCl solution at a potential of −0.2 V (SCE) [67]. Meanwhile, non-monolithic nanoporous Raney copper catalysts were also prepared through a chemical dealloying of $Cu_{35}Zn_{15}Al_{50}$ and $Cu_{50}Al_{50}$ alloys in 14.1 M aqueous NaOH solution or $Cu_{24.8}Zn_{25.2}Al_{48.3}$ in nitric and perchloric acid, and successfully applied in the water–gas shift reaction [68, 69].

In recent years, many researchers have studied the formation of the monolithic CuNPore, a promising high strength/low density material, by the chemical or electrochemical de-alloying of Cu/M (M=Al, Mg, Mn) alloys [44, 70–73]. However, until

2011, the first case of monolithic CuNPore-catalyzed molecular transformation was reported by Jin and Yamamoto [74].

They fabricated the CuNPore catalysts from a homogeneous $Cu_{30}Mn_{70}$ alloy with a thickness of 200 mm under free corrosion conditions [70, 73]. The CuNPore catalyst with tunable nanoporosity was obtained through controlling the dealloying temperature in a mixture of $(NH_4)_2SO_4$ (1 M) and $MnSO_4$ (0.01 M) aqueous solution. XPS analysis showed that the surface of each CuNPore was composed of a mixture of Cu^0, Cu^I, and Cu^{II} species. This CuNPore catalyst exhibited remarkable catalytic properties in Click reaction without using any supports and bases (Eq. 5.2). The catalytic activity of CuNPore catalyst was highly dependent on its ligament (or pore) sizes, and the CuNPore-3 with ligament size about ~40 nm exhibited the highest catalytic efficiency.

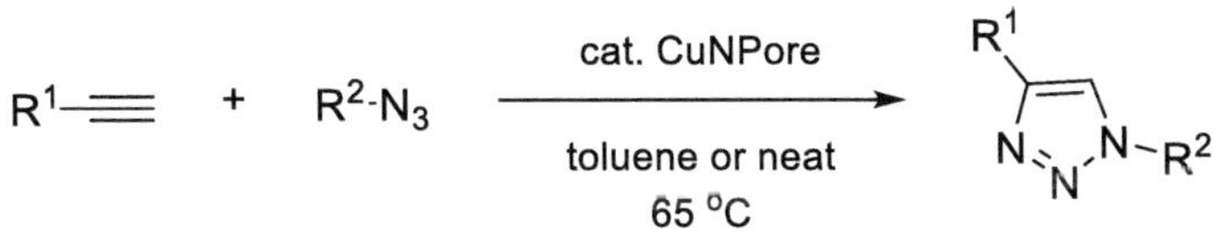

(5.2)

The catalytic activity of CuNPore (2 mol% cat-3) was further investigated with various terminal alkynes and organic azides in toluene (2 M) at 65 °C. The results indicated that not only aromatic and heteroaromatic alkynes but also alkylalkynes underwent the CuNPore-catalyzed Click reaction smoothly, to produce 1,2,3-triazoles in excellent yields. A wide range of functional groups (alkene, ester, and protected amines) were well tolerated under standard conditions.

Under neat conditions, the amount of catalyst could be decreased to 0.1 mol%, which did not have a strong impact on the reaction efficiency, and the TOF was increased up to 5.2 s^{-1} (Eq. 5.3).

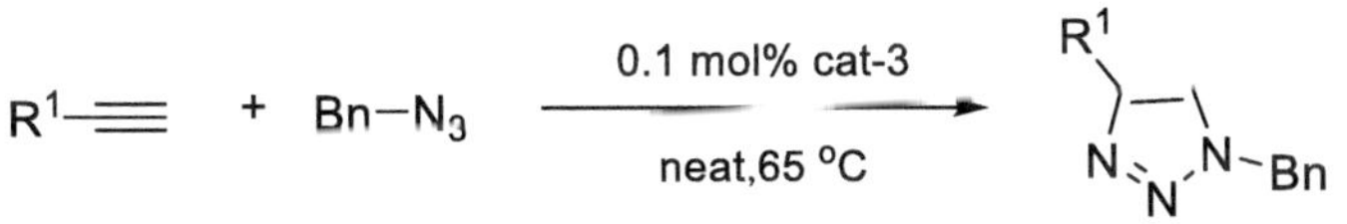

(5.3)

1a R¹ = Ph **2a** **3a** 99% (2 h)
1b R¹ = c-hoxyl **3b** 95% (4 h)

The CuNPore catalyst showed an excellent reusability and was recycled for ten times without any loss of activity. Systematic characterization of the CuNPore catalyst surface, leaching experiments, and the generation of nanostructured copper acetylide indicated that the Click reaction was catalyzed by Cu(I) species on the CuNPore surface.

5.7 Nanoporous AuPd Catalyst

In 2014, Jin and co-workers reported for the first time that AuPdNPore exhibited superior chemoselectivity and high catalytic activity for the direct 1,4-hydrosilylation of conjugated cyclic enones with hydrosilane, which cannot be achieved by using monometallic AuNPore and PdNPore catalysts [75]. In this work, $Au_{20}Pd_{10}Al_{70}$ (at.%) ternary alloy was prepared using a high-frequency induction furnace, and then the AuPdNPore was fabricated by the selective etching of Al from the $Au_{20}Pd_{10}Al_{70}$ (at.%) ternary alloy using NaOH (20%) as the electrolyte at room temperature for 36 h. The scanning electron microscopy (SEM) and transmission electron microscopy (TEM) images of the AuPdNPore showed a bicontinuous nanoporous structure and the curved metallic ligaments with an average size of ~8 nm (Fig. 5.16).

The catalytic properties of AuPdNPore-1 were further examined by the reaction with various conjugated cyclic enones (Table 5.17). The monomethyl- and dimethyl-substituted cyclohexen-2-ones, cyclopenten-2-one as well as cyclohepten-2-one underwent selective 1,4-hydrosilylation sufficiently, producing the desired silyl ethers in good to high yields, while both 3-methyl- and 5-methylcyclohex-2-enones produced a mixture of regioisomers **2b** and **2c**.

The catalytic performances and characterization disclosed that the alloy effect of AuPdNPore plays a crucial role in the reaction efficiency and product selectivity. The proposed mechanism was shown in Scheme 5.40. The chemisorption of hydride on the Pd atom and Si on the Au atom promoted the dissociation of $HSiEt_3$ on the AuPdNPore catalyst. The high affinity of Si for carbonyl oxygen of **1a** induced the selective addition of hydride to the β-position of **1a**, giving the corresponding silyl ether **2a**. The authors speculated that the remaining Al atoms in the AuPdAl alloy modified the electronic and geometric structures of the AuPdNPore catalyst, but Al itself did not directly incorporate in the catalysis.

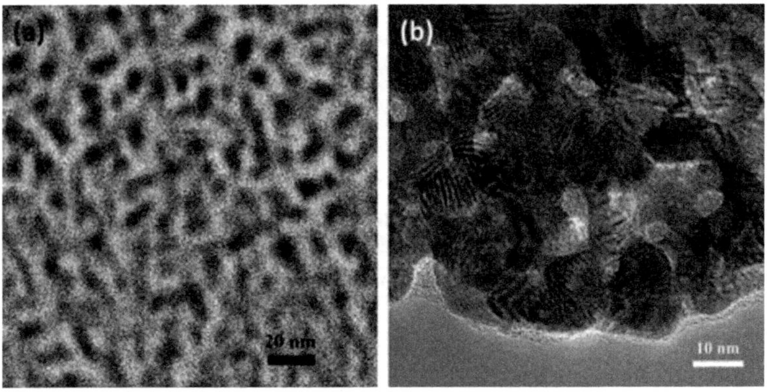

Fig. 5.16 SEM (**a**) and TEM (**b**) images of the AuPdNPore catalyst [75]. Copyright 2014, with permission from Royal Society of Chemistry

Table 5.17 1,4-Hydrosilylation of various conjugated cyclic enones using the AuPdNPore catalyst[a,b] [75]. Copyright 2021, with permission from Royal Society of Chemistry

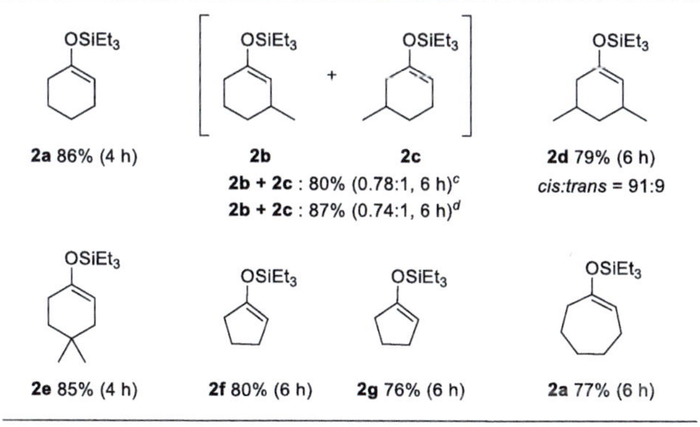

[a] Reaction conditions: conjugated cyclic enones **1** (2.0 mmol), $HSiEt_3$ (2.0 mmol), and AuPdNPore-1 (2.5 mol%) at 100 °C without solvents. [b] Isolated yields of products **2**. [c] 3-Methylcyclohex-2-enone was used as a starting substance. [d] 5-Methylcyclohex-2-enone was used as a starting substance.

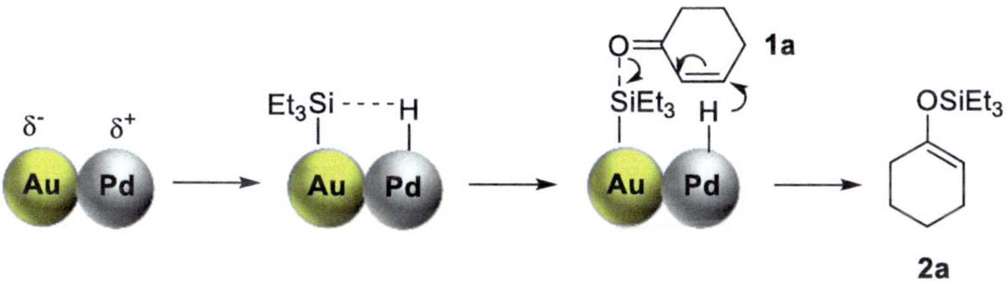

Scheme 5.40 Proposed mechanism for the selective 1,4-hydrosilylation of **1a** [75]. Copyright 2014, with permission from Royal Society of Chemistry

5.8 Nanoporous PtRu Catalyst

In 2010, Xu and Ding et al. prepared the PtRuNPore catalyst by a chemical dealloying of $Pt_{14}Ru_6Al_{80}$, $Pt_{10}Ru_{10}Al_{80}$, and $Pt_6Ru_{14}Al_{80}$ alloys in 2 M NaOH solution for 48 h, and successfully applied them in methanol and CO electro-oxidation [76]. Subsequently, Xu et al. synthesized PtRuNPore catalysts ($Pt_{70}Ru_{30}$NPore, $Pt_{50}Ru_{50}$NPore, $Pt_{30}Ru_{70}$NPore) by chemical dealloying of $Pt_{14}Ru_6Al_{80}$, $Pt_{10}Ru_{10}Al_{80}$, and $Pt_6Ru_{14}Al_{80}$ alloys in dilute hydrochloric acid (Eq. 5.4) [77]. They found that the PtRuNPore catalyst had uniform ligament sizes and controllable bimetallic ratios, and exhibited superior and structure-dependent catalytic activity for the hydrolysis of ammonia borane under an ambient atmosphere at room temperature. Among these PtRuNPore catalysts, $Pt_{70}Ru_{30}$NPore showed the higher catalytic activity than $Pt_{50}Ru_{50}$NPore and $Pt_{30}Ru_{70}$NPore in the hydrolytic dehydrgenation of ammonia borane due to a suitable binding energy for interactions between the catalyst and substrate.

$$H_3NBH_3(aq) + 2H_2O \xrightarrow{\text{catalyst}} NH_4BO_2(aq) + 3H_2(g) \quad (5.4)$$

The hydrolysis activation energy of $Pt_{70}Ru_{30}$NPore was then measured as 38.9 kJ mol^{-1}, which was lower than most of the reported values for various Pt-, Ru-, and other metal-based catalysts. Moreover, $Pt_{70}Ru_{30}$NPore catalyst exhibited good stability in the hydrolysis of ammonia borane due to the unique nanoporous architecture and still showed highly active even after five runs.

5.9 Nanoporous PtFe Catalyst

Zhang and co-workers successfully prepared a soft magnetic PtFeNPore catalyst with a single fcc-FePt phase by electrochemical dealloying of an amorphous $Fe_{60}Pt_{10}B_{30}$ alloy in 0.1 M H_2SO_4 solution [78]. The PtFeNPore catalyst possessed a higher electrocatalytic activity for methanol oxidation reaction in the acid media than that of the Pt/C catalyst. Then, Feng and Bao applied the successfully prepared PtFeNP ore in the chemoselective hydrogenation of halonitrobenzenes to haloanilines under mild reaction conditions (Scheme 5.41) [79].

Firstly, they prepared the PtFeNPore catalyst with an electrochemical dealloying method [78]. The structure and composition of PtFeNPore were characterized through various measurement techniques including scanning electron microscope (SEM), transmission electron microscope (TEM), X-ray photoelectron spectroscopy (XPS) and X-ray absorption spectroscopy (XAFS). The results indicated that the intermetallic compound and iron oxide speciesboth existed in the PtFeNPore catalyst, and the iron oxide species, formed during the dealloying process, enabled the PtFeNPore catalyst to act as a Lewis acid catalyst. The Lewis acidity and nanoporous structure of the PtFeNPore catalyst led to a good catalytic activity and high selectivity. Then, various halonitrobenzenes underwent the hydrogenation reaction smoothly under low H_2 pressure at room temperature, to produce the desired haloanilines in good to excellent yields with excellent selectivity. The PtFeNPore catalyst could be

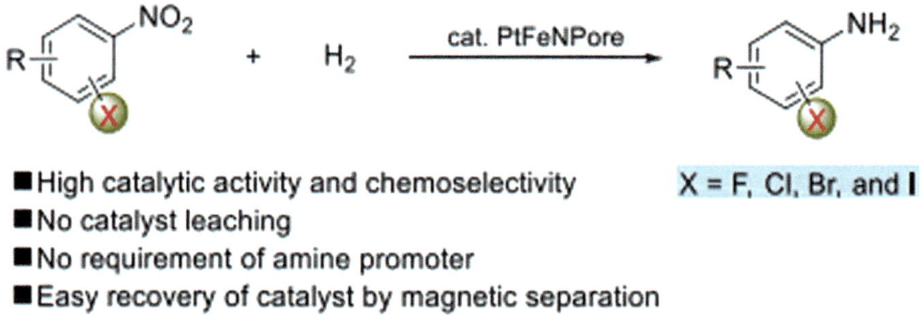

Scheme 5.41 PtFeNPore-catalyzed chemoselective hydrogenation of halonitrobenzenes to haloanilines [79]. Copyright 2021, with permission from ACS Publications

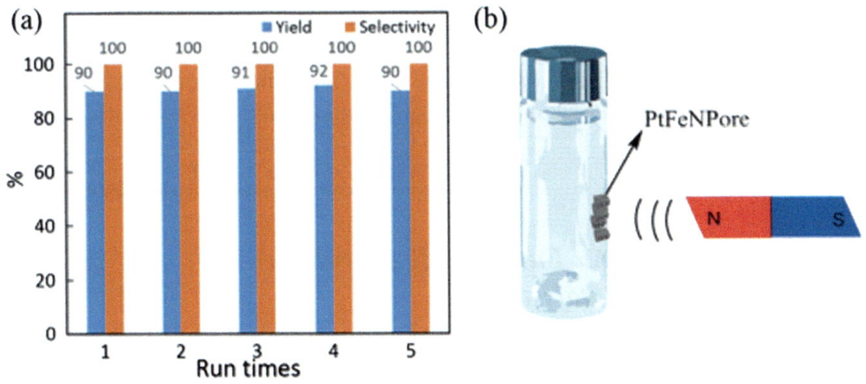

Fig. 5.17 **a** Reusability of PtFeNPore for the hydrogenation reaction of 1b. **b** Recovery of the PtFeNPore catalyst by magnetic separation [79]. Copyright 2021, with permission from ACS Publications

easily recovered through magnetic separation and still had high catalytic activity and excellent chemoselectivity after it was recycled 5 times (Fig. 5.17).

5.9.1 Nanoporous $PtCo/Co_3O_4$ Catalyst

Xu et al. fabricated PtCoNPore/Co_3O_4 catalysts with different Pt/Co ratios by a two-step chemical dealloying of $Pt_{4.5}Co_{14.5}Al_{81}$ alloy and successfully applied them in the hydrolytic dehydrogenation of ammonia borane (Eq. 5.5) [80]. Specifically, $Pt_{4.5}Co1_{4.5}Al_{81}$ alloy slices first underwent the chemical dealloying in 0.5 M NaOH solution for 48 h, to produce the nanoporous PtCo alloy with many Co_3O_4 nanosheets on the surface. And then followed further chemical dealloying in 0.2 M HNO_3 for varying time (0, 2.5, 5, 10, and 20 min) to provide the desired PtCoNPore/Co_3O_4 catalyst with different Pt/Co ratios, respectively. For convenience, these PtCoNPore/Co_3O_4 catalyst with different Pt/Co ratios were named as $Pt_{25}Co_{75}$NPore/Co_3O_4, $Pt_{40}Co_{60}$NPore/Co_3O_4, $Pt_{55}Co_{45}$NPore/Co_3O_4, $Pt_{70}Co_{30}$NPore/Co_3O_4, and $Pt_{90}Co_{10}$NPore/Co_3O_4, respectively.

$$NH_3BH_3(aq) + 2H_2O(l) \xrightarrow{catalyst} NH_4BO_2(aq) + 3H_2(g) \quad (5.5)$$

The catalytic performance of PtCoNPore/Co_3O_4, PtNPore, and Co_3O_4 were investigated in the hydrolytic dehydrogenation of ammonia borane. The results indicated that the PtCoNPore/Co_3O_4 catalysts exhibited much better catalytic performances than that of PtNPore, and Co_3O_4 for the hydrolysis of ammonia borane due to the synergistic effect between Pt and Co. The Pt/Co ratios have evident effects on the catalytic activity of PtCoNPore/Co_3O_4 catalysts, in which, $Pt_{40}Co_{60}$NPore/Co_3O_4 showed the highest catalytic activity, providing a high initial turnover frequency of 131 mol H_2 min^{-1} (mol Pt)$^{-1}$. The interaction between the small amount of oxides and the surface of PtCoNPore alloy could promote the catalytic activity toward the hydrolysis of ammonia borane. And after the fifth run of hydrolysis of ammonia borane, the $Pt_{40}Co_{60}$NPore/Co_3O_4 catalyst retained 65% of the initial catalytic activity.

5.9.2 *Nanoporous Au Functionalized with Ce–TiO$_x$ or Pr–TiO$_x$ Catalyst*

In 2016, Wittstock and co-workers prepared Ce–TiO_x/AuNPore catalyst using the sol–gel coating method [81]. They first synthesized AuNPore by chemical dealloying of AuAg alloy (30% Au and 70% Ag) in 65 wt.% nitric acid for 48 h. Then, the modification of AuNPore with Ce–TiO_2 mixed metal oxides was performed via a sol–gel method following a procedure reported by Li et al. [82]. Metal-oxides with different composition were produced inside the AuNPore material (ligaments and pores ~45 nm) after thermal treatment at 200–500 °C for 2 h. The resulting Ce–TiO_x/AuNPore catalysts exhibited highly catalytic activity for the water–gas-shift reaction in a continuous flow reactor at ambient pressure (Fig. 5.18). As shown in Fig. 18, $Ce_1Ti_2O_x$/AuNPore (with the molar ratio of Ce: Ti = 1:2) showed the highest activity that was nearly twice as high as the activity of all other samples at 300 °C.

Moreover, $Ce_1Ti_2O_x$/AuNPore catalyst exhibited good stability in the water–gas-shift reaction, with only about 10% decrease in catalytic activity after 2 catalytic runs at 450 °C after 20 h under reaction conditions. Raman spectroscopic characterization of the Ce–TiO_x/AuNPore catalysts indicated a dynamic correlation between the

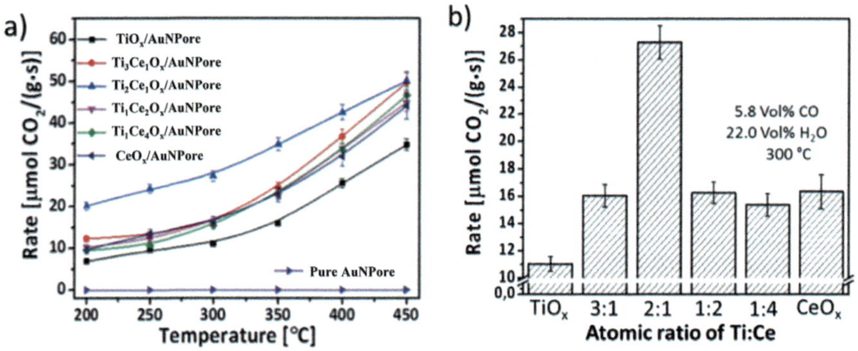

Fig. 5.18 WGS activity ($CO + H_2O \rightarrow CO_2 + H_2$): **a** comparison of the reaction rate of different binary Ti–CeO_x and TiO_2 and CeO_2 single oxide functionalized AuNPore samples (5.8 vol.% CO, 22.0 vol.% H_2O in He, total gas flow 32 mL min^{-1}, $M_{catal} = 6.0 \pm 0.2$ mg, space velocity 320 000 mL $h^{-1}gcat^{-1}$). **b** The bar plot for the catalysts is based on the reaction rates measured at 300 °C (each measurement was repeated at least 4 times with different catalysts) [81]. Copyright 2016, with permission from Royal Society of Chemistry

crystallization (oxygen storage) of the metal-oxides under oxidizing and reducing conditions.

Subsequently, Wittstock et al. applied $Ce_1Ti_2O_x$/AuNPore, CeO_x/AuNPore and TiO_x/AuNPore catalysts for the methanol steam reforming reaction (Eq. 5.6) [83]. Among these catalysts, the CeO_x/AuNPore catalyst showed excellent low-temperature catalytic activity and selectivity within a broad temperature range (250–450 °C). The calculated turn over frequency at 300 °C of the CeOx/AuNPore catalyst was 0.06 s^{-1}. The reforming of methanol was negligible on pure AuNPore or pure metal oxides. The SEM and Raman scattering results showed that ceria and tiania–ceria mixed oxides could more effectively stabilize the porous structure of AuNPore during the methanol steam reforming reaction up to 500 °C. Meanwhile, the water–gas-shift reaction was not observed in the methanol steam reforming reaction pathway at lower than 350 °C. Neither was any CO detected during in situ FTIR spectroscopy nor did the addition of CO during steady-state steam reforming of methanol tests at 300 °C led to higher activity.

$$CH_3OH + H_2O \rightarrow 3H_2 + CO_2 \quad \Delta H^{\theta}_{298} = 49.4\,\text{kJ mol}^{-1} \tag{5.6}$$

Parallel investigation by in situ infrared spectroscopy (DRIFTS) revealed that the activation of water and the formation of OH_{ads} are the key to the activity/selectivity of the catalysts. All catalysts could generate sufficient OH_{ads} to prevent complete dehydrogenation of methanol to CO, but only the most active catalysts (e.g., CeO_x/AuNPore) showed direct reaction with formic acid and its decomposition to CO_2 and H_2.

Two different main reaction mechanisms for the steam reforming of methanol were described in Scheme 5.42. The dehydrogenation of CH_3OH to CH_3O_{ads} and further to H_2CO_{ads} and $HCOO_{ads}$ was the dominating reaction pathway for all catalysts (determined by FTIR). And the availability of reactive OH_{ads} was the key to the catalytic activity (determined by FTIR under steady state conditions). While on the surface of the less active TiO_2/AuNPore catalyst, methyl formate could be detected; and on the more active CeO_2/AuNPore catalysts, direct oxidation of formate with OH_{ads} to formic acid which then decomposed to CO_2 and H_2 that appeared to be the dominating reaction pathway.

In 2019, Bäumer and co-workers also prepared AuNPore catalysts functionalized with Pr–TiO_x mixed oxides (PrO_x/AuNPore and $Pr_1Ti_2O_x$/AuNPore) by a sol–gel method for the water–gas shift reaction [84]. Compared with other catalysts (PrO_x/AuNPore, $Pr_1Ti_2O_x$/AuNPore, $Ce_1Ti_2O_x$/AuNPore, CeO_x/AuNPore, and TiO_x/AuNPore), $Pr_1Ti_2O_x$/AuNPore catalyst exhibited the highest catalytic activity and stability both in the low (180 °C for 100 h) and high-temperature range (300 °C for 50 h) for the water–gas shift reaction (Fig. 5.19). But pure AuNPore or bulk Pr–TiO_x mixed oxide alone could not catalyze the water–gas shift reaction. Their studies demonstrated that the good performance of Pr–TiOx/AuNPore catalysts in water–gas shift reaction was attributed to the interaction between the Au substrate and the oxide deposits and that the oxide deposits played an essential

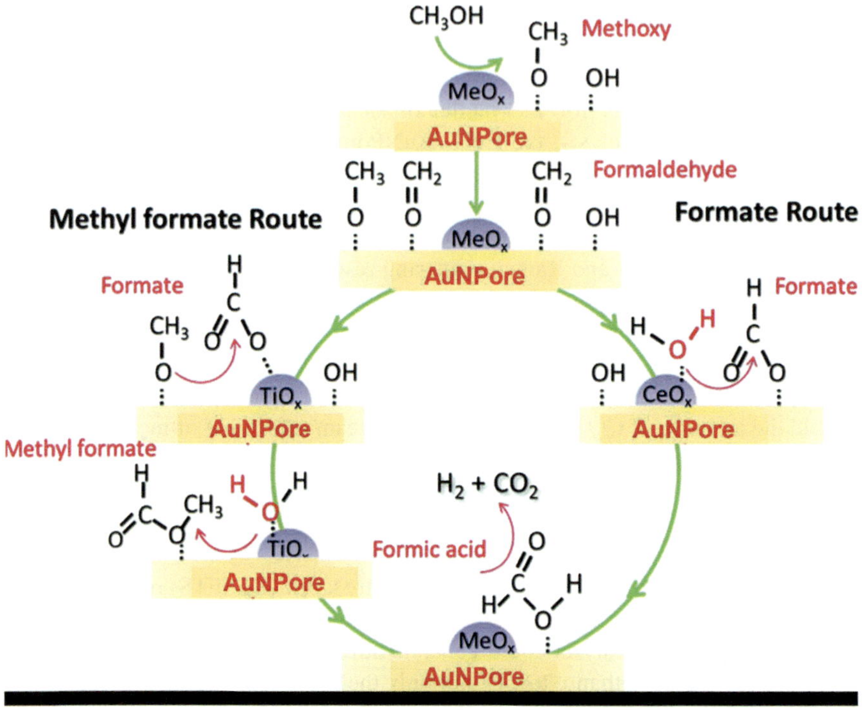

Scheme 5.42 Mechanisms of steam reforming of methanol on an oxide functionalized AuNPore surface [83]. Copyright 2017, with permission from Royal Society of Chemistry

role in the dissociation of water in the catalytic cycle. In addition, the characterizations revealed that the deposits forming mixed oxide Pr–TiOx solid solutions on the stepped surface of AuNPore played a key role in preventing the coarsening of the nanoporous Au network by lowering the free energy of the catalyst.

5.10 Summary and Outlooks

This chapter summarized remarkable catalytic performances of various MNPores catalysts, including Au, Pd, Ag, Ru, Ni, Cu, AuPd, PtRu, PtFe, and PtCo/Co_3O_4 as well as AuNPore functionalized with Ce–TiO_x or Pr–TiO_x catalyst in organic molecule transformations with regard to preparation method, characterization, substrate scope, reusability, leaching resistance, and reaction mechanism. The distinct features of MNPores, such as unique self-supported nanoporous structure, high surface area, robustness, stability, and simple operation for recovery and reuse, made them promising green and sustainable heterogeneous catalysts. However, generally speaking, the catalytic activity of MNPores was not as good as that of

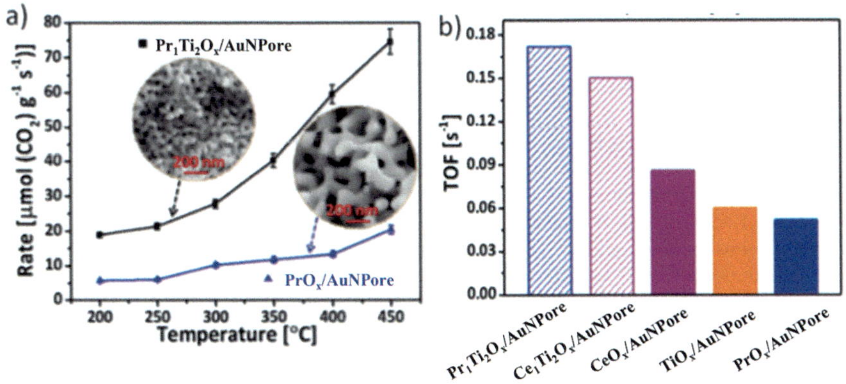

Fig. 5.19 **a** WGS activity of $Pr_1Ti_2O_x$/AuNPore and PrO_x/AuNPore as a function of temperature. Feed-gas mixture: 5.8 vol.% CO, 22 vol.% H_2O, helium as carrier gas, total gas flow 32 mL min^{-1}, $m_{catal} = 6.0 \pm 0.2$ mg, space velocity 320 000 mL h^{-1} $gcat^{-1}$. Insets are SEM images of samples after calcination at 450 °C in helium for 2 h (before catalysis test). **b** Histogram of TOF values at 350 °C determined for $Pr_1Ti_2O_x$/AuNPore and compared with other representative metal oxide-AuNPore based catalysts [84]. Copyright 2019, with permission from Royal Society of Chemistry

MNPs owing to the small surface area of MNPores, which limited their application in more complicated reactions. To date, the reported research works on MNPores catalysis mainly focused on oxidation and reduction reactions. Therefore, designing and preparing multicomponent MNPores catalysts with high surface area has become a great challenge.

References

1. Forty AJ (1979) Corrosion micromorphology of noble metal alloys and depletion gilding. Nature 282:597–598. https://doi.org/10.1038/282597a0
2. Erlebacher J, Aziz MJ, Karma A, Dimitrov N, Sieradzki K (2001) Evolution of nanoporosity in dealloying. Nature 410:450–453. https://doi.org/10.1038/35068529
3. Zielasek V, Jürgens B, Schulz C, Biener J, Biener MM, Hamza AV, Bäumer M (2006) Gold Catalysts: nanoporous gold foams. Angew Chem Int Ed 45:8241–8244. https://doi.org/10.1002/anie.200602484
4. Wittstock A, Neumann B, Schaefer A, Dumbuya K, Kübel C, Biener MM, Zielasek V, Steinrück HP, Gottfried JM, Biener J, Hamza A, Bäumer M (2009) Nanoporous Au: An unsupported pure gold catalyst. J Phys Chem C 113:5593–5600. https://doi.org/10.1021/jp808185v
5. Moskaleva LV, Röhe S, Wittstock A, Zielasek V, Klüner T, Neyman KM, Bäumer M (2011) Silver residues as a possible key to a remarkable oxidative catalytic activity of nanoporous gold. Phys Chem Chem Phys 13:4529–4539.https://doi.org/10.1039/c0cp02372h
6. Wittstock A, Wichmann A, Biener J, Bäumer M (2011) Nanoporous gold: a new gold catalyst with tunable properties. Faraday Discuss 152:87–98. https://doi.org/10.1039/c1fd00022e
7. Xu C, Su J, Xu X, Liu P, Zhao H, Tian F, Ding Y (2007) Low temperature CO oxidation over unsupported nanoporous gold. J Am Chem Soc 129:42–43. https://doi.org/10.1021/ja0675503
8. Xu C, Xu X, Su J, Ding Y (2007) Research on unsupported nanoporous gold catalyst for CO oxidation. J Catal 252:243–248. https://doi.org/10.1016/j.jcat.2007.09.016

9. Yin H, Zhou C, Xu C, Liu P, Xu X, Ding Y (2008) Aerobic oxidation of D-glucose on support-free nanoporous gold. J Phys Chem C 112:9673–9678. https://doi.org/10.1021/jp8019864
10. Wittstock A, Biener J, Bäumer M (2010) Nanoporous gold : a new material for catalytic and sensor applications. Phys Chem Chem Phys 12:12919–12930. https://doi.org/10.1039/c0cp00757a
11. Wittstock A, Zielasek V, Biener J, Friend CM, Bäumer M (2010) Nanoporous gold catalysts for selective gas-phase oxidative coupling of methanol at low temperature. Science 327:319–322. https://doi.org/10.1126/science.1183591
12. Kosuda KM, Wittstock A, Friend CM, Bäumer M (2012) Oxygen-mediated coupling of alcohols over nanoporous gold catalysts at ambient pressures. Angew Chem Int Ed 51:1698–1701. https://doi.org/10.1002/anie.201107178
13. Kong XM, Shen LL (2012) Selectivity control of ethanol gas-phase oxidation over nanoporous gold. Catal Commun 24:34–37. https://doi.org/10.1016/j.catcom.2012.03.019
14. Fujita T, Guan P, McKenna K, Lang X, Hirata A, Zhang L, Tokunaga T, Arai S, Yamamoto Y, Tanaka N, Ishikawa Y, Asao N, Yamamoto Y, Erlebacher J, Chen M (2012) Atomic origins of the high catalytic activity of nanoporous gold. Nat Mater 11:775–780. https://doi.org/10.1038/NMAT3391
15. Asao N, Ishikawa Y, Hatakeyama N, Menggenbateer, Yamamoto Y, Chen M, Zhang W, Inoue A (2010) Nanostructured materials as catalysts: nanoporous-gold-catalyzed oxidation of organosilanes with water. Angew Chem Int Ed 49:10093–10095. https://doi.org/10.1002/anie.201005138
16. Asao N, Hatakeyama N, Menggenbateer, Minato T, Ito E, Hara M, Kim Y, Yamamoto Y, Chen M, Zhang W, Inoue A (2012) Aerobic oxidation of alcohols in the liquid phase with nanoporous gold catalysts. Chem Commun 48:4540–4542. https://doi.org/10.1039/c2cc17245c
17. Tanaka S, Minato T, Ito E, Hara M, Kim Y, Yamamoto Y, Asao N (2013) Selective aerobic oxidation of methanol in the coexistence of amines by nanoporous gold catalysts: highly efficient synthesis of formamides. Chem Eur J 19:11832–11836. https://doi.org/10.1002/chem.201302396
18. Wichmann A, Bäumer M, Wittstock A (2015) Oxidative coupling of alcohols and amines over bimetallic unsupported nanoporous gold: tailored activity through mechanistic predictability. ChemCatChem 7:70–74. https://doi.org/10.1002/cctc.201402843
19. Bond GC, Louis C, Thompson DT (2006) Catalysis by gold. Imperial College Press, London Chapter 9, p 244
20. McEwan L, Julius M, Roberts S, Fletcher JCQ (2010) A review of the use of gold catalysts in selective hydrogenation reactions. Gold Bull 43:298–306. https://search.ebscohost.com/login.aspx?direct=true&db=a9h&AN=58027481&lang=zh-cn&site=eds-live
21. Yan M, Jin T, Ishikawa Y, Minato T, Fujita T, Chen LY, Bao M, Asao N, Chen MW, Yamamoto Y (2012) Nanoporous gold catalyst for highly selective semihydrogenation of alkynes: remarkable effect of amine additives. J Am Chem Soc 134:17536–17542. https://doi.org/10.1021/ja3087592
22. Yan M, Jin T, Chen Q, Ho HE, Fujita T, Chen LY, Bao M, Chen MW, Asao N, Yamamoto Y (2013) Unsupported nanoporous gold catalyst for highly selective hydrogenation of quinolines. Org Lett 15:1484–1487. https://doi.org/10.1021/ol400229z
23. Takale BS, Tao SM, Yu XQ, Feng XJ, Jin T, Bao M, Yamamoto Y (2014) Exclusive chemoselective reduction of imines in the coexistence of aldehydes using AuNPore catalyst. Org Lett 16:2558–2561. https://doi.org/10.1021/ol500958p
24. Takale BS, Tao S, Yu X, Feng X, Jin T, Bao M, Yamamoto Y (2015) Highly chemoselective reduction of imines using a AuNPore/$PhMe_2SiH$/water system and its application to reductive amination. Tetrahedron 71:7154–7158. https://doi.org/10.1016/j.tet.2014.11.023
25. Chen M, Yan Y, Gebre M, Ordonez C, Liu F, Qi L, Lamkins A, Jing D, Dolge K, Zhang B, Heintz P, Shoemaker DP, Wang B, Huang W (2021) Thermal unequilibrium of PdSn intermetallic nanocatalysts: from in situ tailored synthesis to unexpected hydrogenation selectivity. Angew Chem Int Ed 60:18309–18317. https://doi.org/10.1002/anie.202106515

26. Takale BS, Wang S, Zhang X, Feng X, Yu X, Jin T, Bao M, Yamamoto Y (2014) Chemoselective reduction of α,β-unsaturated aldehydes using an unsupported nanoporous gold catalyst. Chem Commun 50:14401–14404. https://doi.org/10.1039/c4cc07068b
27. Zhao Y, Zhang S, Yamamoto Y, Bao M, Jin T, Terada M (2019) Heterogeneous catalytic reduction of tertiary amides with hydrosilanes using unsupported nanoporous gold catalyst. Adv Synth Catal 361:4817–4824. https://doi.org/10.1002/adsc.201900838
28. Gibson HW (1969) The chemistry of formic acid and its simple derivatives. Chem Rev 69:673–692. https://doi.org/10.1021/cr60261a005
29. Blaser HU, Malan C, Pugin B, Spindler F, Steiner H, Studer M (2003) Selective hydrogenation for fine chemicals: recent trends and new developments. Adv Synth Catal 345:103–151. https://doi.org/10.1002/adsc.200390000
30. Liu X, Li S, Liu Y, Cao Y (2015) Formic acid: a versatile renewable reagent for green and sustainable chemical synthesis. Chin J Catal 36:1461–1475. https://doi.org/10.1016/S1872-2067(15)60861-0
31. Wagh YS, Asao N (2015) Selective transfer semihydrogenation of alkynes with nanoporous gold catalysts. J Org Chem 80:847–851. https://doi.org/10.1021/jo502313d
32. Butt M, Feng X, Yamamoto Y, Almansour AI, Arumugam N, Kumar RS, Bao M (2017) Unsupported nanoporous gold-catalyzed chemoselective reduction of α,β-unsaturated Aldehydes using formic acid as hydrogen source. Asian J. Org. Chem. 6:867–872. https://doi.org/10.1002/ajoc.201700130
33. Butt M, Zhao Y, Feng X, Lu Y, Jin T, Yamamoto Y, Bao M (2019) Unsupported nanoporous gold-catalyzed chemoselective reduction of quinolines using formic acid as a hydrogen source. ChemistrySelect 4:6572–6577. https://doi.org/10.1002/slct.201901309
34. Takale BS, Feng X, Lu Y, Bao M, Jin T, Minato T, Yamamoto Y (2016) Unsupported nanoporous gold catalyst for chemoselective hydrogenation reactions under low pressure: effect of residual silver on the reaction. J Am Chem Soc 138:10356–10364. https://doi.org/10.1021/jacs.6b06569
35. Asao N, Menggenbateer, Seya Y, Yamamoto Y, Chen M, Zhang W, Inoue A (2012) Nanoporous gold-catalyzed [4+2] benzannulation between ortho-alkynylbenzaldehydes and alkynes. Synlett 23:66–69. https://doi.org/10.1055/s-0031-1289527
36. Ishikawa Y, Yamamoto Y, Asao N (2013) Selective hydrosilylation of alkynes with a nanoporous gold catalyst. Catal Sci Technol 3:2902–2905. https://doi.org/10.1039/c3cy00242j
37. Li H, Guo H, Li Z, Wu C, Li J, Zhao C, Guo S, Ding Y, He W, Li Y (2018) Silylation reactions on nanoporous gold via homolytic Si–H activation of silanes. Chem Sci 9:4808–4813. https://doi.org/10.1039/c8sc01427b
38. Chen Q, Zhao J, Ishikawa Y, Asao N, Yamamoto Y, Jin T (2013) Remarkable catalytic property of nanoporous gold on activation of diborons for direct diboration of alkynes. Org Lett 15:5766–5769. https://doi.org/10.1021/ol4028013
39. Chen Q, Zhang X, Su S, Xu Z, Li N, Li Y, Zhou H, Bao M, Yamamoto Y, Jin T (2018) Nanoporous gold catalyzed diboration of methylenecyclopropanes via a distal bond cleavage. ACS Catal 8:5901–5906. https://doi.org/10.1021/acscatal.8b01193
40. Eong Ho H, Ishikawa Y, Asao N, Yamamoto Y, Jin T (2015) Highly efficient heterogeneous aerobic cross-dehydrogenative coupling via C-H functionalization of tertiary amines using a nanoporous gold skeleton catalyst. Chem Commun 51:12764–12767. https://doi.org/10.1039/c5cc04856g
41. Zhao Y, Feng X, Zhang S, Yamamoto Y, Bao M (2020) Hydrodebromination of aromatic bromides catalyzed by unsupported nanoporous gold: heterolytic cleavage of hydrogen molecule. ChemCatChem 12:4951–4957. https://doi.org/10.1002/cctc.202000674
42. Zeng Y, Nishiyama N, Wada T, Louzguine-Luzgin DV, Inoue A (2006) Ni-Rich Ni–Pd–P glassy alloy with high strength and good ductility. Mater Trans 47:175–178. https://doi.org/10.2320/matertrans.47.175
43. Yu J, Ding Y, Xu C, Inoue A, Sakurai T, Chen M (2008) Nanoporous metals by dealloying multicomponent metallic glasses. Chem Mater 20:4548–4550. https://doi.org/10.1021/cm8009644

44. Zhang Z, Wang Y, Qi Z, Zhang W, Qin J, Frenzel J (2009) Generalized fabrication of nanoporous metals (Au, Pd, Pt, Ag, and Cu) through chemical dealloying. J Phys Chem C 113:12629–12636. https://doi.org/10.1021/jp811445a
45. Tanaka S, Kaneko T, Asao N, Yamamoto Y, Chen M, Zhang W, Inoue A (2011) A nanostructured skeleton catalyst: suzuki-coupling with a reusable and sustainable nanoporous metallic glass Pd-catalyst. Chem Commun 47:5985–5987. https://doi.org/10.1039/c1cc10710k
46. Kaneko T, Tanaka S, Asao N, Yamamoto Y, Chen M, Zhang W, Inoue A (2011) Reusable and sustainable nanostructured skeleton catalyst: heck reaction with nanoporous metallic glass Pd (PdNPore) as a support, stabilizer and ligand-free catalyst. Adv Synth Catal 353:2927–2932. https://doi.org/10.1002/adcs.201100358
47. Li Z, Lin S, Ji L, Zhang Z, Zhang X, Ding Y (2014) Nanoporous palladium catalyzed silicon-based one-pot cross-coupling reaction of aryl iodides with organosilanes. Catal Sci Technol 4:1734–1737. https://doi.org/10.1039/c4cy00256c
48. Lu Y, Feng X, Takale BS, Yamamoto Y, Zhang W, Bao M (2017) Highly selective semihydrogenation of alkynes to alkenes by using an unsupported nanoporous palladium catalyst: no leaching of palladium into the reaction mixture. ACS Catal 7:8296–8303. https://doi.org/10.1021/acscatal.7b02915
49. Lu Y, Yamamoto Y, Almansour AI, Arumugam N, Kumar RS, Bao M (2018) Unsupported nanoporous palladium-catalyzed chemoselective hydrogenation of quinolines: heterolytic cleavage of H_2 molecule. Chinese J Catal 39:1746–1752. https://doi.org/10.1016/S1872-2067(18)63151-1
50. Li Z, Zhao D, Luo J (2019) Nanoporous palladium catalyst for the reduction of aromatic nitro compounds with silane/alcohol system as the hydrogen source. Inorg Chem Commun 100:1–5. https://doi.org/10.1016/j.inoche.2018.12.004
51. Cao Y, Li Z, Hou C, Cao X (2020) Nanoporous palladium catalyzed one-pot synthesis N-alkyl amines by hydrogen transfer reaction under mild conditions. Inorg Chem Commun 120:108143. https://doi.org/10.1016/j.inoche.2020.108143
52. Zhao Y, Feng X, Li Y, Zhang W, Yamamoto Y, Bao M (2020) Effect of nanoporous structure on the catalytic activity of nanoporous palladium for hydrogenation of nitro compounds. ChemistrySelect 5:7086–7092. https://doi.org/10.1002/slct.202001550
53. Wang J, Zhou C, Gao Z, Feng X, Yamamoto Y, Bao M (2021) Unsupported nanoporous palladium catalyst for highly selective hydrogenation of carbon dioxide and sodium bicarbonate into formate. ChemCatChem 13:2702–2708. https://doi.org/10.1002/cctc.202100148
54. Wang J, Li S, Wang Y, Feng X, Yamamoto Y, Bao M (2022) Unsupported nanoporous palladium catalyst for N-formylation of amines using CO_2 as a sustainable C1 source. Asian J Org Chem 11:e202200064. https://doi.org/10.1002/ajoc.202200064
55. Lu Y, Wang J, Feng X, Li Y, Zhang W, Yamamoto Y, Bao M (2022) Hydrogenation of nitriles to primary amines catalyzed by an unsupported nanoporous palladium catalyst: understanding the essential reason for the high activity and selectivity of the catalyst. Nanoscale 14:9341–9348. https://doi.org/10.1039/d2nr01722a
56. Ji H, Wang X, Zhao C, Zhang C, Xu J, Zhang Z (2011) Formation, control and functionalization of nanoporous silver through changing dealloying media and elemental doping. CrystEngComm 13:2617–2628. https://doi.org/10.1039/c0ce00900h
57. Li Z, Xu J, Gu X, Wang K, Wang W, Zhang X, Zhang Z, Ding Y (2013) Selective gas-phase oxidation of alcohols over nanoporous silver. ChemCatChem 5:1705–1708. https://doi.org/10.1002/cctc.201200862
58. Li Z, Zhang C, Tian J, Zhang Z, Zhang X, Ding Y (2014) Highly selective oxidation of organosilanes with a reusable nanoporous silver catalyst. Catal Commun 53:53–56. https://doi.org/10.1016/j.catcom.2014.04.009
59. Zhou Q, Yang H, Xu C (2016) Nanoporous Ru as highly efficient catalyst for hydrolysis of ammonia borane. Int J Hydrogen Energ 41:12714–12721. https://doi.org/10.1016/j.ijhydene.2016.05.128
60. Raney M (1927) U.S. Patent 1628190

61. Qi Z, Zhang Z, Jia H, Qu Y, Liu G, Bian X (2009) Alloy composition dependence of formation of porous Ni prepared by rapid solidification and chemical dealloying. J Alloys Compd 472:71–78. https://doi.org/10.1016/j.jallcom.2008.04.01
62. Ito Y, Tanabe Y, Qiu HJ, Sugawara K, Heguri S, Tu NH, Huynh KK, Fujita T, Takahashi T, Tanigaki K, Chen M (2014) High-quality three-dimensional nanoporous graphene. Angew Chem Int Ed 53:4822–4826. https://doi.org/10.1002/anie.201402662
63. Qiu HJ, Kang JL, Liu P, Hirata A, Fujita T, Chen MW (2014) Fabrication of large-scale nanoporous nickel with a tunable pore size for energy storage. J Power Sources 247:896–905. https://doi.org/10.1016/j.jpowsour.2013.08.07
64. Wang T, Ren D, Huo Z, Song Z, Jin F, Chen M, Chen L (2017) A nanoporous nickel catalyst for selective hydrogenation of carbonates into formic acid in water. Green Chem 19:716–721. https://doi.org/10.1039/c6gc02866g
65. Keir DS, Pryor MJ (1980) The dealloying of copper-manganese alloys. J Electrochem Soc 127:2138–2144. https://doi.org/10.1149/1.2129360
66. Pryor MJ, Fister JC (1984) The mechanism of dealloying of copper solid solutions and intermetallic phases. J Electrochem Soc 131:1230–1235. https://doi.org/10.1149/1.2115793
67. Min US, Li JCM (1994) The microstructure and dealloying kinetics of a Cu-Mn alloy. J Mater Res 9:2878–2883. https://doi.org/10.1557/JMR.1994.2878
68. Mellor JR, Coville NJ, Sofianos AC, Copperthwaite RG (1997) Raney copper catalysts for the water–gas shift reaction: I. Preparation, activity and stability. Appl Catal A-Gen 164:171–183. https://doi.org/10.1016/S0926-860X(97)00167-1
69. Mellor JR, Coville NJ, Durbach SH, Copperthwaite RG (1998) Acid leached raney copper catalysts for the water–gas shift reaction. Appl Catal A-Gen 171:273–281. https://doi.org/10.1016/S0926-860X(98)00084-2
70. Hayes JR, Hodge AM, Biener J, Hamza AV, Sieradzki K (2006) Monolithic nanoporous copper by dealloying Mn–Cu. J Mater Res 21:2611–2616. https://doi.org/10.1557/JMR.2006.0322
71. Qi Z, Zhao C, Wang X, Lin J, Shao W, Zhang Z, Bian X (2009) Formation and characterization of monolithic nanoporous copper by chemical dealloying of Al–Cu alloys. J Phys Chem C 113:6694–6698. https://doi.org/10.1021/jp810742z
72. Zhao C, Qi Z, Wang X, Zhang Z (2009) Fabrication and characterization of monolithic nanoporous copper through chemical dealloying of Mg–Cu alloys. Corros Sci 51:2120–2125. https://doi.org/10.1016/j.corsci.2009.05.043
73. Chen LY, Yu JS, Fujita T, Chen MW (2009) Nanoporous copper with tunable nanoporosity for SERS applications. Adv Funct Mater 19:1221–1226. https://doi.org/10.1002/adfm.200801239
74. Jin T, Yan M, Menggenbateer, Minato T, Bao M, Yamamoto Y (2011) Nanoporous copper metal catalyst in click chemistry: nanoporosity-dependent activity without supports and bases. Adv Synth Catal 353:3095–3100. https://doi.org/10.1002/adcs.201100760
75. Chen Q, Tanaka S, Fujita T, Chen L, Minato T, Ishikawa Y, Chen M, Asao N, Yamamoto Y, Jin T (2014) The synergistic effect of nanoporous aupd alloy catalysts on highly chemoselective 1,4-hydrosilylation of conjugated cyclic enones. Chem Commun 50:3344–3346. https://doi.org/10.1039/c3cc49524h
76. Xu C, Wang L, Mu X, Ding Y (2010) Nanoporous PtRu alloys for electrocatalysis. Langmuir 26:7437–7443. https://doi.org/10.1021/la9041474
77. Zhou Q, Xu C (2016) Nanoporous PtRu alloys with unique catalytic activity toward hydrolytic dehydrogenation of ammonia borane. Chem Asian J 11:705–712. https://doi.org/10.1002/asia.201500970
78. Ou S, Ma D, Li Y, Yubuta K, Tan Z, Wang Y, Zhang W (2017) Fabrication and electrocatalytic properties of ferromagnetic nanoporous PtFe by dealloying an amorphous $Fe_{60}Pt_{10}B_{30}$ alloy. J Alloys Compd 706:215–219. https://doi.org/10.1016/j.jallcom.2017.02.203
79. Geng Y, Chen C, Gao Z, Feng X, Liu W, Li Y, Jin T, Shi Y, Zhang W, Bao M (2021) Unsupported nanoporous platinum-iron bimetallic catalyst for the chemoselective hydrogenation of halonitrobenzenes to haloanilines. ACS Appl Mater Interfaces 13:23655–23661. https://doi.org/10.1021/acsami.1c02734

80. Zhou Q, Xu C (2017) Nanoporous PtCo/Co_3O_4 composites with high catalytic activities toward hydrolytic dehydrogenation of ammonia borane. J Colloid Interf Sci 508:542–550. https://doi.org/10.1016/j.jcis.2017.08.054
81. Shi J, Mahr C, Mangir Murshed M, Zielasek V, Rosenauer A, Gesing TM, Bäumer M, Wittstock A (2016) Versatile Sol–Gel coating for mixed oxides on nanoporous gold and their application in the water gas shift reaction. Catal Sci Technol 6:5311–5319. https://doi.org/10.1039/c5cy02205c
82. Li W, Wang F, Feng S, Wang J, Sun Z, Li B, Li Y, Yang J, Elzatahry AA, Xia Y, Zhao D (2013) Sol–Gel design strategy for ultradispersed TiO_2 nanoparticles on graphene for high-performance lithium ion batteries. J Am Chem Soc 135:18300–18303. https://doi.org/10.1021/ja4100723
83. Shi J, Mahr C, Murshed MM, Gesing TM, Rosenauer A, Bäumer M, Wittstock A (2017) Steam reforming of methanol over oxide decorated nanoporous gold catalysts: a combined in situ FTIR and flow reactor study. Phys Chem Chem Phys 19:8880–8888. https://doi.org/10.1039/c6cp08849
84. Shi J, Wittstock A, Mahr C, Mangir Murshed M, Gesing TM, Rosenauer A, Bäumer M (2019) Nanoporous gold functionalized with praseodymia–titania mixed oxides as a stable catalyst for the water–gas shift reaction. Phys Chem Chem Phys 21:3278–3286. https://doi.org/10.1039/c8cp06040a

Chapter 6
Conclusions and Outlook

Abstract Nanostructured catalysts are of great interest from a basic science point of view as technological applications. The above four chapters introduced the preparation, characterization and application of nanoparticle catalysts, nano-cluster catalysts, single-atom catalysts, and nanoporous metal catalysts in organic synthesis. It can be found that nanostructured catalysts have broad application prospects in the field of organic synthesis. Among them, the nanoparticle catalysts, single-atom catalysts, and nanoporous metal catalysts belong to heterogeneous catalysts; while, the unsupported nano-cluster catalysts belong to homogeneous catalysts. These catalysts have different advantages and disadvantages in organic synthesis.

Nanostructured catalysts are of great interest from a basic science point of view as technological applications. The above four chapters introduced the preparation, characterization and application of nanoparticle catalysts, nano-cluster catalysts, single-atom catalysts, and nanoporous metal catalysts in organic synthesis. It can be found that nanostructured catalysts have broad application prospects in the field of organic synthesis. Among them, the nanoparticle catalysts, single-atom catalysts, and nanoporous metal catalysts belong to heterogeneous catalysts; while, the unsupported nano-cluster catalysts belong to homogeneous catalysts. These catalysts have different advantages and disadvantages in organic synthesis. Table 6.1 gives a comparative summary of the essential characteristics of the four kinds of nanostructured catalysts. The characteristics of the four kinds of catalysts may complement each other.

The catalytic organic reaction with nanostructured catalysts may involve diffusion of the reactants to the catalyst surface, physical adsorption and chemical adsorption on the active sites, surface reaction, desorption of products, and products diffusion from the surface, etc. Although the unsupported nano-cluster-catalyzed reactions belong to homogeneous catalysis, the organic molecule transformations occurred also on the catalyst surface. Therefore, the leaching of active species may be one of the main reasons for the decrease of catalyst activity. In addition, agglomeration or structural changes also have an important effect on the reduction of catalyst activity.

M. Bao et al., *Nanometal Catalysis in Organic Synthesis*, Molecular Catalysis 3,
https://doi.org/10.1007/978-981-97-4573-9_6

Table 6.1 Comparison of four kinds of Nanostructured Catalysts

	Nanoparticle catalysts	Single-atom catalysts	Nanoporous metal catalysts	Unsupported nano-cluster catalysts
Active centers	Only surface atoms	All metal atoms	Only surface atoms	Only surface atoms
Selectivity	Medium	High	High	High
Mass transfer limitation	Present	Present	Present	Practically absent
Catalyst separation	Easy	Easy	Easy	Difficult
Modification possibility	High	High	Low	High (by ligand/ stabilizer)
Thermal stability	High	High	High	Low
Catalyst recycling	Easy	Easy	Easy	Difficult
Reaction phase state	Heterogeneous	Heterogeneous	Heterogeneous	Homogeneous
Temperature range of reaction	Wide (even >200 °c)	Wide (even >200 °c)	Wide (even >200 °C)	Narrow (<200 °C)
Applicability	Wide (liquid-phase and gas-phase reaction)	Wide (liquid-phase and gas-phase reaction)	Wide (liquid-phase and gas-phase reaction)	Limited (liquid-phase reaction)
Preparation of catalyst	Complex	Complex	Simple	Simple
Cost of catalyst losses	Low	Low	Low	High

At present, the selectivity, including chemoselectivity, regioselectivity, diastereoselectivity, and enantioselectivity, obtained by using nanostructured catalysts is usually not as high as that obtained by using metal complex homogeneous catalysts.

In the future, novel catalyst design strategies need to be proposed to prepare more stable nanostructured catalysts with high activity and selectivity; looking forward to more industrial applications of nanostructured catalysts for synthesis of high value-added fine chemicals such as medicine, pesticides, and functional materials.

GPSR Compliance

The European Union's (EU) General Product Safety Regulation (GPSR) is a set of rules that requires consumer products to be safe and our obligations to ensure this.

If you have any concerns about our products, you can contact us on ProductSafety@springernature.com

In case Publisher is established outside the EU, the EU authorized representative is:

Springer Nature Customer Service Center GmbH
Europaplatz 3
69115 Heidelberg, Germany

Batch number: 10726532

Printed by Printforce, the Netherlands